A Maple Approach to CALCULUS

SECOND EDITION

John T. Gresser

Upper Saddle River, NJ 07458

Executive Editor: George Lobell
Supplement Editor: Melanie Van Benthuysen
Assistant Managing Editor: John Matthews
Production Editor: Wendy A. Perez
Supplement Cover Manager: Paul Gourhan
Supplement Cover Designer: PM Workshop Inc.
Manufacturing Buyer: Ilene Kahn
Cover Image: Antonio Martinelli

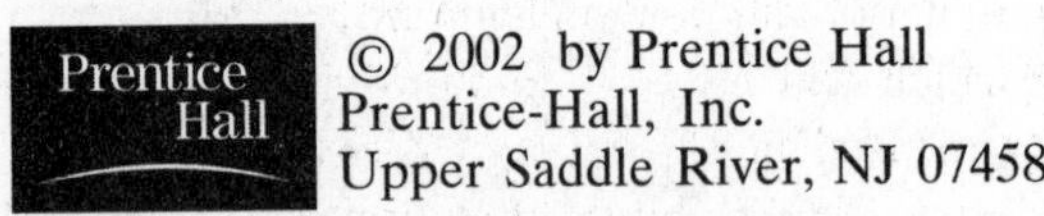

Printed in the United States of America

10 9 8 7 6 5 4

ISBN 0-13-092014-2

Pearson Education Ltd., *London*
Pearson Education Australia Pty. Ltd., *Sydney*
Pearson Education Singapore, Pte. Ltd.
Pearson Education North Asia Ltd., *Hong Kong*
Pearson Education Canada, Inc., *Toronto*
Pearson Educacíon de Mexico, S.A. de C.V.
Pearson Education—Japan, *Tokyo*
Pearson Education Malaysia, Pte. Ltd.
Pearson Education, *Upper Saddle River, New Jersey*

Preface

The last decade of the twentieth century was an exciting and interesting time for the study and use of mathematics. At the dawn of the new millennium, mathematics has undergone a technological change which will have a profound effect on the way mathematics is studied, understood and used in the future. What is driving all this change is the introduction and widespread availability of computer software programs that allow one to compute and manipulate mathematical symbols on a computer screen in much the same way that these symbols are computed or manipulated with pencil and paper. Several such programs, generally referred to as Computer Algebra Systems (CAS) or symbol processors are commercially available. One of the most elaborate of these is the program called Maple that will be used in this manual.

One of the strengths of Maple is that it is so easy to use. To be sure, there are some special Maple language peculiarities that one must learn before Maple can be used to solve problems, but these are held to a minimum. To a large extent, Maple mimics the ordinary language of mathematics. With no more than an hour or two worth of effort, one can begin to use Maple to solve mathematical problems. Virtually all of the words used in mathematics are commands or data structures in the Maple language. Think of a mathematical command, and you will likely find it in Maple. If you simply guess at the spelling (usually an abbreviation) of the Maple command, there is a good chance that your guess will be correct. If not, you will certainly be close enough to find the command in Maple's great on-line help, which will be discussed shortly.

With Maple, there is no longer a need to focus attention on the difficulties of carrying out a mathematical computation. The user becomes the director of the production, the creative force which drives the solution, while Maple does all the work. Problems which were previously inaccessible because of computational complexity, can now be solved using Maple. Just the freedom to concentrate on the creative aspects of a problem rather than the computational aspects, makes some problems much more approachable.

Mathematicians, indeed, all scientists, are naturally skeptics. Their first reaction to any claim is to ask for a proof, and then they are inclined to look for mistakes and counter examples when they study the work of their colleagues. It is exceedingly important that students of science and mathematics acquire this attitude of skepticism. Is the answer reasonable? Is it correct? Is it unique? Are there easier or alternative ways to get the answer? Is there a counter example? These are questions we should always ask, but in the past there was such a heavy price to pay to answer these questions that students typically became accustomed to accepting as correct the outcome of any computation or argument. Understandably, few of us were willing to verify the correctness of an answer with yet another lengthy calculation.

With Maple, on the other hand, it is usually easy to establish the correctness of a solution by an alternate calculation of some kind. Consequently it is now easier to be a skeptic, and it is just as important to question the validity of a calculation now as it ever was before. Maple, computers, and human beings are all capable of giving inappropriate answers, incomplete answers, misunderstood answers, and even wrong answers. Maple provides an excellent opportunity for students of science and mathematics to question their work and to acquire, in the process, this all important skeptical attitude at an early stage in their careers.

Computer Preliminaries

This manual, revised and updated to cover **Maple 6 and Maple 7**, is written for the Macintosh operating system. Maple for the **Windows operating system is similar enough** that very little difficulty should be experienced in modifying this manual to the Windows operating system. Rather than risk confusion, however, by discussing both operating systems, it is written entirely in terms of just one operating system.

We assume a passing familiarity with a computer using one of these operating systems. Excellent on-line tutorials are available on most computers. Two hours or so spent on such a tutorial should be sufficient preparation for anyone with no computing experience. In particular, we assume that the reader knows how to: 1) open up applications and files, 2) create folders and files and save files, 3) select, click, double click, and drag, 4) use menu-bars.

Maple needs a minimum of 24MB of RAM to operate, but 32 MB RAM is preferred, and even more RAM may be needed for memory intensive operations like 3-dimensional graphing. As you begin to demand more of Maple, consider the benefits of giving Maple more RAM to operate, if it is available on your computer.

Calculus and Maple

This manual is not meant to be a self-contained calculus text, but rather a supplement to a regular calculus text. Topics are covered in a way which is meant to parallel the sequence of topics in a fairly typical calculus text. New Maple commands are discussed as they arise in the process of solving problems in calculus.

In Chapter 2, the primary focus of attention is to learn basic Maple by using it to do problems in algebra, trigonometry, and introductory differential calculus. Starting with the integral in Chapter 3, emphasis shifts back to the study of calculus, where it remains for the rest of the manual. As you will see, Maple, as involved as it may be, is also straightforward enough that it will come to be understood quite naturally while attention is paid primarily to calculus. It takes some time, however, to adjust to this merger of mathematics, computers, and Maple. Chapter 2 is meant to nudge us gently in this direction. Introductory differentiation is well suited for more intense Maple activity, but the opportunity to use Maple in this way will come soon enough.

There are several advantages to be gained by postponing the use of a Computer Algebra System until the second calculus course, and this manual is ideally suited for such a program. Hand held graphing calculators are still a mainstay in the mathematical curriculum, and the first calculus course provides a good opportunity to get better acquainted with this excellent tool. By the second course, students who are continuing their study of mathematics are usually ready for more serious work, and their numbers are small enough to make computer lab work more manageable.

Chapter 2 provides a quick tour through the usual topics of elementary differentiation, so it can be used as a review of first semester calculus while learning basic Maple. The review by itself could be seen as an advantage. This chapter is important as a vehicle to learn the fundamentals, and it should be covered by everyone with no previous Maple experience. This chapter also provides an opportunity to get used to the practice of thinking about mathematics with a keyboard and monitor rather than with pencil and paper.

Of course, the best way to learn Maple is not just to read about it, but to use it to do problems. Many of the problems at the end of each chapter are meant to provide routine experience in using Maple commands. **Routine problems from your main calculus text can also be used as Maple practice problems.** You can, with such problems, **look up their answers in the appendix of your main text** to verify that Maple commands are being used correctly.

Problems are, however, also designed to foster an attitude of skepticism and experimentation. The importance of adopting a skeptical scientific attitude has already been discussed. Answers are, by design, not provided in the back. **How do you know** that you have the correct answer? **Is there another** solution to the equation that is being solved? Is there an interesting feature to a graph under consideration that is **too small to be seen** in a window or that occurs outside the window being used? Problems in this manual sometimes create **unexpected, incomplete, or occasionally wrong answers**. In the first few problem sets, you will usually be alerted to look for unexpected results, but eventually, such warnings will not be supplied. **Whenever possible or appropriate, you should supply evidence that your answer is correct**. One of the goals of this manual is to develop a good, skeptical, scientific attitude.

Maple can be used to gain a deeper understanding of calculus by focusing complete attention on an issue of calculus rather than on a computation. Some of the problems are designed with this in mind.

Without question, Maple can be used to enhance problem solving skills. Just imagine the creative freedom that you will have when you can focus all of your energy on ideas rather than computations. Maple makes more substantial and interesting mathematical problems accessible, and this might be its most important contribution to mathematical education. One should strive to get as much experience as possible in doing problems of this sort. Problems designed to improved problem solving skills are included in the exercise sets along with all of the other problems.

Many of the exercise sets have additional problems, labeled **"Projects,"** which are somewhat more involved. They range in difficulty from being just longer and more interesting versions of ordinary problems, to being quite difficult. They should be accessible without outside background reading. These problems are designed to enhance problem solving skills by making use of not only current topics under discussion, but, occasionally, a wide variety of previously discussed topics as well. At least some of them are presented in a playful way, and are meant to be enjoyed, as well as to enlighten.

These projects, however, should be tackled with some discretion as well. Using Maple for the first few times is an interesting, but very different way of doing mathematics, and it takes some time to get accustomed to it and to take advantage of all the opportunities Maple presents to the user. When you are ready to take on the issue of "putting it all together," you are encouraged to work on the projects that appear after the exercise sets at the ends of the chapters.

How to Read this Manual

It is the nature of computer languages that certain words, phrases, sentences, and punctuations have to be read quite carefully, symbol by symbol. We have made an attempt to emphasize all such structures with **bold face** type. **When you encounter these items, read them with great care**. In addition, whenever a Maple command is introduced for the first time, it also appears in bold face type.

The ordinary text of this manual will be separated from the input-output statements which would appear on a computer under Maple in the following manner.

****

(Begin Maple work session.)

Text statements made during Maple work sessions will be enclosed in sharp symbols.#

(End of Maple work session.)

****

The two separators are slightly different. The first one will be used to mark the start of a new Maple session, and the second will be used to mark the end of a Maple session. The Maple environment reproduced between these separators is exactly the same, both in content and appearance to that which would appear on a computer screen under Maple. Ordinary text that appears in this space reserved for Maple will be enclosed in sharp (#) symbols. Maple ignores everything between sharp symbols, so these comments will have no effect on Maple computations whether or not they appear on a computer during a Maple work session. Actually, Maple has better ways to enter text (material ignored by Maple), but this manual is not printed in color, and so sharp symbols are used as an easy means of identifying statements of this type. Maple still does ignore statements enclosed in sharp symbols, but you are encouraged to **avoid using these symbols on your own Maple worksheets**. Use Maple's text environments instead.

The ($\underline{*****}$) and ($\overline{*****}$) separators are our preferred means of separating the ordinary text of this manual, especially lengthy discourse, from the space reserved for Maple, but they can only be used (given the above interpretation) if we are allowed to turn our computer on at ($\underline{*****}$) and off at ($\overline{*****}$). This means that if the same active Maple worksheet is meant to continue before and after such textual material, then this material must be enclosed in sharp (#) symbols. We will follow this practice regardless of how long the discourse between sharp symbols becomes.

Explanations of new commands are given just once, when they are first introduced, so if chapters or sections are skipped, it would help to skim over the material that was skipped, looking for and reading the material on the introduction of new commands. Since new commands appear in bold face type when they are first introduced, this should be relatively painless.

This material can be read and understood without a computer, but surely the best way to use it is to work on a computer at the same time. Enter the same (or similar) expressions on your computer as you read the manual, so you can experience the results first hand.

Some Closing Thoughts

While Maple is fairly easy to learn, it is a wide-ranging language with hundreds of commands effecting many subjects of mathematics. No attempt has been made to make this manual a complete study of Maple. **You are encouraged to explore on your own, and to seek help frequently with Maple's on-line help**. Adopt the attitude that **any "mathematical, or logical word" under consideration is a word that can be found in some form, somewhere in Maple**.

Exercises should be presented in an organized, thoughtful and readable manner. Remember that **Maple is a very good word processor**, and it should be used frequently in the exercises. Projects, in particular, should be done with great care paid to presentation. Think of a project as a term paper, or as a report which is going to your supervisor at work. The same matters of presentation should play a role in creating a paper on any subject—including mathematics.

A pencil-and-paper strategy session can be a useful way to start a Maple work session. It should be a strategy session, however, and not a complete pencil-and-paper solution, unless, of course, such a solution is desired. Remember that Maple will do all of the computations for you. Think of a pencil-and-paper strategy session as a flow chart, where steps are organized, notation is devised, etc.

And finally, above all, remember that **Maple was created by human beings**. Human beings make mistakes. Maple makes mistakes, infrequently, perhaps, but mistakes nevertheless. The same is true for all other software packages. Maple is under continuous improvement, but it will always have some potential for error. Human beings are, of course, a more common source of mistakes. A small mistake in an input statement can have enormous consequences. Even without an input line mistake, a Maple computation can be misinterpreted, or used improperly by us with grave consequences.

Work with Maple should be a partnership between a human being and a machine, not a thoughtless ride on a machine. The way to avoid wrong answers is to know mathematics well enough to see the warning signs when mistakes have been made, or when Maple is misbehaving. **Make it a practice to verify that answers are correct, especially if an answer "looks" doubtful.** Frequently, answers can be verified very quickly.

In order to keep the main story line simple, brief, and readable, this manual will not always participate in this practice of verifying answers. Make it a habit, when reading this manual, and when doing your own work to **check Maple's performance.**

Acknowledgments

I am indebted to the staff of Prentice Hall for its help in preparing this book. Special thanks are offered to George Lobell, Executive Editor, for his encouragement and support; Melanie VanBenthuysen, Supplement Editor; and Wendy A. Perez, Production Editor. I am especially grateful to Bill Salvatore, who proofed the second edition of this manuscript. His careful attention to the smallest detail improved my manuscript immeasurably. Finally, special thanks go to all of my students, past and present and future, who were, are now, or will be a part of my Maple based calculus program. Their cooperation and sense of humor kept and will continue to keep this project moving forward.

———★★★———

Solving problems with Maple, especially more involved problems, can be a rich and rewarding mathematical experience. I hope that this manual and its problems meet with your approval. My e-mail address is included below, because good text books are a community effort, and your comments and suggestions for improvements would be greatly appreciated.

John T. Gresser
Department of Mathematics and Statistics
Bowling Green State University
Bowling Green, OH 43403
jgresse@bgnet.bgsu.edu

To Pamela, Morgan, and Nathan
for their patience
during this writing project

and

to Captain Ralph
off on another adventure.

Contents

To The Teacher

A Maple Approach to Calculus is intended to be used as a supplement to a main text. There is a significant change in philosophy between Chapter 2 (basic Maple learned by a quick tour through introductory differentiation) and the rest of the manual where the focus is more on calculus than on software. This matter was discussed in the Preface as well.

Whether Maple is introduced in the first course or the second, students still need time to adjust to the rigors of merging mathematics, computers, and Maple. There are certain advantages to postponing the introduction of Maple until the second course (a more mature audience is one reason), and this supplement is ideally suited for such courses. By working their way through Chapter 2, students can quickly learn basic Maple and adjust to this computer environment, by reviewing familiar topics from the previous course. While students work on Chapter 2, somewhat on their own, a regular second calculus course can proceed almost at a normal pace. In a fairly short amount of time, students will be ready to use Maple in their second course activity.

If Maple is introduced in the first course, then this adjustment period must still be built into the course. One way to do this is to depend initially on a more traditional development of calculus, and then slowly introduce Maple into the course as it progresses. For such a program, Chapter 2 might well be an acceptable way to start a Maple based calculus program. Certainly it lacks an in depth study of the elementary properties of the derivative, but it allows calculus to proceed at a fairly normal pace, and it gives students time to adapt to a computer based program.

The mathematical community has responded to the issue of calculus reform in a variety of ways. The reform process is still evolving, but it does not appear to be evolving towards one uniform way to teach calculus. Nevertheless, it is hard to deny that computers could (or should) play at least some minimal role in calculus, if not a maximal role, in all of these schemes. Time, unfortunately, is a major player in this game.

Calculus classes have typically been packed with so much material that there has been little or no room left to add new material. If a computer laboratory component is added to a calculus class, certainly classroom time has to be made available to deal with this addition. It is probably true that some topics have to be dropped, or the level of expectation has to be relaxed on some topics, in order to make room for the additional demands made by computer activity. Therein lies the problem. How do we make time for computer work in calculus?

Time plays a role in another way as well. Mathematicians are extremely busy people. Adding a computer component to a calculus class means taking time out of an already packed work schedule to learn new software, and more important, to learn how to use it in a class room setting. The influence of a computer can produce an overwhelming change in the way mathematics is presented, and this can create a demand for more time than a teacher is able to provide for a course. How are mathematicians suppose to make time for all of this change?

A software manual may supply all of the necessary computer background, and a main text may supply all of the pertinent mathematical material. If, however, there is nothing connecting the two books, then a teacher has to spend significant class room time and energy to explain the connection. A student may ask, "Why did I do this problem on a computer? I pushed a button and got this answer. Now, what's the point?"

This manual will certainly not eliminate these problems, but, hopefully, it can soften them somewhat. I believe that it is possible to include a fairly significant computer component to a calculus class with only minor alterations to whatever method is being used to teach calculus. More dramatic reform might be encouraged by some, but then again, each strategy has its advantages and disadvantages.

This manual is mainly about calculus. It contains a fair amount of intuitive—hopefully readable and student oriented—mathematical material in addition to Maple material. These same ideas may well appear in a main calculus text, but they also appear in the manual to provide a connection between mathematics and Maple. Student questions such as those raised above may be answered by reading the manual. If so, then a teacher can spend most of the class room time on strictly mathematical ideas. (Maple, however, frequently offers a slick way to present an idea.) The computer component of the course can then be handled with a command to, "Read the manual, and do problems,...."

Some classroom lab time is probably essential. In my 5 credit (introductory Maple) calculus course, the first 3 days of the semester are spent in the lab devoted exclusively to Maple. This is a very intense time for my students, but they are fresh, they are not yet burdened by the demands of their other courses, and a great deal of Maple is learned in three days. After that, classroom time is devoted almost entirely to mathematics, with slightly less than one day per week spent in the lab for the rest of the semester. Maple frequently enters into classroom discussion, but usually in a mathematical way.

In my 3 credit follow-up Maple based, multivariate calculus course, no class room time is spent in the lab. My classroom activity is almost exclusively mathematical. Maple issues are almost always discussed in some mathematical context. Students are told to read the manual and to do certain problems. I make myself available on a regular basis for lab activity outside of class.

I confess that my students get frustrated and need human encouragement. Much of this frustration is a communication problem they have with computers. Human beings can understand content, sometimes in spite of what is said to them. A computer (lacking a human ability to interpret) can only understand exactly what it is given. But this is a plus, is it not? What a great way to enforce precise thinking and communication!

Except for the computer component, calculus at my home institution is taught in a very traditional way. A traditional book (read big) is used, and this approach has left a clear impact on the structure of this manual. As you can see, topics are covered in much the same way they would be covered in a traditional calculus course.

In spite of the traditional bent, this manual should also be usable in courses which are more reform oriented. Computers are, after all, one of the principal features of this movement. As a supplement, this manual can simply be regarded as a list of topics and problems, and so the choice of topics and their order of presentation does not have to be strictly adhered to.

In fact, regardless of the main text you use and whether your approach to teaching calculus is traditional or more reform oriented, you might wish to avoid a cover to cover study of this manual. A cover to cover study would be nice, but it would require somewhat of a commitment to the Maple based approach. If material is skipped, new Maple commands that are introduced in the skipped material will be missed. It is reasonably important to become acquainted with most of these commands, even if material is skipped, because they are introduced only once in detail. New Maple commands appear in bold face type, when they are first introduced, so it should be easy to skim through the skipped material and pick up the new commands. If a command is missed in this way, it will be noticed later when the command is used again. If the command use is not clear from context, it should be a simple matter to find the introduction of the command in the missed section or look it up in Maple's on-line help file.

In my courses, students need to spend at least two or three hours per week outside of class in the lab in order to finish lab assignments. They are permitted (almost encouraged) to discuss mathematics with fellow class members. So much learning takes place during these sessions that I am only slightly concerned about how much of their work is entirely their own. In order to make their hard work a bit more tolerable, I assign, whenever possible, the problems in this manual that have been written in a playful way. Students are quick to pick up the playful spirit, and grading their lab assignments can be an enjoyably funny experience.

Lab assignments are turned in over the campus network and graded electronically. It takes a while to adjust to grading in this way, but there are some real advantages. I open a blank Maple Worksheet to use as my comment file, and then open student lab assignments, one at a time. (It helps to change the font setting on this comment-worksheet to produce text that looks very different from the fonts used on student lab assignments.) I write all of my comments on this comment-worksheet, and then paste comments onto student lab assignments. Because of the similarity of student mistakes, it takes little time before the same comments are being copied and pasted into several different student work files, and this speeds up the grading process considerably. This practice also encourages detailed comments, since I know they only have to be typed once.

I hope this manual becomes a useful addition to your calculus program. I have included my e-mail address in the Preface. Your comments would be appreciated.

Chapter 1

Some Preliminaries

This chapter contains critical initial details about using Maple that are important for you to become aware of at an early stage of development. Try to pick up as many of the key points as possible here. As you work your way through the first few sections of chapter 2, **return frequently to this preliminary chapter** (especially if you encounter difficulties) to review the important issues raised here. **At certain critical points in Chapter 2, you will be reminded to return to this preliminary Chapter.**

When Maple is opened, the top of your computer screen will be filled with menu items, buttons, tool bars, and so on. For the most part, the menu-bar is straightforward; its use will gradually become a part of your routine as you spend time with Maple and experiment with these accessories. Our first objective is to learn how to move around in Maple's environment and how to enter and process information. Another critical matter involves the way information is processed on a computer screen and the way previously saved files are loaded and processed. Along the way, some of the menu-bar items will be discussed, and we will finish the chapter with an introduction to Maple's on-line help. The **Help Menu will be a fundamental tool in our discovery of Maple.**

1.1 Maple Worksheets

A Maple **work session** is defined to be the activity that takes place between the time that the Maple application is opened and the very next time it is terminated. Frequently, it is useful to quit Maple and then immediately open it up again to continue working on the same document. Thus, one could sit at a computer terminal and pass through several work sessions, all the while working on the same document.

Maple, like most other computer applications, can be activated by opening a new Maple file or a previously saved Maple file. Maple is activated and a new file is created in the same way as any other application, by double-clicking the application icon. A blank sheet (window) will appear on the computer screen after Maple is loaded. This sheet has the appearance and essential functionality of a typical word processor (and much more), with a "Maple prompt" (>) appearing next to a square left bracket in the upper left-hand corner, followed by a blinking cursor indicating the point where work is to begin. The role played by the symbol (>) will be explained later. It appears to be a "greater than symbol," but **it is not, and the "greater than" symbol cannot be used as a Maple prompt.**

If, on the other hand, Maple is activated by opening a previously saved Maple file, then all of the information from that file will appear on the computer screen after Maple is loaded. Actually, although this information appears on the computer screen, it has **not yet been entered into Maple's computational environment** (more on this important point later). Any file opened in this way, whether it is a new blank file or a previously saved file, is called a **worksheet. Several worksheets can be opened at the same time.**

At this point, open the Maple application by double-clicking on the Maple icon. The purpose of this exercise is not to do any mathematics, but rather to experience, first-hand, some of the idiosyncrasies of Maple that can affect your work rather quickly.

At the blinking cursor, type (**2+3;**) after the Maple prompt (**>**) and hit the **enter key** (not the return key). **Do not forget the semicolon** (;). When the blinking cursor reappears, type (**b:=17;**) after the next Maple prompt and again hit the enter key. Be careful with the punctuation. Do not forget the semicolon (;), and use (:=), not just (=). Follow this by entering, in the same way, (**b:=8*A;**) after a Maple prompt. The assignment operator (:=) will be discussed in due course, but, for now, at least, we have enough on our computer screen so that we can discuss Maple's behavior.

In the current Maple work session, the letter ***b*** **now has a value of 8*A*.** To see this, type (**b;**) after the last Maple prompt and press the **enter key**. Maple will respond with "8*A*." Our computer screen should look as follows, except that the **left-hand square brackets** appearing on your computer screen are not shown below. **They will be omitted throughout this manual.**

****

```
> 2+3;
```

$$5$$

```
> b:=17;
```

$$b := 17$$

```
> b:=8*A;
```

$$b := 8\,A$$

```
> b;
```

$$8\,A$$

****

The letter A has no value—Maple treats it as an unknown (**possibly complex-valued**) number. The letter b initially had a value of 17, but that was simply erased and replaced by $8A$ after its last assignment. It is important to realize that Maple is **case sensitive**. The letters A and a are treated as absolutely different objects by Maple.

Pay attention to what we do next. Move the mouse pointer back to the input statement (> $b := 17$;), and **click** anywhere between the Maple prompt **>** and the closing (;), or immediately after the closing (;). The blinking cursor will move to this location. Press the enter key. Notice that the corresponding output statement disappears and then reappears, but nothing else below this point changes. What is the value of b now? To find out, move the mouse pointer down to the very last input statement (> b;), click the mouse, and again press the enter key. The last time we checked the value of b it turned out to be $8 * A$. Now we see that its value is 17, and our computer screen looks as follows.

****

```
> 2+3;
```

$$5$$

```
> b:=17;
```

$$b := 17$$

```
> b:=8*A;
```

$$b := 8\,A$$

```
> b;
```

$$17$$

****

This simple example gives us a way to view an important facet of Maple's behavior. As you can see, we can move back and forth on a Maple worksheet, entering input statements in one order and then another (with or without changes in the input statements). This is a common and effective way of working with Maple. When we go back and reactivate an input statement in this way, the corresponding output statement disappears and then reappears (with an appropriate change if the input statement was changed), but nothing below this point appears to change on the computer screen. Of course, Maple knows, internally, whether changes have been made, and subsequent work will reflect these changes.

In summary, **Maple pays no attention to the location of an input statement on a worksheet.** All that matters is the **sequence in which input statements are entered relative to the passing of time.**

This means that a Maple worksheet might not make sense if it is read from top to bottom, as a paper document would be read. It could make sense to us, if we remember the sequence of clicks and enterings, but it would not make sense to another reader. Naturally, if a Maple file is being prepared for another reader, it will eventually have to be presented in the order in which statements are entered before it is given to the reader. The usual cut-and-paste activity can be used to prepare a document in this way.

The Maple file we are creating (or any other Maple file) consists of three types of statements, which appear on the computer screen in a variety of font types and colors. (Font types and colors are all subject to change by the user. Maple's default settings are used in the following.)

i) **Input statements** contain the information that is entered into Maple's computational environment. Input statements appear in **red type** (on a color monitor), and each input statement must **begin with a "Maple prompt" (>)** and **end with a "semicolon" (;) or "colon" (:)** (more on this later).

ii) **Output statements** are the results computed by Maple by acting on input statements. They appear in **blue type**.

iii) **Text statements** are ignored by Maple. Their purpose is to supply the human reader with any other information that might be useful. Text statements **should be used extensively** in order to motivate, explain what one is doing, and otherwise make a document more readable, interesting, and enjoyable to a reader. Text statements appear in **black type**. It is worth mentioning that Maple is, in fact, a very good scientific word processor. We will explore some of Maple's tools for creating more involved mathematical text in the exercise sets, but not until later, after we have had time to become familiar with more critical issues.

A Maple prompt, the input statements (one or more) linked to this prompt, and the corresponding output statements (one or more) are all enclosed in a group by a square bracket on the left-hand side of the work sheet. This group is called an **execution group**.

It is possible to change from one statement type to another by using the Menu Bar, buttons at the top of the Maple window, or keyboard commands. Much of the time, however, Maple will conveniently choose the statement types automatically.

As we said above, Maple ignores "text" statements, which appear in black type. Maple also ignores everything between "sharp" symbols # ... #, regardless of the statement type used. They are no longer used much for this purpose, but this manual is not printed in color, so we will use sharp symbols to identify all "text" statements appearing in Maple work areas.

1.2 Opening Old Worksheets

When you work with Maple, you will want to save your work, quit, and then return either immediately or at some later date to do additional work on the same file. This can be a puzzling experience unless you understand how Maple files are loaded.

In order to experience this first-hand, let us save the simple Maple file we just created. Files are saved in the same way for all applications. Pull down the File

Menu and click on Save. You will be asked to name your file (just call it "junk"), and to select a place to put it.

Now quit Maple. Don't just close the worksheet, but actually quit the application as well. Then double-click on the file we just named "junk" and wait for Maple to open. Several other files could be opened at the same time. Without doing so, let us imagine, for the sake of argument, that we have, indeed, opened several files or worksheets. All of our previous work will show on these worksheets. One of these worksheets might say that $b := 17$, and another might say that $b := \sqrt{1-x^2}$. Which worksheet carries the current value of b, or are both values correct, each on its own worksheet?

The answer is that b has no current value. Test this by placing the blinking cursor somewhere on the last input statement ($> b;$) of the file we called "junk" (use the mouse or some other means), and press enter. Maple will respond with the answer "b," which simply means that b has no current value.

When we first open a Maple file, all of our previous work will show on the computer screen, but **nothing is entered into Maple's computational environment**. If there are values on the old worksheet that we still need in order to continue our work, **all of the appropriate input statements must be reactivated** by allowing the blinking cursor to pass through each of these input statements (in the appropriate order, of course) and pressing the enter key for each one.

Furthermore, **Maple offers us only one computational environment**. We can open several worksheets, but they must all coexist in the same computational environment. **There can only be one current value for** b (or any other name, for that matter). If we move the mouse pointer back to the input statement ($> b := 8 * A;$), click and enter, then the **current value of b is $8A$ on all of the worksheets, regardless of what appears on the worksheets.**

Actually, it should be said that on <u>some</u> computers, Maple can be configured so that each worksheet is an isolated computational environment. In this manual, however, Maple will never be configured in this way.

1.3 Frequent Saving

One of the first suggestions made in virtually any computer application is to **save your work frequently**—at least every 10 to 15 minutes. Such a practice will prevent much of your work from being lost if a problem is encountered. Earlier versions of Maple occasionally quit unexpectedly, or crashed for some unknown reason. If this happens only a few minutes after the saving of a file, then very little work will be lost. On the other hand, it is easy to get so absorbed in your work that an hour or more passes without saving it. If Maple then quits unexpectedly, or if you have a computer problem or a power failure, much of your work will be lost. The new version of Maple appears to be much more stable, but this practice of saving is always a good idea.

Maple can be configured to save automatically every n minutes, where n is a selected number. With Maple running, pull down the **Options** Menu and click on **AutoSav**.

1.4 Maple's Menus and Buttons

In this section, we pay some attention to the menu items, buttons, and tool bars that appear at the top of your computer screen under Maple. Most of them are straightforward enough that you will naturally become familiar with them as you work with Maple and experiment with these accessories.

Maple can be set up so that the **return** key and the **enter** key function in different ways, with the enter key used for initiating a Maple computation and the return key used simply as a carriage return key. This can be a real convenience. In the default version of Maple, the two keys are equivalent, unless an option to treat them separately is selected. Pull down the **Edit Menu** to **Preferences ...** to select that option if you wish.

Pull down the **View Menu** and select the top three items: **Tool Bar, Context Bar, Status Line**. Selecting the last of these can be very useful. It can be used to explain the function of all the other menu items and buttons. The Status Line is a narrow horizontal window running across the bottom of your computer screen. It shows the amount of memory that has been used, although this is not one of our immediate concerns. It also provides an **instantaneous explanation of the function of a menu item, or a button**. To see this, move the mouse pointer to any button, or to an item in any one of the menus. **Hold the mouse button down–don't let go**! Notice the explanation that appears in the Status Line. **Now slide the mouse pointer off the button or menu item without letting go of the mouse button**. This moves the point off the button or menu item without activating its use. By using this approach, the function of each button and menu item can be discovered. Here are a few buttons and menu items that will be of immediate use.

The Maple prompt (>) looks like a "greater than" symbol, but it is not, and Maple will not recognize that symbol as a prompt. Most of the time, Maple will insert a legitimate prompt in the appropriate place for you, but there are many times when you must take the initiative. Notice that **Maple prompt button [|>] appears in the Tool bar**. It will insert a new execution group, along with a prompt, after the group where the blinking cursor is located. Frequently, you will want to **place a Maple prompt ahead of** where the blinking cursor is located. Pull down the **Insert Menu**, select **Execution Group** and then select the subitem.

Maple can correct some input mistakes, typically unmatched parentheses and punctuation problems. Such mistakes occur frequently, even to experienced Maple users. When you experience mistakes of this sort, **press the button [√]** (the check mark is red) in the lower left-hand group of the tool bar. If Maple can, it will correct the mistake.

The **left and right sectioning buttons** in the tool bar (just to the right of the Maple prompt button [|>], and the **Section and Subsection items in the Insert Menu** can be used to create and remove sections. Sectioning Maple worksheets is not essential, but it can be a real convenience to **collapse or expand sections of a work sheet by clicking a button**. When a section is created, a button appears at the beginning of the material in the newly created section. The space just to the right of the section button is reserved for the section's title. The text entered in this region will automatically be black, large and boldface. The button appears as a [+] **button** when the section is collapsed, but as a [-] **button** when the section is opened up.

Press the [T] button in the Menu-Bar to create unprocessed (black) text, which is ignored by Maple's computational engine. The menu-bar changes considerably when this button is clicked. Text statements can be inserted at the end of an input statement after the closing semicolon. An entire execution group can be turned into a text group by clicking the [T] button while the cursor is within an empty input line. The Maple prompt (>) disappears and the execution group becomes a text group. Text statements should be used extensively to explain what you are doing and to make your work more readable to others. Unprocessed mathematical symbols can also be inserted within a text group by clicking the [Σ] button in the menu-bar. The mathematical typesetting group is placed at the location of the blinking cursor. This is, however, a more advanced topic we are not quite ready to discuss. Make a mental note to explore this feature later.

Most of the features appearing at the top of your computer screen under Maple will be left to your own discovery. **Use the information in the Status Line** to learn about Maple's accessories.

1.5 Maple's On-Line Help

This introductory chapter could hardly end without a word about the on-line help available from the menu-bar. The help apparatus is excellent! This manual can offer very little in the way of guidance, except to say ... **use it**! If you are puzzled about how Maple is behaving or about how to use a command, or if you are searching for a command to use in your work, help is only a click away. Maple's help is good enough, and easy enough to access, that, when help is needed, it might serve as your preferred first line of help, over and above human help. The main purpose of this short section is to entice you into exploring the help file and to offer just a few tips.

Pull down the **Help Menu** and spend a moment in a random exploration of some of the files. **Click on any one of the many "colored" words in any help file, and a help section on that word will open up**. These colored words are **hyperlinks**, and they provide a useful way of moving through help files. You can even create your own hyperlinks, although this is not an immediate or even secondary concern. If you wish to find out how, pull down the **Help Menu**, click on **Full Text Search...**, type *hyperlinks*, and follow your instincts.

Suppose you are in the process of using a Maple command and,for whatever reason, you are puzzled by Maple's response. Place the blinking cursor at the beginning of (or somewhere inside) the command name, then open up the help menu and click on **Help on Context**. The information appearing at the top of the page that opens up will explain the nature of the command in very precise language. Because of the careful language, it can sometimes be hard to read. Scroll down to the **bottom of the help page where the examples appear**. Frequently, just a glance at the examples will supply all the help that is needed.

If you are unsure of, how to spell a command name, or if you have only vague notions about unfamiliar tools that might possibly be useful in solving a problem, it is frequently beneficial to simply **browse through a list of command names** to see if such tools exist and to find their names and spellings. Such a list is available. It can be accessed in many ways. For example, pull down the **Help Menu** and click on **Topic Search....** Type **"Index"** in the topic field, click on index, and click the **[OK]** button to get a broad range of items. Click the hyperlink **functions** (it's colored) to get **a list of Maple functions.** All of the names in this long list are hyperlinks, so you can click on any one of them to get detailed help on that command. **It is easy to forget how to get to this list of commands, so try to remember the word "Index".**

Take some time, especially during your early Maple experiences, to browse through the help file. Many interesting features of Maple can be discovered by continuing this practice throughout your work.

The Beginning

There are many other general, introductory features of Maple that have not been discussed, but we have many chapters yet to cover, and we are anxious to begin our mathematical adventure. At this point, we can discard the file named "Junk" that we created earlier by dragging it to the trash. We are ready to experience the mathematical power of Maple.

Chapter 2

Basic Maple

This introductory material might be more readily understood if you open the Maple application (by double-clicking the Maple icon) and repeat the same calculations on your own worksheet.

Every Maple input statement must begin on a **new line** with a **Maple prompt** (>), as shown below. It has the appearance of a "greater than" symbol, but **Maple will not recognize the "greater than" keyboard character as a legitimate prompt**. The Maple input prompt will **usually appear automatically**, but it can be entered deliberately, either before or after the current location of the cursor. **Click the button in the tool bar that has the appearance [>] of the Maple prompt**, and a Maple prompt along with a new execution group will appear after the cursor. A message, which explains the function of this button, can be seen in the Status Line if you move the mouse pointer to the [>] button and hold the mouse button down. To insert a new execution group **before the cursor**, pull down the **Insert Menu** and do the obvious. Notice the keyboard equivalents.

With that introduction, we begin our first Maple computations. An input statement may contain several separate calculations, as we shall demonstrate shortly, but our first few calculations will be simpler in nature.

2.1 Maple as a Calculator

As you will see, Maple can be an ordinary calculator. Multiplication is denoted by (*) and exponentiation by (^) .

****

```
> 7*(8-5)^2;
```

$$63$$

```
> 2*7+6/2;
```

$$17$$

****

Every Maple input statement must end with a semicolon (;) or colon (:). **If (;) is used**, Maple will compute the input statement and **display the result. If (:) is used**, Maple will compute the input statement, but the **result will not be displayed.** To simplify our discussion, we will only use the semicolon to represent both of these closing punctuations, unless there is a specific reason for using a colon.

To initiate a particular Maple computation, **place the blinking cursor anywhere between its input prompt (>) and its closing (;), or place the cursor immediately after the closing (;)**, and then **press the enter key. Any input statement can be selected for computation, not necessarily the last one.** The choice depends on the **position of the blinking cursor. Pressing the return key rather than the enter key might or might not initiate a Maple computation.** It depends on whether this preference is selected. Pull down the **Edit Menu** and click on **Preferences...** to see or change the current preference. Using the Return Key as carriage return only can be a real convenience, but the choice is up to the user.

After the calculation is performed and the output is displayed (or omitted if a colon ends the statement), Maple automatically inserts another input prompt on a new line and sets the line up for another input statement.

Notice from the two examples above that, in Maple, the **multiplication symbol (*) must be inserted.** This is universally the case in all Maple statements. Maple **never assumes multiplication in the absence of the multiplication symbol.**

Maple follows the usual rules for the order of computations, and parentheses have the same meaning in Maple as they do on paper. In particular, negative numbers must be enclosed in parentheses in basically the same circumstances in Maple as on paper. With that said, we perform a few more calculations, paying particular attention to parentheses. Notice that, in the first calculation, Maple tries to show us the source of our mistake.

****

```
> 80*2^-3;
Error, '-' unexpected
> 80*2^(-3);
```

$$10$$

```
> 5/(4*6);
```

$$\frac{5}{24}$$

```
> 5/4/6;
```

$$\frac{5}{24}$$

```
> -5/4*6;
```

$$\frac{-15}{2}$$

```
> 5/(4/6);
```

$$\frac{15}{2}$$

****

Notice that the output is in the form of a fraction that has been reduced to lowest terms. The answers are not expressed as decimals. This is characteristic of a symbol processor, and it affects all Maple output in a very significant way. **Maple will always be exact unless you allow it to approximate answers with a decimal.** In the following, this is demonstrated in several ways.

The mathematical constant π **is spelled Pi.** The letter **P must be capitalized.** When π is entered in the form "pi" with a small-case letter "p," it appears as the Greek letter π, but it represents only a variable or arbitrary constant. **It has the same look as the constant $\pi = 3.14159\ldots$ but it carries no assigned value.**

Also notice that the **arguments of the trigonometric functions** are assumed to be in **radian measure.**

****

```
> 5/sqrt(2);2/sqrt(2);
```

$$\frac{5}{2}\sqrt{2}$$

$$\sqrt{2}$$

```
> cos(Pi/4);cos(Pi/8); cos(3);
```

$$\frac{1}{2}\sqrt{2}$$

$$\frac{1}{2}\sqrt{2+\sqrt{2}}$$

$$\cos(3)$$

****

As you can see, Maple will carry out as much of the computation as it can under the constraint that the output must represent an exact answer. The value of $\cos(\pi/4)$

is irrational, but algebraic, and so its value is represented algebraically. The value of cos(3) is transcendental. It is not only irrational, but nonalgebraic as well, and so Maple has no way of expressing its value exactly except as cos(3). Initially, this realization may be disconcerting, but eventually this will be realized as a powerful feature of Maple. Decimal numbers and approximations play an important role in mathematics, and, as we will see, in Maple as well, but mathematics is also a study of symbols and of the exact relationships among its symbols.

Also notice, in this last set of calculations, that Maple input prompts and closing delimiters do not have to be paired with one another in a simple way. **One input prompt** can serve as a prompt for **several closing (;)**. When the blinking cursor is placed in or at the end of an input statement, and the enter key is pressed, Maple starts at the input prompt (>), and looks for the first (;). This causes Maple to perform a calculation and display (or omit) the result. Maple then continues to parse the input statement, looking for the next (;), which causes Maple to make the next calculation, and so on, until the computation corresponding to the last (;) is performed or another input prompt (>) is encountered. The computational process stops at the next input prompt (>) if one already exists on the worksheet.

We will return in a moment to discuss the mechanism for triggering decimal approximations,but first let us take a mathematical break and discuss some of the more basic software issues that were raised in the preliminary chapter.

At some point, you will want to save your work and quit Maple. Saving a Maple file is the same as saving any other file in the Macintosh operating system. If you need help, consult the manual that came with your computer. **When you save a file, be sure to give it a name that is peculiar to just you, so that your file does not have the same name as someone else's file**. Even if you do not plan to quit, **your work should be saved frequently**, so that you have a copy if something should go wrong with the software or hardware. This is a common practice that is used with all software **to avoid losing work**. After you have done some work, save your file, then quit Maple by pulling down the **File Menu** and clicking on **Quit**.

If you have not yet read Chapter 1 carefully, it would be a good idea to study that chapter before you resume your work with Maple. Working with old Maple files can be a confusing experience unless you understand how Maple files are activated. Pay particular attention to Section 1.2. When you are ready to resume your work, open your previously saved file by double-clicking on its icon. The next topic is decimal approximations.

There are two important ways to get Maple to approximate answers in decimal form. Most of the time we will use a command-based approach, which we will discuss shortly. The other way is to have a decimal number appear somewhere in the input statement. We discuss this approach first. Maple makes the reasonable assumption that even one decimal number in an expression compromises the integrity of a pure symbolic answer, and so there is no significance represented by the answer other than as a decimal approximation. Compare the answers above with the following.

****

```
> 5/sqrt(2.0);2/sqrt(2.);cos(3.);
```

$$3.535533907$$

$$1.414213563$$

$$-.9899924966$$

****

Maple will compute with **as many digits of accuracy** as you wish. The default setting is 10. (Some versions of Maple are limited to at most 100 digits of accuracy.) Maple looks for the value of an internal variable named **"Digits"** (Notice the capital D) and then computes and displays decimal answers with that many digits of accuracy. The value of the variable Digits can be changed with an input statement by using the **assignment operator** (:=). This operator will be discussed in greater detail shortly, and so, for now, we will simply use it without comment.

****

```
> Digits:=30;
```

$$Digits := 30$$

```
> 5/sqrt(2.0);cos(3.);
```

$$3.53553390593273762200422181053$$

$$-.989992496600445457271572794731$$

```
> Digits:=10:
```

****

Rather than slow our calculations down with needless accuracy, we finished the above work session by returning "Digits" to its default setting of 10 digits. Notice that here is a place where output is unneeded, and so we used a (:).

As we mentioned, Maple also has a command-based approach for initiating decimal approximations. In the Maple language, the symbol **f...** (**f** for floating decimal point) is attached to some command names to change the commands from those that perform exact computations to those that perform decimal approximations. The commands **solve( )** and **eval( )** mean what you would expect them to mean, whereas **fsolve( )** causes Maple to approximate the solution to an equation with a decimal, and **evalf( )** causes Maple to approximate the value of an expression by a decimal. Whether the **f** is placed before or after the command name seems to depend primarily on the appearance of the name.

****

```
> evalf(sqrt(2));
```

1.414213562

```
> cos(3);
```

cos(3)

```
> evalf(%);
```

−.9899924966

****

Notice the role played by the **percent sign (%)** in the last input statement. The last Maple output value is denoted by (%), the second to last output value is denoted by (%%), and the third to last output is denoted by (%%%).

The evalf( ) command in the form **evalf(,n)**, where **n is an integer**, can be used to **control the number of digits of accuracy in a single computation without changing the value of the Maple name Digits**. With the value of Digits set at its default setting 10, observe the following.

****

```
> evalf(sqrt(2),4);evalf(Pi,15);evalf(Pi);
```

1.414

3.14159265358979

3.141592654

****

We continue our discussion about Maple's role as a calculator by displaying a few calculations. Some of them might surprise you. Remember that Maple will always be exact unless you allow it to approximate.

Also notice below that **Maple ignores anything between sharp (#) symbols**. This is one of the old ways Maple had to display comments on a work sheet. They are usually not used any longer, but colors cannot be used in this manual to set off text from input statements in the space reserved for Maple, and so we will continue to use sharps for this purpose. For the most part, this practice will be reserved for short comments; occasionally, they will have to be used for lengthy comments as well. The meaning of the separator lines that begin and end a Maple session was given in the Preface. If it is important for the same work session to continue on either side of comment, then we will have no alternative to enclosing the comment in sharp symbols, regardless of how long it is.

****

```
> 2^50;
```

$$1125899906842624$$

```
> 80!;
```

71569457046263802294811533723186532165584657342365752\
577109445058227039255480148842668944867280814080000000\
0000000000000

These are not decimal approximations. We asked for exact calculations, not decimals. The backslash at the end of the first line of output above indicates a break, with the display continuing on the next line.

```
> abs(-7); abs(sin(4));
```

$$7$$

$$-\sin(4)$$

```
> sqrt(-2);
```

$$I\sqrt{2}$$

Maple will compute and display complex-valued answers just as it does real-valued answers. The **symbol "I"**(expressed as a capital letter only) is reserved by Maple to represent the **imaginary number** $\sqrt{-1}$, and it should not be used for any other purpose.#

```
> (17+2*5)^(1/3)/(32^(1/5)+1)^2;
```

$$\frac{27^{1/3}}{(32^{1/5}+1)^2}$$

```
> simplify(%);
```

#Don't use evalf(%) here. It would give a decimal approximation.#

$$\frac{1}{3}$$

****

The command **simplify( )** can be used in a variety of situations. To find out more about it, **consult Maple's on-line Help File**. If the cursor is at the beginning or somewhere inside the the word "simplify," just pull down the Help Menu and click on **Help on Context**. Try this now, and explore the consequences.

We end this section with a calculation demonstrating really astonishing behavior on Maple's part. It turns out that Maple has good reason for its behavior, and this will lead to a **new command** to overcome the difficulty.

****

```
> a:=(-27)^(1/3);
```

$$a := (-27)^{1/3}$$

```
> a:=simplify(a);
```

$$a := \frac{3}{2} + \frac{3}{2} I \sqrt{3}$$

#Instead of the obvious real-valued answer, $a = -3$, Maple has given us a complex number. Is this answer even correct? In response to this question, we ask Maple to compute a^3. If the above answer is correct, then we should get $a^3 = -27$. The command **expand( )** is introduced for the first time to finish up this calculation.#

```
> a^3;
```

$$(\frac{3}{2} + \frac{3}{2} I \sqrt{3})^3$$

```
> expand(%);
```

$$-27$$

****

Evidently, the above answer is correct, but it is certainly a disappointment. As long as we plan to live in a real-valued world, we should still insist that -3 is the only acceptable answer for the cube root of -27. Maple, on the other hand, is quite inflexible about this matter and will refuse to grant us our request. It turns out that Maple has a good reason for its behavior, although it is one rather hard to explain why until after the completion of a more advanced course in complex variables.

Maple treats all **roots of negative numbers** in a similar fashion. **They will always be complex valued.** An even root of a negative number *should* be complex valued, but surely we want an odd root of a negative number to be real (and negative), even though there are complex-valued roots as well. One way to get Maple to cooperate is to make a practice of using $-(-x)^{1/n}$ instead of $x^{1/n}$ in our Maple calculations whenever x is negative and n is an odd integer. This is not altogether a bad feature of Maple. After all, we know that there are quite a few symbolic complications to working with roots of negative numbers. The usual rules of manipulating exponents frequently fail for negative bases, even with pencil-and-paper techniques. Maple's behavior here should serve as an warning to us that we need to exert more caution in dealing with roots of negative numbers. As an interesting footnote, earlier versions of Maple (before Release 3) did not have this difficulty in computing roots of negative numbers. However, the formulas it used to create such nice clean real-valued roots of negative numbers also caused major mistakes to be made in a broad range of other mathematical computations. All things considered, this change in subsequent versions of Maple has been a welcome improvement.

Fortunately, Maple has a command for **extracting the real root of a real number**, if it exists. **The command surd(x,n) computes the real-valued nth root of a real number x, if it exists.** If $n \geq 4$ is an even integer and x is a negative real number, it returns $\sqrt[n]{|x|}\,(-1)^{(n-1)/n}$. To explain the value of the term $(-1)^{(n-1)/n}$ would take us beyond the scope of this book. Suffice it to say that it is a **complex-valued expression** that has no place in a real-valued setting.

If you are younger than 100 years old, you might be puzzled by the choice of words used for this command. One English dictionary defines the word as *"an irrational number like $\sqrt{2}$."* Is there a historical connection between the words *"surd"* and *absurd*? We leave the question unanswered—it will have served its purpose, if this command name is remembered in the future.

****

```
> surd(-8,3),surd(-7,-3),surd(-243,-5);
```

$$-2,\ -\frac{1}{7}\,7^{2/3},\ \frac{-1}{3}$$

```
> evalf(surd(-7,-3));
```

$$-.5227579587$$

```
> surd(-25,2);
```

$$5\,I$$

```
> surd(8,1/2);

Error, (in surd) numeric 2nd argument must be an integer
```

****

For simple calculations such as those appearing above, the surd( ) command can easily be avoided, but it will be an essential command as soon as we move beyond simple numerical evaluation.

2.2 Assigned and Unassigned Names

A **name** or **string** in Maple is any letter followed by zero or more letters, digits, and underscores (lower- and upper-case letters are distinct). Spaces are not allowed; an underscore should be used instead. The maximum length of a name is 499 characters. If names are unassigned, they are simply treated as variables or as unknown real (or complex) numbers.

To assign a value to a name, we use the **assignment operator (:=)**. Be sure to use (:=) and not just (=). Absolutely **anything in Maple can be represented by a name.** Using names is a real convenience, because complicated statements can be referred to again by simply entering their names. Equally important, carefully chosen names carry (human) meaning, and this can help to motivate what we do, making our work easier to create and easier to explain. As a consequence, **names should be used frequently.**

There is more to the assignment operator (:=) than might be evident at first glance. Suppose a and b are assigned or unassigned names (or expressions involving other assigned or unassigned names). When Maple reads the assignment $a := b$, it fully evaluates the right-hand side and leaves the left-hand side unevaluated. The fully evaluated right-hand side is then assigned to the unevaluated name a, replacing any previous value of a, if there was one, in the process. In particular, **a name does not have to be unassigned before a new value is assigned to it.** The name **a can even include the name a**, although this **must be done cautiously.** If **a already has a value**, then an assignment such as

```
> a:=a+1;
```

is quite acceptable. If the previous value of a was 10, then the right-hand side evaluates to 11, and this becomes the new value of a. **If a is unassigned**, however, then, in its attempt to evaluate the right-hand side fully, Maple would look for, and find, a value for a back in the same statement. This forces Maple into **an infinite loop**, something to be avoided.

****

```
> (p12q + 2*x+7*george)^2;
```

$$(p12q + 2\,x + 7\,george)^2$$

```
> poly:=x^2+5*x-9;
```

$$poly := x^2 + 5\,x - 9$$

```
> a:=(6+8)^2/4;
```

$$a := 49$$

```
> eq:=x=4;
```

$$eq := x = 4$$

#To see the significance of these assignments, notice the following.#

```
> eq;
```

$$x = 4$$

```
> sqrt(a*poly)/25;
```

$$\frac{7}{25}\sqrt{x^2 + 5x - 9}$$

```
> x;
```

$$x$$

#Here is an interesting point. Notice that ***x* is unassigned**. It does not have the value 4. The equation $x = 4$ and the assignment $x := 4$ are treated very differently by Maple.

As you proceed with a Maple work session, you are likely to encounter situations where you **attempt to use a letter or name as a variable (an unassigned name) only to discover that had you assigned some value to that letter or name in some previous problem.** The most direct way to **return a name to its unassigned status** is to enter the input statement **name:='name'**. For example, the name poly was assigned to a quadratic above. To return this to its unassigned status, enter the following.#

```
> poly:='poly';
```

$$poly := poly$$

```
> poly;
```

$$poly$$

****

Enclosing a name in single quotes (') prevents evaluation of the name. Besides their use in unassigning names, single quotes can be used in a variety of situations to prevent or postpone evaluation of a name.

Quitting Maple and then opening the same worksheet again has the effect of returning all user-defined names to unassigned status.

Once a file is opened again, its contents will be visible on the screen, but nothing will be entered into Maple's computational environment.

This **important issue** was raised in Chapter 1. It could help to read the appropriate material in Chapter 1 carefully at this time.

In order to focus your attention on this emptiness, it sometimes helps to pull down the **Edit Menu** and click on the command **Remove Output Areas.** While this step is optional, it is a useful way to keep track of what is entered into Maple during the current (new) work session. Whether output areas are removed or not, to enter the input statements that are needed again, place the blinking cursor inside each input statement and press the enter key.

Quitting Maple and then opening the same worksheet has the added advantage of **recovering all of the spent memory** set aside for computations. This practice of quitting and then restarting is the standard way of dealing with memory problems on worksheets involving memory-intensive computations.

The command **restart** has the same effect as quitting Maple and then opening up the same worksheet, except that **spent memory is not recovered.** To use this command, simply enter

```
> restart;
```

in a Maple work session. It is a useful way of returning all user-defined names to unassigned status. In Maple 7, there is a **Restart button** in the upper right-hand corner of the Menu Bar.

2.3 Functions and Expressions

Students of mathematics are reminded frequently, when discussing functions, to appreciate the difference between the symbol f, which denotes the "whole idea" of the correspondence, and the symbol $f(x)$ that denotes the value of the function f at x. Maple has the same understanding of the symbols f and $f(x)$, but it does not have the human ability to translate one into the other when a symbolic mistake is made. **If these symbols are misused, Maple will give answers that are inappropriate and can look strange.**

The easiest way to **create a <u>function</u> is to use the symbol (->)** (a hyphen followed by a greater-than symbol) in the manner shown in the following example.

<u>************************************</u>

```
> f:=x->5*x^2+1;
```

$$f := x \to 5\,x^2 + 1$$

#The letter x is said to be a **local variable.** If a letter, such as x, is used in this way ($x-$ >), as an argument of a function, then x **is treated (locally) as an unassigned letter, regardless of whether x had, or did not have an assigned value previous to this input statement.** If x had a previously assigned value, it will **continue to have that same value after** this input statement.#

```
> f(x);f(t);f(x^2);f(p+q);f(3);(f(1))^2;
```

$$5\,x^2+1$$

$$5\,t^2+1$$

$$5\,x^4+1$$

$$5\,(p+q)^2+1$$

$$46$$

$$36$$

#Another way to create a <u>function</u> is to use the procedure command **proc( )**.#

```
>  g:=proc(x) x^2 end;
```

$$g := \mathbf{proc}(x)\, x^2\, \mathbf{end}$$

```
>  g(x);g(cos(y));
```

$$x^2$$

$$\cos(y)^2$$

#A **procedure must terminate with the word "end"** as indicated in the definition of the function g above (and of course a semicolon). The proc( ) command is one of the **most versatile commands in the Maple language.** It can be used in a large number of situations other than the creation of real-valued functions. The entries between the opening and closing symbols in proc( )......end; can consist of a large sequence of input statements separated by semicolons. We will use this command in the next chapter when we discuss *Simpson's Rule.*

With more convenient methods for defining <u>functions</u> available at our finger tips, the proc( ) command could be of limited use for the creation of most normal real-valued functions. Its use is usually reserved for more involved statements. On the other hand, it is versatile enough that we might choose to use it when the name of some other Maple command is forgotten. We will discuss this matter further in a moment. #

An expression is another Maple data type, which closely resembles a function but is **very different**. Notice carefully the differences between the definition of f above and of h in the next input statement.#

```
> h:=x^2+5*x-8;
```

$$h := x^2 + 5\,x - 8$$

#This is a perfectly legitimate (and frequently used) way to name an **expression**, but it **does not define a function.**#

```
> h(x);
```

$$\mathrm{x}(x)^2 + 5\,\mathrm{x}(x) - 8$$

#As you can see, Maple doesn't know what you mean by $h(x)$, it only knows what you mean by h. Maple has given this a valid interpretation, but it is not what we want it to mean and not worth discussing here. Contrast this with the following assignment.#

```
> p(x):=x^2+5*x-8; p(3);
```

$$\mathrm{p}(x) := x^2 + 5\,x - 8$$

$$\mathrm{p}(3)$$

#Maple allow us to make this assignment, but it is **a fatal mistake** to do so. Maple understands what we mean by $p(x)$ (it's just a name), but notice that $p(3)$ is just an unassigned name without a value.#

****

It would help to establish some terminology. By an **expression (expr)**, we mean any Maple structure that evaluates to a real or complex number if all unassigned names are given real or complex values. By an **equation (eqn)**, we mean a statement of equality, involving the symbol (=), where the left- and right-hand sides are expressions. By a **function (fcn)**¡ we mean a function in its *classic mathematical sense only*. Actually, regardless of the techniques used to define it, **a function is always ultimately a Maple procedure.** With this understanding, expressions, equations, and functions are all distinctly different. The names f and g above represent functions, and $f(x)$, $g(x)$, and p represent expressions.

Expressions are undoubtedly the most common Maple objects. The easiest way to **change an expression into a function** is to use the **unapply command.**

****

```
> p:=x^2+5*x-8;
```

$$p := x^2 + 5\,x - 8$$

```
> p:=unapply(%,x);
```

$$p := x \to x^2 + 5\,x - 8$$

****

Initially p was an expression, but after the unapply command, it became a function. The word "unapply" might seem to be a poor choice of words for this command, but it is not. If f is a **function**, we **apply** f to x to get the **expression** $f(x)$, and so it is reasonable language to say that if *expr* is an **expression**, we **unapply** *expr* to get the **function itself.**

If f and g are Maple names of functions, then $f + g$, $f - g$, $f * g$, f/g, **and** f^n (where n is a real number) all represent the expected functions in Maple. The **composition of f by g is denoted by $f@g$.**

To evaluate these functions at a point x, the function symbols must be enclosed in parentheses. Surely this is to be expected. To evaluate $f + g$, or $f * g$ at x, it would not make notational sense to write $f + g(x)$ or $f * g(x)$. To **evaluate these functions** in Maple or in traditional mathematics, we would **use parentheses** and write instead

$$(f \pm g)(x) = f(x) \pm g(x), (f * g)(x) = f(x)g(x), (f/g)(x) = f(x)/g(x),$$
$$(f\hat{\ }n)(x) = (f(x))^n, (f@g)(x) = f(g(x))$$

In ordinary mathematical text, we are accustomed to seeing an expression like $\sin^2(x)$. It may be tempting to enter this in a Maple expression as sin^ 2(x), but this will not work. In view of the above, it is clear that Maple understands the meaning of the function sin^ 2, but, in order to evaluate this function at x, we must enter the expression as (sin^ 2)(x). Of course, writing sin(x)^2 will also work, as will (sin(x))^2, although in the latter case the extra parentheses are unnecessary.

The next Maple work session contains a sampling of these ideas. Notice, in particular, that $f@g$, and $g@f$ are different functions.

****

```
> f:=x->7*x^2+3*x-9; g:=x->sin(sqrt(x));
```

$$f := x \to 7\,x^2 + 3\,x - 9$$

$$g := x \to \sin(\sqrt{x})$$

```
> h:=f+g;h(x);(f*g)(x);
```

$$h := f + g$$

$$7\,x^2 + 3\,x - 9 + \sin(\sqrt{x})$$

$$(7\,x^2+3\,x-9)\sin(\sqrt{x})$$

```
> (f@g)(x);
```

$$7\sin(\sqrt{x})^2+3\sin(\sqrt{x})-9$$

```
> (g@f)(x);
```

$$\sin(\sqrt{7\,x^2+3\,x-9})$$

```
> (f/g)(a*t+b);
```

$$\frac{7\,(a\,t+b)^2+3\,a\,t+3\,b-9}{\sin(\sqrt{a\,t+b})}$$

```
> h:=(f*g)^2;
```

$$h := f^2\,g^2$$

```
> h(2);
```

$$625\sin(\sqrt{2})^2$$

```
> f^2(x);(f)^2(x);(f^2)(x);f(x)^2;
```

$$f^2$$

$$f^2$$

$$(7\,x^2+3\,x-9)^2$$

$$(7\,x^2+3\,x-9)^2$$

```
> sin^2(x);(sin)^2(x);(sin^2)(x);sin(x)^2;
```
#The first two expressions are not well defined.#

$$\sin^2$$

$$\sin^2$$

$$\sin(x)^2$$

$$\sin(x)^2$$

#Does Maple understand the first two of the above sine expressions? Not according to our preliminary discussion! Observe the following calculations.#

```
> sin^2(Pi/4);(sin)^2(Pi/4);(sin^2)(Pi/4);sin(Pi/4)^2;
```

$$\sin^2$$

$$\sin^2$$

$$\frac{1}{2}$$

$$\frac{1}{2}$$

Finally, we introduce the new Maple command **piecewise( )**, which is worth remembering. We can readily see what the three arguments represent, if we look at the output for $f(x)$ following the input definition.

```
> f:=x->piecewise(x<2,x^2,1-3*x);
```

$$f := x \to \text{piecewise}(x < 2,\, x^2,\, 1 - 3\,x)$$

```
> f(x);
```

$$\begin{cases} x^2 & x < 2 \\ 1 - 3\,x & \textit{otherwise} \end{cases}$$

```
> f(t);
```

$$\begin{cases} t^2 & t < 2 \\ 1 - 3\,t & \textit{otherwise} \end{cases}$$

$$\begin{cases} x^4 & x^2 < 2 \\ 1 - 3\,x^2 & \textit{otherwise} \end{cases}$$

```
> f(x^2);
```

$$\begin{cases} x^4 & x^2 < 2 \\ 1 - 3\,x^2 & \textit{otherwise} \end{cases}$$

The next evaluation is interesting.#

```
> f(cos(x));
```

$$\begin{cases} \cos(x)^2 & \cos(x) < 2 \\ 1 - 3\cos(x) & \textit{otherwise} \end{cases}$$

#What has happened here? Doesn't Maple know that $\cos(x) < 2$? Actually, this brings up an issue that can haunt us from time to time throughout our Maple activity. Maple assumes that unassigned variables may be **complex-valued**, and it turns out, strangely enough, that $\cos(x) < 2$ is not always true if x is allowed to be complex-valued.

From time to time we will have to use the **new Maple command assume( )** in order to impose a variety of assumptions on variables.#

```
> assume(x,real);
> f(cos(x));
```

$$\cos(x)^2$$

 # The assumptions made with the assume( ) command remain in force for the duration of the work session, unless the associated names are unassigned or there is a call to restart. In this sense, the assumptions, like assignment itself, are permanent. Additionally, to remind us of the assumptions placed on a name, Maple can attach a trailing tilde to any such name. In the above application of the assume( ) command, the name x could subsequently appear in the form x˜. If this is found to be annoying, the **tilde can be removed.** Pull down the **Options** Menu, select **Assumed Variables**, and make the obvious selection.

In Maple 7, there is another interesting way to make assumptions. Let us remove the assumptions on x and try again to evaluate $f(\cos(x))$. #

```
> x:='x':
```

After an unsuccessful attempt at evaluating the expression, we could write the following interesting input statement:#

```
> f(cos(x)) assuming x::real;
```

$$\cos(x)^2$$

 # This assumption is not permanent. It is in effect for only this one computation. This can be quite a convenience. We can try a range of assumptions until we find one that works. Because the assumptions are not permanent, we will not have to undo then before we move on to other assumptions. **Notice the punctuation.** To save space, we will not comment further, but this very useful command should be looked up in the **Help** Menu. Sadly, the "assuming" facility is **not available in Maple 6.**

The piecewise( ) command can be used to define a function that uses **several different formulas in several different intervals.** The statement

>f:=piecewise($cond_1, result_1, cond_2, result_2, result_3$);

for example, defines an expression f which has the value $result_1$ if $cond_1$ is true. **If $cond_1$ is false and $cond_2$ is true,** f has the value $result_2$. Otherwise, f has the value $result_3$. Notice that the middle condition assumes the first condition is false.

In the next example, **notice how the condition $x \leq -2$** as opposed to $x < -2$ **is entered into a Maple imput statement.**#

```
> g:=x->piecewise(x<=-2,-x,x<2,x^2,3*x);
```

$$g := x \to \text{piecewise}(x \le -2,\ -x,\ x < 2,\ x^2,\ 3\,x)$$

```
> g(x);
```

$$\begin{cases} -x & x \le -2 \\ x^2 & x < 2 \\ 3\,x & \textit{otherwise} \end{cases}$$

If the conditions used to evaluate g are not clear enough from the output statement for $g(x)$ above, the next evaluations can be used to remove any lingering doubt.#

```
> g(-2.1);g(-2);g(-1.9999);g(0.8);g(2);g(2.1);
```

$$2.1$$

$$2$$

$$3.99960001$$

$$.64$$

$$6$$

$$6.3$$

****

2.4 Algebra

Maple has a large supply of commands that can be used to manipulate algebraic expressions and equations. There are too many commands to cover all of them in this chapter, and so only a sample of basic commands will be introduced. Others will be discussed in future chapters. Use your experience in algebra to predict the nature of other commands and to guess their spelling. **Explore the Help File**. Pull down the **Help Menu**, click on **Topic Search...**, type **"Index"** in the search field, and click on the hyperlink functions. Exhibited below are a few brief examples using the commands **factor( ), expand( ), subs( ), normal( ), numer( ), denom( ), solve( ), fsolve( ).**

****

```
> ex1:=(x-1)*(x+4)^3*(x^2+8);
```

$$ex1 := (x-1)\,(x+4)^3\,(x^2+8)$$

```
> ex2:=expand(ex1);
```

$$ex2 := x^6 + 44\,x^4 + 11\,x^5 + 104\,x^3 + 224\,x^2 + 128\,x - 512$$

```
> ex3:=factor(ex2);
```

$$ex3 := (x-1)\,(x+4)^3\,(x^2+8)$$

```
> ex4:=subs(x=cos(t),ex2);
```

$$ex4 := \cos(t)^6 + 44\cos(t)^4 + 11\cos(t)^5 + 104\cos(t)^3 + 224\cos(t)^2 + 128\cos(t) - 512$$

****

There is another way to make this last substitution. By **making the assignment $x := \cos(t)$**, the expression **ex2 would automatically take on the value ex4.** No other command would be needed, and no assignment beyond ex2 would be necessary. The assignment for x would be **permanent**, however, until x is reassigned. The **importance of the subs( )** command is that the substitution is carried out **without making the assignment $x := \cos(t)$**. The variable x remains unassigned.

In the next example, we use the substitution command to make more than one substitution. You will see that this must be done thoughtfully.

****

```
> ex5:=(2*x*a+3*y^2*b)^3;
```

$$ex5 := (2\,x\,a + 3\,y^2\,b)^3$$

```
> ex6:=subs(x=y,y=2,ex5);
```

$$ex6 := (4\,a + 12\,b)^3$$

#It is important to understand the **order of substitution** in the subs( ) command, so let us explain how Maple performed the last calculation. Maple **first** replaced each x in ex5 by y. When this was **finished**, it **then replaced** each y in ex5 by 2. The substitutions were done sequentially, leftmost first. Until this order of substitution is fully appreciated, it is easy to make substitution mistakes. Contrast this with the following substitution.#

```
> ex7:=subs({x=y,y=2},ex5);
```

$$ex7 := (2\,y\,a + 12\,b)^3$$

#With **set brackets** around both substitutions, Maple does **all of the substitutions simultaneously.**#

```
> ex8:=5+3*x+(2*x^2*x+1)/(x^2-4);
```

$$ex8 := 5 + 3\,x + \frac{2\,x^3 + 1}{x^2 - 4}$$

```
> ex9:=normal(ex8);
```

$$ex9 := \frac{5\,x^2 - 19 + 5\,x^3 - 12\,x}{x^2 - 4}$$

```
> top:=numer(ex9); bottom:=denom(ex9);
```

$$top := 5\,x^2 - 19 + 5\,x^3 - 12\,x$$

$$bottom := x^2 - 4$$

```
> eq:=x^3-4*x^2-25*x+28=0;
```

$$eq := x^3 - 4\,x^2 - 25\,x + 28 = 0$$

```
> solve(eq,x);
```

$$1,\ -4,\ 7$$

```
> s:=%;
```

$$s := 1,\ -4,\ 7$$

#The output from this solve command (which we then named s) is an important and very common type of Maple structure known as a **sequence.** A **sequence is any open-ended listing** (finite or infinite) of **Maple objects separated by commas.** By open-ended we mean that the sequence is not enclosed in parentheses, brackets, or so on. This is an important distinction. Enclosures would create other Maple structures, as we shall see. **Maple recognizes the order of the entries in a sequence,** and the **individual entries of a sequence can be accessed by using square brackets** as follows. (We gave the sequence a name so we could do this conveniently.)#

```
> s[3];
```

$$7$$

```
> a:=(s[2])^2;
```

$$a := 16$$

****

Recall that the letter "f" is attached to some commands to prompt a decimal approximation. As we pointed out earlier in this chapter, Maple will always compute exact answers unless it is allowed to approximate with decimals. Solutions to equations can frequently be very complicated to express exactly, and in these situations, a simple decimal answer is often appreciated. Try solving the next equation exactly, and you will surely agree. Only the approximate decimal solution is entered here.

****

```
> fsolve(x^3+7*x^2-3*x-6=0,x);
```

$$-7.298407943, -.7696860788, 1.068094022$$

#The last equation could have been solved exactly, but the solution would have looked awkward. Now consider the consequence of trying to solve the next equation exactly.#

```
> solve(x-cos(x)=0,x);
```

$$\mathrm{RootOf}(\_Z - \cos(\_Z))$$

#Surely there is a solution to this equation for some x between 0 and $\pi/2$. A simple plot would reveal this fact, but plotting is not discussed until the next section. There is no need for a plot anyway. The left-hand side, $x - \cos(x)$, of the equation is a continuous function that clearly is negative (below the x-axis) at $x = 0$ and clearly is positive (above the x-axis) at $x = \pi/2$. Consequently, it must be 0 (cross the x-axis) somewhere between $x = 0$ and $x = \pi/2$.

The reason for the strange output in the solve command is that we asked for an **exact solution**, and, although it exists theoretically, neither Maple nor mathematics itself has a means of expressing the answer exactly. **Maple will simply not give an answer if it cannot do exactly what it is asked to do.** Here is another reason for allowing Maple to make decimal approximations. It is worth mentioning that "most equations" are complicated enough that they cannot be solved exactly. This is not Maple's fault, but the nature of mathematics itself. As a consequence, the fsolve command plays a very important role in Maple computations.

Take a second look at Maple's response to the above solve( ) command. Strange as it may look, there is a reason for Maple's notation. Whenever Maple needs to take the initiative and introduce its own unassigned names, it uses names like: $\_A$, $\_B, \ldots, \_Z$. Maple feels, and rightly so, that no human would use names like that, and so it will not be introducing names that are already owned by us. Now that makes sense!#

```
> fsolve(x-cos(x)=0,x);
```

$$.7390851332$$

#The solve command can also be used to solve systems of equations.#

```
> solve({3*x+5*y=8,7*x+4*y=2},{x,y});
```

$$\{x = \frac{-22}{23}, y = \frac{50}{23}\}$$

****

Set brackets { } play a similar role in many Maple commands. (We are referring here to the set brackets in input statements, not output statements.) In the case above, the first argument of the solve command is the place for an equation, and the second argument is the place for an unknown. When we have **two or more objects to place in the same argument** of a command, the usual Maple policy is to **group them together with set brackets.** This is an **important issue,** which plays a role in many input statements.

2.5 Plotting Graphs

Certainly one of the distinct advantages of using any Computer Algebra System is the relative ease of graphing functions. We will have ample occasion to take advantage of a large number of exciting graphing tools found in Maple. At this time, we merely introduce the most basic graphing command, **plot( )**, by graphing below first one expression, and then several functions. In the second plot, notice how set brackets { } play a role similar to the role they played in the last solve command.

After the plots appear in the worksheet, move the mouse pointer to one of the plots and click the mouse. A rectangle will appear around the plot, and the tool and menu-bar at the top of the computer screen will change. Later, when we do 3-dimensional plots, the tool and menu-bar will change dramatically. While in this mode, click some of the buttons and notice their consequence on the plot. **Pull down the View Menu, and activate the Status Line.** We explained on page 6 how this can be used to explain the various functions of buttons and menu items. In particular, **notice the [1:1] button**, which can be used to provide a view of the graph with the **same scale on both axes.**

****

```
> plot(x^2+x*sin(5*x),x=-3..3);
```

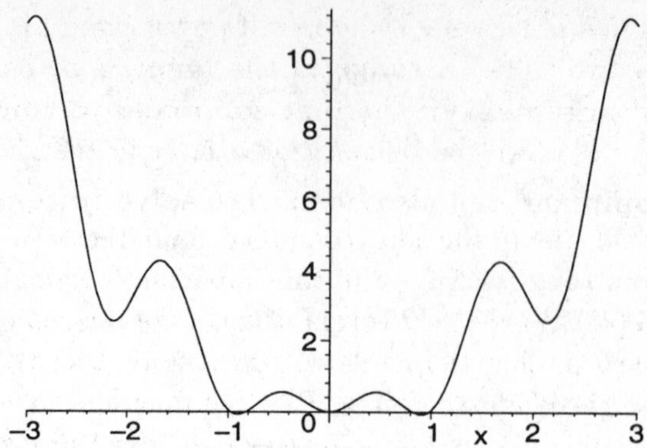


```
> f:=x->x^2;g:=x->80+8*x;h:=x->80-8*x;
```

$$f := x \to x^2$$

$$g := x \to 80 + 8\,x$$

$$h := x \to 80 - 8\,x$$

```
> plot({f(x),g(x),h(x)},x=-10..10);
```

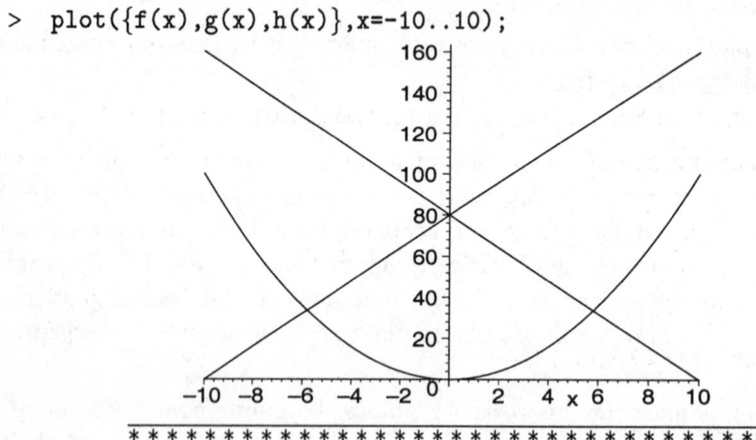

****

The second argument in each of the plot statements above is an example of another Maple structure used in a wide variety of situations. **A range in Maple is an object of the form, $x = a..b$,** where x represents any variable (unassigned name), and a and b are given real numbers. Many commands in Maple have arguments calling for ranges, and so this is an **important new structure of general interest.**

The first argument in the plot( ) command must be an **expression in one variable.** Actually, some variation is allowed here, and it will be discussed in due time, but the **first argument cannot be an equation.** In other words, the first argument cannot include the equality symbol (=).

Unless Maple is told otherwise, it **chooses its own scale on the vertical axis,** as it did in the last two plots. **A range of the form, *c..d*, can be inserted as an optional third argument in the plot command to control the scale on the vertical axis.** This will be demonstrated later in the chapter as our study continues.

Much can be said about the plot command, and these two examples fall far short of giving a complete description of this important command. As you become acquainted with Maple over the next several chapters of this manual, take advantage frequently of Maple's on-line Help File to learn more about the plot command. Experiment and use the command often. Plotting functions is no longer something to be avoided. It should be your first response to many problems. With Maple, functions are so easy to graph, and the graphs provide us with such a wealth of information, that there is no longer a reason to avoid them.

Now that the plot command has been introduced, we can continue our discussion of certain complications regarding fractional powers. We mentioned earlier that Maple has decided to accept a complex-valued interpretation for $x^{1/n}$ when $x < 0$ and $n \geq 3$ is an odd integer. This may be undesirable, but it is necessary in order to avoid even more serious problems in manipulating expressions involving radicals. Some problems, such as evaluating $(-8)^{1/3}$, are easily dealt with by using $-8^{1/3}$ instead of $(-8)^{1/3}$ in a Maple expression. Other problems are not so easily handled, and plotting $f(x) = x^{1/n}$ over an interval that includes both positive and negative values of x is certainly one of them. Here is a situation where the surd( ) command, introduced on page 17, is more critical.

In order to plot $f(x) = x^{1/n}$ on a negative interval, it is necessary to define this function with the surd( ) command.

****

```
> f:=x->surd(x,3);
```

$$f := x \to \mathrm{surd}(x, 3)$$

```
> plot(f(x),x=-10..10);
```

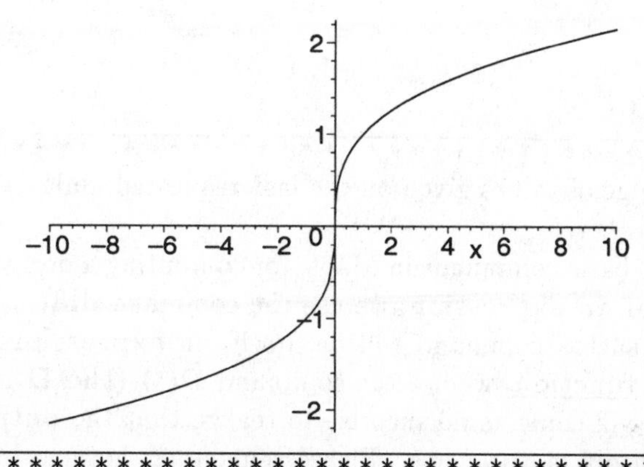

****

2.6 The Differential Calculus

Calculus is the study of limits, and Maple will compute both full limits and one-sided limits with the command **limit( )**. Just as with "Pi," "I," and a few other names, Maple has also reserved the name **infinity**. Like all other Maple assigned names, care should be taken to avoid assigning any other value to this name. Here are some examples.

****

```
> limit((x^2+5*x-6)/(2*x^3+7),x=4);
  limit(sin(x)/x,x=0);
```

$$\frac{2}{9}$$

$$1$$

```
> limit(tan(x),x=Pi/2);limit(tan(x),x=Pi/2,left);
  limit(tan(x),x=Pi/2,right);
```

$$\mathit{undefined}$$

$$\infty$$

$$-\infty$$

```
> limit((a*x+b)/(c*x+d),x=-infinity);
  limit(sin(x),x=infinity);
```

$$\frac{a}{c}$$

$$-1..1$$

****

Notice the range of values given on the last requested limit. This limit does not exist.

There are two basic commands in Maple for computing a derivative. To compute the **derivative of an expression**, we use the command **diff( )**. Not surprisingly, the output from such a command will be, itself, an **expression.** To compute the **derivative of a <u>function</u>**, we use the command **D( ) (the D must be capitalized)**. Again, it will come as no surprise to realize that the **output from such a command is, itself, a function**. Please remember, however, that we are using the words expression and function very carefully. If f is a function of x, then $f(x)$ is an expression, not a function.

****

```
> diff(x^2*cos(x),x);
```

$$2\,x\cos(x) - x^2\sin(x)$$

#Maple differentiates with respect to the variable placed in the second argument of the diff( ) command. Any other unassigned name appearing in the expression is treated as an arbitrary constant.#

```
> f:=(a*x^2+b*x+c)/(p*x^2+q);
  g:=t->tan(5*t^3+4);
```

$$f := \frac{a\,x^2 + b\,x + c}{p\,x^2 + q}$$

$$g := t \to \tan(5\,t^3 + 4)$$

```
> fp:=diff(f,x);
```

f is an expression, as is the output fp.#

$$fp := \frac{2\,a\,x + b}{p\,x^2 + q} - 2\,\frac{(a\,x^2 + b\,x + c)\,p\,x}{(p\,x^2 + q)^2}$$

```
> gp:=diff(g(t),t);
```

g(t) is an expression, as is the output gp.#

$$gp := 15\,(1 + \tan(5\,t^3 + 4)^2)\,t^2$$

#The names *fp* and *gp* are just names, and any other names would do just as well. They are, however, appropriate names for derivatives, if we think of them as representing *f-prime* and *g-prime* respectively. We started with a function g above; by applying the diff( ) command, we obtained an expression *gp* rather than a function. If this is a disappointment, it is easily corrected by using the unapply( ) command.#

```
> gp:=unapply(gp,t);
```

$$gp := t \to 15\,(1 + \tan(5\,t^3 + 4)^2)\,t^2$$

#This changes gp from an expression to a function. As we mentioned above, there is another way to get *gp* directly as a function without first passing through an expression.#

```
> gp:=D(g);
```

$$gp := t \rightarrow 15\,(1 + \tan(5\,t^3 + 4)^2)\,t^2$$

****

As you can see, we get a function directly. To evaluate the derivative of g, at $x = 5$, for example, we have only to enter $gp(5)$ in a Maple input statement. Notice that, when we used the D( ) command, we did not have to specify the variable as a second argument. This is entirely appropriate if we simply consider the mathematical significance of the choice of letter used to define a function. We used the letter "t," but using any other letter would have defined the same function g. Mathematically, the defining equation for the function g can be thought of essentially as meaning

$$g(\,) = \tan(5(\,)^3 + 4).$$

Looking at the function g in this way, you can see why we did not have to specify a variable in the command D(g).

Higher order derivatives are handled as follows.

****

```
> diff(x^3*sin(x),x,x,x);   # The third derivative.#
```

$$6\sin(x) + 18\,x\cos(x) - 9\,x^2\sin(x) - x^3\cos(x)$$

****

This notation can be used on all higher order derivatives, but clearly the notation is awkward for derivatives of large order. Recall that we mentioned earlier that **a sequence is an open-ended list of objects separated by commas.** This is exactly what is needed here. Maple has several ways to **generate a sequence**, and the approach that works best in this situation is one involving the **Maple operator ($)**. We plan to use it in only a minor way at this time, and so complete details of this operator will not be provided now. At some point, this operator should be explored in the Help File. First, note the effect of the next input statement, and then it will be clear how to use it to compute higher order derivatives.

****

```
> x$10;
```

$$x,\ x,\ x,\ x,\ x,\ x,\ x,\ x,\ x,\ x$$

```
> diff(x^20,x$10);
```

$$670442572800\,x^{10}$$

****

Maple can handle differentiation on a much more abstract level. **As we mentioned earlier, Maple treats all unassigned names as constants except for**

the name (variable) it differentiates with respect to. In view of this, it is not surprising that, in the next Maple work session, the answer to the first input statement is zero. In the second computation, even though $y(x)$ is unassigned, **Maple interprets the parentheses to mean that $y(x)$ is an unassigned expression that depends on x.**

****

```
> diff(y,x);
```

$$0$$

```
> diff(y(x),x);
```

$$\frac{\partial}{\partial x}\,\mathrm{y}(x)$$

****

Because of this ability to interpret abstract symbols, Maple can perform implicit differentiation.

Example 2.1 *Suppose that $y = y(x)$ is defined implicitly by the equation*

$$xy^2 + 5x^2y + y^3 = 4x^3 + 7.$$

Find the derivative $y'(x)$ in terms of x and of y.

Imagine how you would solve this problem with pencil and paper. Maple can be used to solve the problem in basically the same way. This is a good way to review the basic principle of implicit differentiation. After we are finished, we will introduce an **interesting new Maple command for getting the answer directly.** It's still a good learning experience to go through all of the details, so let's proceed in spite of the promise of an easy command.

****

```
> eq:=x*y^2+5*x^2*y+y^3=4*x^3+7;
```

$$eq := x\,y^2 + 5\,x^2\,y + y^3 = 4\,x^3 + 7$$

```
> eqx:=subs(y=y(x),eq);
```

$$eqx := x\,\mathrm{y}(x)^2 + 5\,x^2\,\mathrm{y}(x) + \mathrm{y}(x)^3 = 4\,x^3 + 7$$

```
> deq:=diff(eqx,x);
```

$deq :=$

$$\mathrm{y}(x)^2 + 2\,x\,\mathrm{y}(x)\,(\frac{\partial}{\partial x}\mathrm{y}(x)) + 10\,x\,\mathrm{y}(x) + 5\,x^2\,(\frac{\partial}{\partial x}\mathrm{y}(x)) + 3\,\mathrm{y}(x)^2\,(\frac{\partial}{\partial x}\mathrm{y}(x)) =$$
$$12\,x^2$$

```
> yp:=solve(deq,diff(y(x),x));
```

$$yp := \frac{-\mathrm{y}(x)^2 - 10\,x\,\mathrm{y}(x) + 12\,x^2}{2\,x\,\mathrm{y}(x) + 5\,x^2 + 3\,\mathrm{y}(x)^2}$$

```
> yp:=subs(y(x)=y,yp);
```

$$yp := \frac{-y^2 - 10\,x\,y + 12\,x^2}{2\,x\,y + 5\,x^2 + 3\,y^2}$$

#Here is the **new Maple command, implicitdiff( )**, which has the power to take us from the very first equation *eq* to the final answer in just one step. The second argument in this command is the dependent variable y, and the third argument is the independent variable x. As an added measure, we compute, as well, the second derivative of y with respect to x. **This command is worth remembering!**#

```
> yp:=implicitdiff(eq,y,x);
```

$$yp := \frac{-y^2 - 10\,x\,y + 12\,x^2}{2\,x\,y + 5\,x^2 + 3\,y^2}$$

```
> ypp:=implicitdiff(eq,y,x,x);
```

$$ypp := -6(14\,y^5 + 148\,x^5 - 27\,y^4\,x - 8\,y^3\,x^2 - 298\,x^3\,y^2 - 21\,y\,x^4)\Big/(188\,x^3\,y^3 + 125\,x^6 + 27\,y^6 + 285\,x^4\,y^2 + 171\,x^2\,y^4 + 150\,x^5\,y + 54\,x\,y^5)$$

****

Several commands in Maple have **inert forms**, which are obtained by **capitalizing command names.** Inert commands are identical to active commands, except that **computations are postponed** until they are initiated by a **value( )** command—more on this later. One such command is the **inert differentiation command Diff( )**, and there are many others. Notice the effect the capital letter has on the following. As you can see, it can be used to create visually pleasing text, but we will find many other uses in the future for the inert forms of this and many other commands.

****

```
> Diff(sin(x),x)=diff(sin(x),x);
```

$$\frac{\partial}{\partial x}\sin(x) = \cos(x)$$

****

2.7 Solving Problems in Maple

Much remains to be done in subsequent chapters, but, with this brief introduction, we already have a sufficient knowledge of Maple to be able to solve a large number of interesting problems.

Example 2.2 *Plot the function*

$$f(x) = x^6 - 189x^5 - 2192x^4 - 1736x^3 + 27207x^2 - 1547x + 29300.$$

Supply evidence that you have the complete graph (that there are no hidden turns, which are not displayed).

****

```
> f:=x->x^6-189*x^5-2192*x^4-1736*x^3+27207*x^2-1547*x+29300;
```

$$f := x \to x^6 - 189\,x^5 - 2192\,x^4 - 1736\,x^3 + 27207\,x^2 - 1547\,x + 29300$$

```
> plot(f(x),x=-10..10);
```

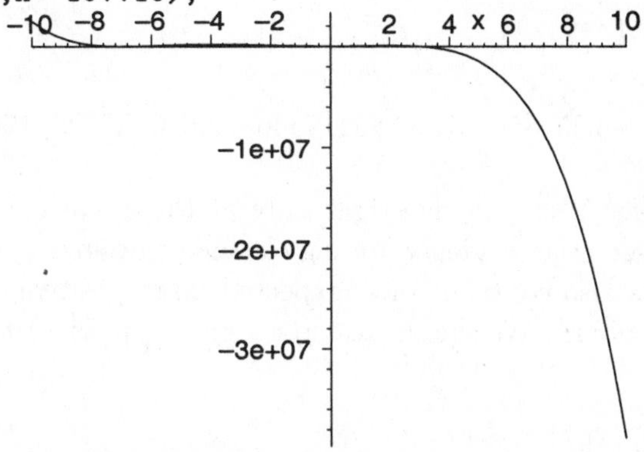

#This is disappointing but not surprising. The values of this polynomial get large very quickly, and Maple responds to this by placing a very large scale on the vertical axis. If there are any peculiarities to this graph, they are clearly hidden by the large scale. The vertical axis can be scaled as desired by inserting a range as an optional third argument in the plot command. For example, to scale the vertical axis between -100 and 100 we would write the following.#

```
> plot(f(x),x=-10..10,-100..100);
> #Output omitted.#
```

#As you will see by experimenting with different vertical ranges, the graph doesn't begin to assume a classic shape until a vertical range from -100000 to 100000 is used. In the next Maple work session, this vertical range is fine-tuned still further, by trial and error, to produce a graph that appears to show all of the "twists and turns" of the function. At the same time, we must avoid using too large a range, or we will just produce the same "flat" graph that we started with.#

```
> plot(f(x),x=-10..10,-100000..220000);
```

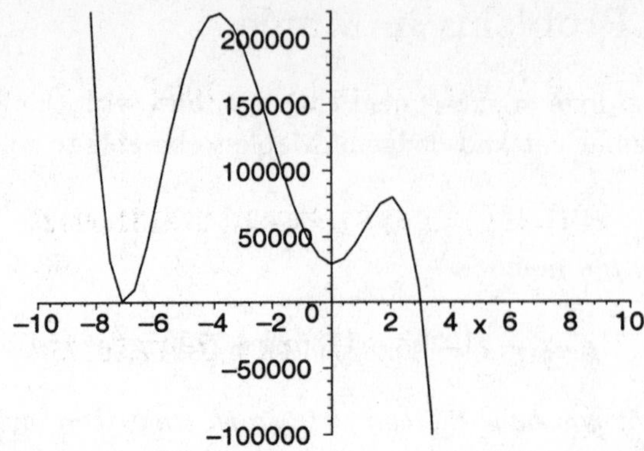

#Do we have a complete graph? How do we know? **Make it a habit to be a skeptic**. One way to answer such questions is to use the differential calculus. There are other ways as well.#

```
> fp:=diff(f(x),x);
```

$$fp := 6\,x^5 - 945\,x^4 - 8768\,x^3 - 5208\,x^2 + 54414\,x - 1547$$

```
> s:=fsolve(fp=0,x);
```

$$s := -7., -3.857850491, .02851173464, 2.013437798, 166.3159010$$

#The derivative fp can change signs only at these five points. By looking at the last plot, we see clear evidence for sign changes occurring at the first four of these points, but s also contains an unexpected large positive value of x where a sign change could occur. We should test the sign of fp for some value of x larger than $s[5]$.#

```
> subs(x=200,fp);
```

$$337658561253$$

#This positive value implies that the graph is increasing for x larger than $x = s[5]$, and so there is **one last hidden turn in the graph which is not shown** in the above plot. Any further graphing using the plot command is probably not going to be very effective. Maple's desire to interpret the large vertical scale "scientifically" rather than "artistically" is hard to overcome. Perhaps the best way to finish this graph is to evaluate just the relative maximums and relative minimums. Surely this is easy to do. The words "RelMax" and "RelMin" are just names—not Maple commands. #

```
> RelMin1:=[s[1],f(s[1])];
  RelMax1:=[s[2],f(s[2])]; RelMin2:=[s[3],f(s[3])];
  RelMax2:=[s[4],f(s[4])]; RelMin3:=[s[5],f(s[5])];
```

$$RelMin1 := [-7., -100.]$$

$$RelMax1 := [-3.857850491, 219131.0002]$$

$$RelMin2 := [.02851173464, 29277.96775]$$

$$RelMax2 := [2.013437798, 80099.28445]$$

$$RelMin3 := [166.3159010, -.4571023679\,10^{13}]$$

****

Example 2.3 *Find all solutions (as decimal numbers) to the system of equations* $y = 2x^2 + 5x - 17,\ 3x + 5y = 34$.

****

```
> eq1:=y=2*x^2+5*x-17;eq2:=3*x+5*y=34;
```

$$eq1 := y = 2\,x^2 + 5\,x - 17$$

$$eq2 := 3\,x + 5\,y = 34$$

```
> fsolve({eq1,eq2},{x,y});
```

$$\{x = 2.322902094,\ y = 5.406258744\}$$

#Are we done? Are there other solutions? Again, be a skeptic. One way to gain further insight would be to plot the two curves. Maple **can plot equations** in two variables, but **not with the plot command**. Rather than introduce more commands, we proceed as follows.#

```
> y1:=solve(eq1,y);y2:=solve(eq2,y);
```

$$y1 := 2\,x^2 + 5\,x - 17$$

$$y2 := -\frac{3}{5}\,x + \frac{34}{5}$$

```
> plot({y1,y2},x=-6..6,-20..20);
```

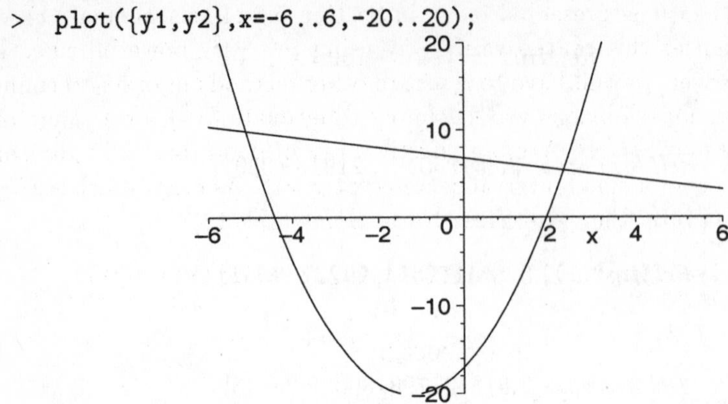

#We can see that there is another solution that the fsolve command has not yet provided. To help the solve( ) command find it, place the mouse pointer inside the plot and click, so that the rectangular border around the plot is activated. Place the **mouse pointer close to this new point of intersection and click.** You will see the **coordinates of the point appear in the left-hand corner of the tool bar.** Using this approach, we find that another solution is approximately $x = -5$, $y = -9$. Armed with this information, we can **help Maple give us the solution we wish to find.** Make note of this **important additional input.**#

```
> fsolve({eq1,eq2},{x,y},x=-6..-4,y=8..10);
```

$$\{x = -5.122902094, y = 9.873741256\}$$

****

Example 2.4 *Compute the following limit, and verify the limit through alternate means.*

$$\lim_{x \to 1} \frac{2\cos(x-1) + x^2 - 2x - 1}{(x-1)^4}$$

****

```
> h:=x->(2*cos(x-1)+x^2-2*x-1)/(x-1)^4;
```

$$h := x \to \frac{2\cos(x-1) + x^2 - 2\,x - 1}{(x-1)^4}$$

```
> L:=limit(h(x),x=1);
```

$$L := \frac{1}{12}$$

#Is the answer correct? Some limits have obvious values, but this is not one of them. The numerator and denominator of $h(x)$ both go to 0 as $x \to 1$, and so the limit cannot be evaluated purely by observation. The best way to answer the question is to use *L'Hôpital's Rule*, which might or might not be presented in a first calculus course. When it is presented later, in Section 5.6 of this manual, we will show how limits such as this can be verified by using this very powerful rule. For the time being, however, we will have to resort to other methods in order to confirm the limit value. The most obvious way is simply to evaluate $h(x)$ for a range of x values that are close to $x = 1$. A more effective way of doing basically the same thing is to plot $h(x)$ over a small interval centered at $x = 1$. As x approaches $x = 1$, are the points $(x, h(x))$ on the graph close to the point $(1, 1/12)$? #

```
> evalf(L); evalf(h(1.1)); evalf(h(1.01)); evalf(h(1.001));
```

$$.08333333333$$

$$.08331000000$$

$$.1000000000$$

$$0$$

```
> plot(h(x),x=.5..1.5);
```

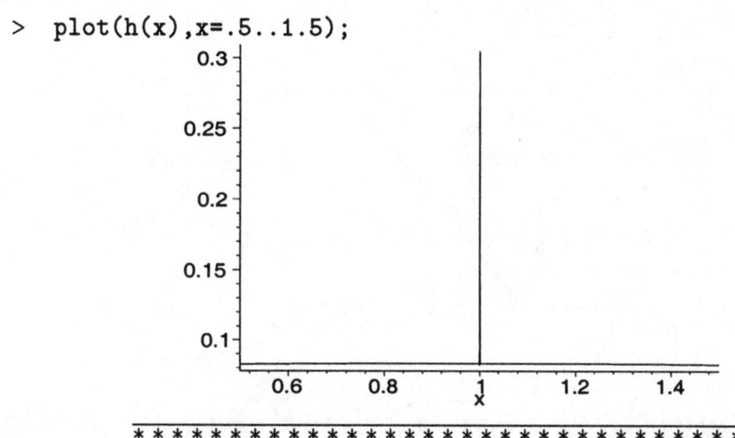

****

Something is clearly wrong here! It is not at all obvious, but the limit value $L = 1/12$ happens to be correct. Our attempt to confirm this, however, has certainly failed. The fault lies not with Maple, but with unavoidable problems in dealing with decimal approximations. When decimal arithmetic is performed on a fraction of the form A/B where both A and B are small, round-off errors become significant. They are unavoidable! We can ask Maple for more accuracy, by increasing the value of **Digits**, but, regardless of how large a value we assign to Digits, all we have to do is choose x close enough to $x = 1$ and decimal arithmetic on the value of $h(x)$ will fail to be accurate. We will not be able to confirm the value of this limit until Section 5.6.

In spite of this problem, the method above frequently will confirm the value of a limit. Because of the ever-present possibility of round-off errors, however, it is just not a very reliable way to confirm a limit value. Other methods sometimes work. Certainly, all of the methods that were used to compute limits in a pencil-and-paper environment can be used as well in a Maple work session, to confirm the value of a limit. Until *L'Hôpital's Rule* is established, however, our ability to confirm Maple's limit calculations is somewhat restricted.

Our next and final example in this chapter is admittedly a bit involved. Think of it as an example whose purpose is to demonstrate Maple's problem-solving power. If we keep our minds on the ideas and let Maple take care of the calculations, then perhaps we can avoid getting bogged down in the computational details. **Maple lets us focus all of our energy on ideas.** Nevertheless, we should not turn a blind eye to the calculations—we can at least glance at the calculations to see whether they look reasonable.

Example 2.5 *An Air Traffic Controller is monitoring the positions of two airplanes on a radar screen. Both are flying at the same altitude. At a certain moment, the passenger jet is 75.4 miles South of the airport and traveling North at 547 miles per hour. At the same moment, the air force fighter jet is 126 miles Northeast of the airport and traveling Southwest at 714 miles per hour. Neither plane will be landing at the airport, and both will be flying over the airport with no change in speed or direction. How close do the planes get to each other?*

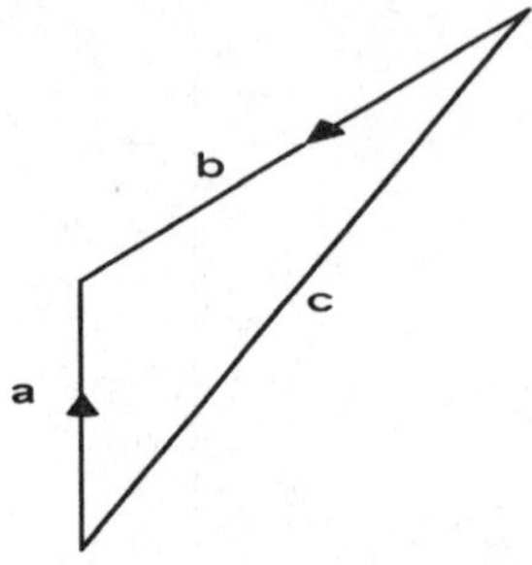

The *Law of Cosines* gives us the equation $c^2 = a^2 + b^2 - 2ab\cos(\alpha)$. The angle α is 135°. We let t be time in hours with $t = 0$ corresponding to that "initial moment," and we wish to minimize c as a function of t.

****

```
> a:=75.4-547*t;b:=126-714*t;

  c:=sqrt(a^2+b^2-2*a*b*cos(135*Pi/180));
```

$$a := 75.4 - 547\,t$$

$$b := 126 - 714\,t$$

$$c := \sqrt{(75.4 - 547\,t)^2 + (126 - 714\,t)^2 + (75.4 - 547\,t)\,(126 - 714\,t)\,\sqrt{2}}$$

```
> c:=unapply(c,t);
```

$$c := t \to \sqrt{(75.4 - 547\,t)^2 + (126 - 714\,t)^2 + (75.4 - 547\,t)\,(126 - 714\,t)\,\sqrt{2}}$$

#As the planes pass over the airport, the values of a and b will turn from positive to negative values. This situation needs to be considered because the minimum distance between the planes could occur after one of the planes passes the airport. At first glance it is not clear that the *Law of Cosines* stated above is still valid for negative values of a and/or b. However, a moment's thought will convince you that as exactly one of the variables becomes negative, and then as they both become negative, the angle α changes from 135° to 45° and then back to 135°, introducing in the process just the right combination of negative signs to make the formula valid for all values of a and b. The key idea here is that $\cos(135^o) = -\cos(45^o)$.#

```
> cp:=diff(c(t),t);
```

$$cp := \frac{1}{2}(-262415.6 + 1618010\,t - 547\,(126 - 714\,t)\,\sqrt{2} - 714\,(75.4 - 547\,t)\,\sqrt{2}) \Big/ \sqrt{(75.4 - 547\,t)^2 + (126 - 714\,t)^2 + (75.4 - 547\,t)\,(126 - 714\,t)\,\sqrt{2}}$$

```
> close_time:=fsolve(cp=0,t);
```

$$close\_time := .1601443758$$

```
> distance:=c(close_time);
```

$$distance := \sqrt{284.6986395 - 142.2024064\,\sqrt{2}}$$

```
> distance:=evalf(%);
```

This is the closest distance in miles.#

$$distance := 9.142979153$$

#To verify this answer we plot the distance function $c(t)$. Clearly the minimum distance would occur before one hour.#

```
> plot(c(t),t=0..1);
```

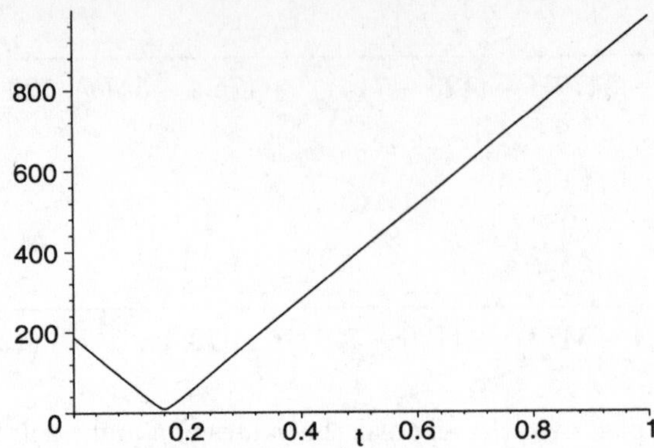

#To see where the planes will be at this time, we evaluate.#

```
> A:=evalf(subs(t=close_time,a));B:=evalf(subs(
  t=close_time,b));
```

$$A := -12.19897356$$

$$B := 11.6569157$$

****

These answers indicate that the passenger jet will have passed the airport and the fighter jet will still be approaching the airport at the moment of closest approach.

2.8 Exercise Set

New Maple Commands in this Chapter

assume( )	assuming	D( )	denom( )	diff( )
Diff( )	evalf(,n)	evalf( )	expand( )	factor( )
fsolve( )	implicitdiff( )	limit( )	normal( )	numer( )
plot( )	proc( )	simplify(,*options*)	simplify( )	solve( )
subs( )	surd(,n)	unapply( )	value( )	

Remember the following format for creating a function f.

```
> f:=x->expression in x;
```

Click and hold on a menu button, and an explanation of the button's function will appear at the bottom of the screen in the Status Bar. Hold the button down and slide the cursor off to avoid activating the button.

Maple's sectioning tools are easy to use and can be a real convenience in organizing worksheets, especially longer worksheets. After you finish a problem, **select (highlight) the entire problem, then press the right Section button in the Tool Bar** (The second of the two buttons just to the right of the Maple prompt [|>] button). The highlighted material will be indented and a [-] section button will appear above it. The space to the right of the [-] section button is reserved for the name of the section. Text in this region will automatically be black, large, and boldface. Click the [-] section button to compress the output of the section (it turns into a [+] button). Click the [+] section button to expand the section.

To avoid interference from previous Maple activity, start new work with the **Restart command**. In Maple 7, there is a Restart button in the upper right-hand corner of the menu-bar.

1.Evaluate the following expressions exactly.
$a) 3 \times \frac{6\times7^2-8}{5} - 2 + \frac{14}{6\times5+8}$ $\quad b) \frac{8}{5}\times 3 - \frac{8}{5\times3}$ $\quad c)(-2^4 + \frac{1}{5^2})^{-3}$
$d) \frac{\frac{5}{3}}{14} - \frac{6}{\frac{2}{3}}$ $\quad e) \frac{\sqrt[3]{-8}}{\sqrt{25}} + 81^{1/4}$ $\quad f) 7\cos^3(\pi/6) + \frac{\tan(\pi/12)}{5}$

2. Evaluate the following expressions as decimals with 10 digits of accuracy.
$a) \frac{\sqrt{43.7}}{5^2}$ $\quad b) 8\sqrt{\cos(13°)}$
$c) \frac{(73)^{2/3}+17}{(-3)^{1/5}}$ $\quad d) \sin^2(-3\pi/7)$

3. Evaluate the expressions in problem 2 as decimals with 3 digits of accuracy.

4. Evaluate the expressions in problem 2 as decimals with 15 digits of accuracy.

5. Evaluate the following expressions, and express the answers in simplest form. (Do not express them as decimals.)
$a)(2 + \sin(\pi/3))^6$ $\quad b) \frac{(27)^{2/3}+5}{6\times(8+\frac{1}{3})}$ $\quad c)(\tan(\pi/8) + 2\cos(3))^2$

6. Graph the function $f(x) = \frac{3x+7}{x^2+5}$. Verify that you have a complete graph. Find the x-intercepts and the coordinates of the relative maximums and minimums.

7. Graph the function $f(x) = x^4 - 50x^3 - 28x^2 - 280x - 642$. Verify that you have a complete graph. Find the x-intercepts and the coordinates of the relative maximums and minimums.

8. Simplify the following rational function. How do you know that the answer Maple gives you is in simplest terms?

$$f(x) = \frac{x^5 + 14x^4 + 30x^3 - 235x^2 - 506x + 1456}{2x^5 - 5x^4 - 85x^3 + 336x^2 - 72x - 504}$$

9. Compute the following limits if they exist in a finite or infinite sense. Verify each answer by some means using an algebraic approach whenever appropriate.
$a) \lim_{x\to4}(\frac{3x^2-17x+20}{5x^2-19x-4})$ $\quad b) \lim_{x\to3^+}(2x^2\cot(\pi x))$ $\quad c) \lim_{x\to4}(\frac{2-\sqrt{x}}{x-4})$
$d) \lim_{x\to0}(\frac{5x-\sin(5x)}{x^3})$ $\quad e) \lim_{x\to\infty}(\frac{7x^2+3x-9}{2x^2+8x+5})$

10. Use the diff( ) command to compute $f'(4)$ for $f(x) = x^2\sin(\pi x)$. Repeat the computation, using the D( ) command. What are the advantages and disadvantages of each? If you use the diff( ) command, how can you turn your output into a function? Compute $f'(4)$ by using the definition of a derivative as the limit of a certain difference quotient.

11. Use Maple to generate the rule $f'(x) = 10x^9$ for $f(x) = x^{10}$, and then verify the rule by using the definition of a derivative as the limit of a certain difference quotient. Also verify the value of the limit algebraically.

12. Compute the derivatives of the following functions, where a denotes an arbitrary constant. Use the diff( ) command, then repeat the computation, using the D( ) command. What are the advantages and disadvantages of each? If you use the diff( ) command, how can you turn your output into a function?

$a)f(x) = x^2\sqrt{a^2 - x^2}$ $b)g(w) = \sqrt{1+\sqrt{1+\sqrt{1+w^2}}}$

$c)H(t) = \sec^5(\frac{t^2}{t^2+a^2})$ $d)S(X) = X\sin^2(X) + X\cos^2(X)$

In part d), think about what this answer should be and force Maple to give you that answer.

13. Compute $\frac{d^8y}{dx^8}$ for $y = \tan(x)$. Is your answer what you expected? (Recall the rule for differentiating the tangent function.) Make the answer look more like what you would expect by using the subs( ) command to interject the secant function into the answer. The simplify( ) command won't quite work here. Try it and see what happens.

14. Use implicit differentiation and the diff( ) command to find the slope of the tangent line to the graph of the equation $x^3y^2 + 4xy = 85 - (2x^4 - 8y)^2$ at the point on the curve corresponding to $x = -5$. Use the point with the smallest y-coordinate. How do you know there isn't a smaller one that Maple simply failed to give you? **Check your answer** for the slope, with Maple's **implicitdiff( )** command. It would be nice (and reassuring) if we had a graph of the equation to look at, but we have not yet discussed the command for plotting equations. This problem can be done just as well without a graph.

Use Maple's implicitdiff( ) command to compute the second derivative of y with respect to x at the same point.

15. Use Maple to generate a Product Rule for differentiating the product of three differentiable functions f, g, and h. Recall how to set up an unassigned name that depends on another variable.

16. Use Maple to generate a Chain Rule for the composition $p(x) = f(g(h(x)))$ of three differentiable functions f, g, and h. Recall how to set up an unassigned name that depends on another variable.

17. Find the absolute maximum and absolute minimum of $f(x) = x^4 + 2x^3 - 76x^2 - 242x + 100$ on the closed interval $[-9, 9]$. Do this problem without graphing. Use a plot to verify your results only after you are finished.

18. At noon, a passenger ship is 100 miles South of an oil tanker. The passenger ship travels due North at 35 miles per hour, while the oil tanker travels due East at 17 miles per hour. At what rate is the distance between the ships changing at 2:30 pm of the same day?

19. Radar at an airport spots a passenger jet traveling toward the airport at a constant elevation 25,000 feet. Radar determines that, when the line-of-sight distance between the jet and radar is exactly 12 miles, this line-of-sight distance is decreasing at the rate of 370 miles per hour. What is the ground speed of the jet at this moment? (Recall that 1 mile = 5280 feet.)

20. A rectangular pool-deck complex is being built for a hotel. The pool must have a water surface area of 4,000 square feet, and the pool must be enclosed on three sides by a 10-foot-wide deck and on the fourth side (the side facing the building) by a 30-foot-wide patio. Find the dimensions that minimize the total area of the pool-deck complex.

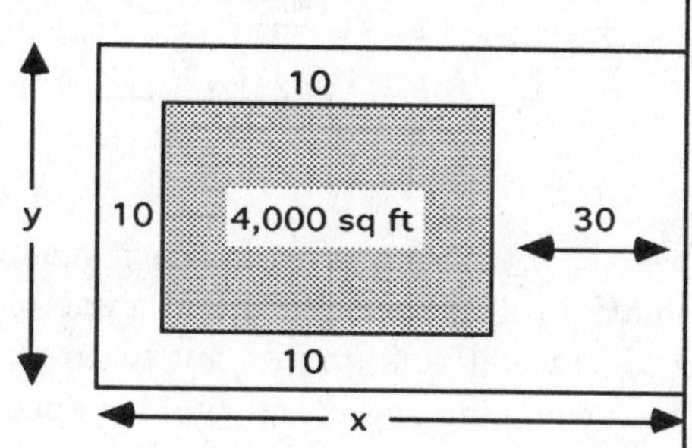

21. A computer company can expect to sell

$$q = 1080 - \frac{1}{9375}p^2 + \frac{8}{75}p \quad (1000 \le p \le 3500)$$

computers per day if the selling price of a computer is p in dollars What should the selling price be in order to maximize total revenue from sales? How many computers can it expect to sell at that price? What will the revenue be from sales? Suppose that the total cost (overhead, labor, and material) of manufacturing q computers per day is

$$C = 1470 + 1300q + (q - 1000)^2.$$

What should the selling price be in order to maximize profit? How many computers can it expect to sell at that price? What will the profit be?

Project: Swimming Coach

Suppose that you are the head coach for a large team of athletes who all plan to compete in a combination swimming/running event. The event begins on an island 2 miles out from a straight shore line. The end of the race is 10 miles down the shoreline, as shown in the picture below. Each competitor in the race will swim to some *"transfer point"* along the beach and then run from that point to the finish line. The choice of a transfer point is entirely up to each athlete. Simply by choosing the appropriate transfer point, an athlete may decide to run the whole 10 miles or decide to completely avoid running, and any transfer point between (and including) these extremes is allowed.

As the head coach, it is your responsibility to chose the best possible transfer point for each member of your team. Your only concern is that each athlete turn in his/her best total time. (Typical head-coach attitude, right?) You know the swimming and running speeds of each member of your team, but these speeds change from day to day, and so what you need is a means of determining, quickly, and

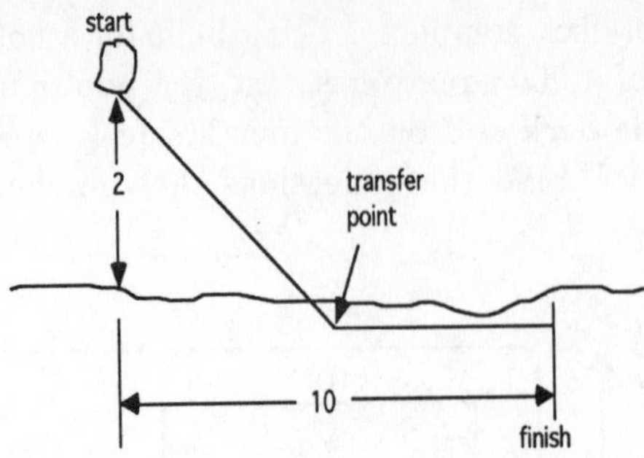

just before the race begins, a transfer point for each member of your team. The Department of Mathematics will graciously lend you one of its portable computers (and Maple) to take to the race, if you promise not to drop it in the water. Suppose we let s and r denote the swimming speed and running speed in mph for any one of your athletes. Find expressions that depends on r and s and will quickly determine the transfer point and total race time for that athlete. (This is why they hired a mathematician as head coach.) Good luck!

Project: Road Construction

Towns A and B are separated by a river running basically east-west, and plans are being made to connect the towns by a road and a bridge over the river.

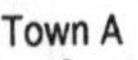

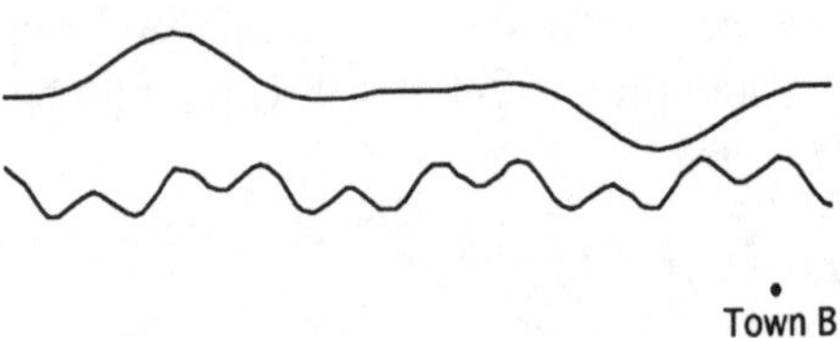

To locate everything precisely, a north-south and an east-west axis are placed over the region with units in kilometers (km). The origin is located at some point in the river as can be seen in the picture below. According to these axes, Town A is located 15 km west and 10 km north of the origin and Town B is located 17 km east and 6 km south of the origin. The bridge can be built at any point along the river, but it must cross the river due north-south. To save money, it is decided to build the road straight from Town A to the point P on the north side of the bridge, and straight again from the point Q on the south side of the bridge to Town B. (To simplify matters, assume that the road/bridge complex intersects the river bank only at P and Q.)

Cost, however, is a major complication. The north side of the river is hilly, expensive farm land, and it will cost \$500,000 per km to build the road. The bridge itself will cost \$2,000,000 per km to build. The south side of the river is

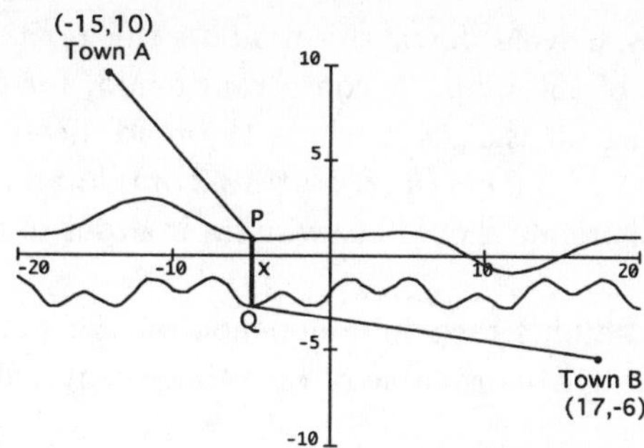

flat, undeveloped government land, and this part of the road can be built for only $190,000 per km. By trial and error, it is determined that the curves representing the northern and southern boundaries of the river can be described reasonably well by the equations

$$y = 1 + 2\cos(x/4)\sin(x/7)^3, \text{ and } y = -2 + \cos(x/2)\sin(x),$$

where the northern boundary is listed first.

At what point along the river should the bridge be built in order to minimize the cost of the whole project?

Project: Best Race Time

A swimming–running–biking contest is to take place in a region described by the map shown in the figure. Map coordinates are in kilometers. The shoreline is defined by the curve $y = \frac{x^2}{10} + 2x^{\frac{1}{3}}$. A road defined by the equation $x = 3$ runs south from the shore line. The contest begins on a small island out in the lake located at the point with map coordinates (-3,3). Contestants must swim to a point of their own choosing on the shoreline. From that point, they must run due east to the road, where their assistant will be ready with a bicycle.

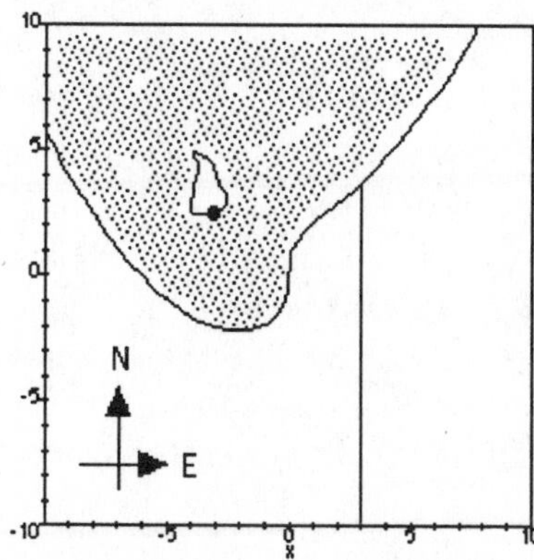

The race continues by bicycle down the road to the point with map coordinates (3,-10) at the bottom of the map. A contestant could, for example, avoid running altogether by swimming for the point where the road meets the shoreline.

(i) Ralph can swim at 3.9 km/h, run at 12.2 km/h, and bike at 34.4 km/h. At what position on the shoreline should he swim for in order to turn in his best personal performance?

(ii) Observe how Ralph's race time depends on the point (x, y) he chooses on the shoreline, by creating a short table of race times centered around his best choice for x.

(iii) This problem can be done in a more general way. Let s, r, and b represent the swimming, running, and biking velocities, respectively, of any contestant in the race. Determine formulas both for the total race time and for the "best" point (x, y) to swim to on the shore, in terms of s, r, and b. Check to see whether your formula agrees with part *(i)*. Select at least one other set of values for s, r, and b and give the results.

Chapter 3

Basic Integration

3.1 Area and the Definite Integral

The definition of a definite integral as the limit of its Riemann sums is so important to mathematics and its applications that we begin this chapter by using Maple to gain some intuitive insight into this concept. Even if we downplay the many mathematical reasons for understanding this definition thoroughly, we are still left with some very important applied reasons.

As long as it is known that the integral of f on $[a, b]$ exists, its value can be determined by looking at a restricted, more manageable class of Riemann sums. In this case, if

$$a = x_0 < x_1 < \cdots < x_n = b$$

is the partition obtained by dividing $[a, b]$ into n subintervals of equal length

$$\Delta x = \frac{b-a}{n},$$

and if

$$x_0 \le c_1 \le x_1, x_1 \le c_2 \le x_2, \ldots, x_{n-1} \le c_n \le x_n,$$

then

$$\int_a^b f(x)dx = \lim_{n\to\infty} \sum_{k=1}^{n} f(c_k)\Delta x. \tag{3.1}$$

Typically, the points $c_j (j = 1, 2, \ldots, n)$ are chosen to be left endpoints, right endpoints, or midpoints of their respective subintervals, but any scheme for determining these points is acceptable. Furthermore, if f is continuous on $[a, b]$, then it is integrable on $[a, b]$, and so this simplified setting is usually sufficient. Applied problems, in particular, are usually continuous in nature.

Frequently, when problems from the real world are formulated into mathematics, they take the form of the right-hand side of (3.1). A complicated problem might not be solvable "as a whole," but if we partition the problem up into a large number of parts, we might conclude that the "whole" problem is just the sum of its parts.

We might be able to estimate each of the small parts easily, to get approximations $P_1, P_2, \ldots P_n$. Then a good approximation for the whole problem would be

$$\sum_{k=1}^{n} P_k.$$

If the approximations get better as n (the number of subdivisions) gets larger, then the answer to the original problem would be

$$Answer = \lim_{n\to\infty} \sum_{k=1}^{n} P_k. \tag{3.2}$$

Without a knowledge of the definition of a definite integral, we would be left with a formidable problem. Frequently, however, (3.2) can be made to look like (3.1), so, by using the definition of an integral, we can identify our problem as an integral. At this point, we have a well-understood problem, which can easily be evaluated.

The Maple commands we need to study this definition are not available until we load them. Maple has a large number of specialized **packages** of commands. Maple is so big, that it could not possibly load everything automatically, so when a worksheet is opened, it makes available only a list (a very large list) of the most common and frequently used commands. To see what packages are available, pull down the **Help Menu**, open up **Topic Search...**, type **"Index"** in the topic field, and click on the hyperlink **packages**.

The **packages** in Maple **are loaded** by using the **with( ) command**, and the package we wish to use now is called the **student package**. Remember that, with a colon rather than a semicolon, Maple output display is suppressed. Once you become familiar with the names of the commands in the package that you plan to use, you can use (:) and suppress displaying the list of names.

****

```
> with(student);
```

[*D*, *Diff*, *Doubleint*, *Int*, *Limit*, *Lineint*, *Product*, *Sum*, *Tripleint*, *changevar*, *combine*, *completesquare*, *distance*, *equate*, *extrema*, *integrand*, *intercept*, *intparts*, *isolate*, *leftbox*, *leftsum*, *makeproc*, *maximize*, *middlebox*, *middlesum*, *midpoint*, *minimize*, *powsubs*, *rightbox*, *rightsum*, *showtangent*, *simpson*, *slope*, *trapezoid*, *value*]

****

Consider the definition of an integral as it applies to the problem

$$\int_1^6 (9 - x^2)dx.$$

This integral can be approximated by a Riemann sum obtained by dividing the interval [1, 6] into, say 20 subintervals of equal length, by using, let's say, right endpoints to establish (±)heights of rectangles and by computing the sum of the (±)areas of all the rectangles. To draw the corresponding picture, we use the command **rightbox( )**. Similar commands **leftbox( )** and **middlebox( )** are also available.

Before we use these commands in the next round of computations, we introduce another Maple command that is of more general interest. The command **value( )** can easily be confused with the commands eval( ), and evalf( ), which, as names, seem to be equivalent. **The command value( ) is used to evaluate a Maple expression whose value has been directly or indirectly suppressed.** It is sometimes useful to set up a Maple computation, but postpone its evaluation until a later point in the development of the problem. We will demonstrate how to suppress an evaluation later in this manual. The value( ) command may be used infrequently, because of its limited purpose, but it can find its way into a broad range of problems. In the next round of calculations, it is used to evaluate an expression whose value is suppressed by Maple.

****

```
> with(student):
> f:=x->9-x^2;
```

$$f := x \to 9 - x^2$$

```
> rightbox(f(x),x=1..6,20);
```

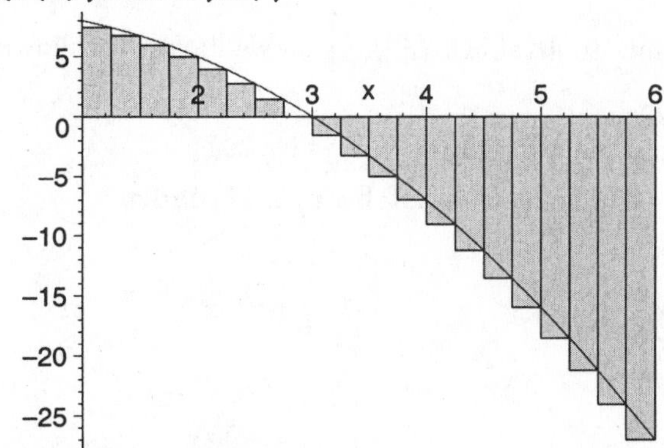

#To see the corresponding Riemann sum expressed in Σ-notation, use the command **rightsum( )**. To get a numerical value, follow this with the command **value( )**. When you do this, notice how the areas of rectangles below the x-axis naturally subtract from the areas of rectangles above the x-axis. From the above picture, we should expect a negative answer.#

```
> rightsum(f(x),x=1..6,20);
```

$$\frac{1}{4}\left(\sum_{i=1}^{20}(9-(1+\frac{1}{4}i)^2)\right)$$

```
> value(%);
```

$$\frac{-995}{32}$$

#Frankly, although the command rightsum( ) supplies an answer, it requires so little input that it might not contribute to a deeper understanding of the definition of an integral, which is the whole point of this exercise. At any rate, experiment with the leftbox, rightbox, and middlebox commands for different values of n (n =number of rectangles). Notice that for n large, it doesn't visually seem to make much difference whether we use left endpoints, right endpoints, or midpoints to establish ($\pm$)heights of rectangles.

To find Riemann sums, it is much more instructive to put them together ourselves. To do this, we use the **sum( ) command**. This is another basic Maple command that we will be using frequently in the future.

If we divide the interval $[1, 6]$ into 20 subintervals of equal length

$$h = \frac{6-1}{20}$$

then the partition points are

$$1 < 1+h < 1+2h < 1+3h < ... < 1+20h = 6.$$

Using right end points to establish ($\pm$)heights we have the following.#

```
> h:=(6-1)/20;
  approx_value:=sum(f(1+j*h)*h,j=1..20);
```

#the letter j (the index) must be unassigned.#

$$h := \frac{1}{4}$$

$$approx\_value := \frac{-995}{32}$$

#For polynomial functions, we can carry this analysis out to its ultimate conclusion and actually take the limit of the Riemann sums as $n \to \infty$. Divide the interval $[1, 6]$ into n subintervals of equal length $h = (6-1)/n$. This gives the partition points

$$1 < 1+h < 1+2h < \ldots < 1+nh = 6.$$

The corresponding Riemann sum is computed next. Maple assumes, given the location of n in the argument of the sum( ) command, that this unassigned letter represents a positive integer.#

```
> h:=(6-1)/n;s:=sum(f(1+j*h)*h,j=1..n);
```

#Maple actually evaluates this sum for an arbitrary integer n.#

$$h := \frac{5}{n}$$

$$s := 40\,\frac{n+1}{n} - 25\,\frac{(n+1)^2}{n^2} + 25\,\frac{n+1}{n^2} - \frac{125}{3}\,\frac{(n+1)^3}{n^3} + \frac{125}{2}\,\frac{(n+1)^2}{n^3} - \frac{125}{6}\,\frac{n+1}{n^3} - \frac{40}{n}$$

```
> exact_integral_value:=limit(s,n=infinity);
```

$$exact\_integral\_value := \frac{-80}{3}$$

****

Can you see, by observation, that this last limit value is correct? Just look at the formula for s.

It is now possible to gain a real insight into the definition of an integral. Try another function and experiment. Draw some left, right, and middle Riemann sums. Evaluate some of them by using the sum( ) command. (Use the commands leftsum( ), rightsum( ), middlesum( ) only to check your answer.) Take the limit as n (the number of rectangles) goes to infinity. (Maple cannot always compute the limit, but it can when $f(x)$ is a polynomial.)

As a final matter, take note of the issue raised about the index in the sum( ) command. **In the input statement**

```
> sum(f(j),j=1..100);
```

the index j must be unassigned. One way to avoid using this command, after j has been inadvertently assigned a value, is to use this command in the following form instead.

```
> sum(f('j'),'j'=1..100);
```

We have used this punctuation before. A letter enclosed with single quotes is simply left unevaluated. Frankly, you may find it a lot easier to simply use the command without the punctuation. If you ever try to use the sum( ) command after assigning a value to the index j, rest assured that **Maple will tell you about the mistake.** It can then be easily corrected with an input statement like

```
> j:='j':
```

3.2 The Fundamental Theorem of Calculus

According to the *Fundamental theorem of calculus*, if f is continuous on the interval $[a,b]$, then the function F defined by

$$F(x) = \int_a^x f(t)dt, \tag{3.3}$$

is differentiable on $[a,b]$, and $F'(x) = f(x)$ for all x in $[a,b]$. As a consequence of (3.3), if G is any antiderivative of $f(x)$, then

$$\int_a^b f(x)dx = G(b) - G(a) = G(x)\mid_a^b . \tag{3.4}$$

Formula (3.3) is important for reasons other than the fact that it gives us formula (3.4), which is the most basic formula for evaluating definite integrals. Formula (3.3) also implies that every continuous function has an antiderivative, and it gives us a formula for it. This is of considerable consequence in its own right, and so we begin by using Maple to study F.

The command for evaluating a definite integral is **int( )**. This is one of many Maple commands that has an interesting **inert form—Int( )**. Remember that **inert forms are accessed by capitalizing the first letter of the command name.** Recall the function $f(x) = 9 - x^2$ from our Riemann sum example.

****

```
> Int(9-x^2,x=1..6)=int(9-x^2,x=1..6);
```

$$\int_1^6 9 - x^2\,dx = \frac{-80}{3}$$

****

Example 3.1 *Let* $f(x) = \sqrt{2+\sin^3(x)}$. *Find an antiderivative* F *of* f, *and verify. Then evaluate* $F(5)$ *as a decimal.*

It turns out that the antiderivative is quite complicated, and it is hard to express in any way other than as an integral by using the *Fundamental Theorem of Calculus.* We can use any number as the lower limit.

****

```
> f:=x->sqrt(2+sin(x)^3);
```

$$f := x \to \sqrt{2+\sin(x)^3}$$

```
> F:=x->int(f(t),t=1..x);
```

$$F := x \to \int_1^x \mathrm{f}(t)\,dt$$

```
> F(x);
```

#Notice that Maple **cannot evaluate the integral, and so it returns just the integral symbol itself.** #

$$\int_1^x \sqrt{2+\sin(t)^3}\,dt$$

```
> F(5);
```

#Maple can't evaluate the above integral either, so it returns just the symbol.#

$$\int_1^5 \sqrt{2+\sin(t)^3}\,dt$$

It can, however, compute the derivative of $F(x)$.#

```
> diff(F(x),x);
```

$$\sqrt{2+\sin(x)^3}$$

#This shows that $F(x)$ is an antiderivative of $f(x)$ and confirms the *Fundamental Theorem of Calculus* for this example.

Maple cannot find the exact value of $F(x)$ for any x (other than $x = 1$), but a decimal approximation for $F(x)$ can be found for any x. The definition of an integral as a Riemann sum is itself a numerical approximation technique, one that doesn't depend on a search for an antiderivative. There are other techniques introduced later in this chapter that work much better. We will discuss Maple's numerical approximation command for integrals shortly. For now, let us be brief. Decimal values for $F(x)$ are available in a very straightforward way. #

```
> evalf(F(5));
```

$$5.654170993$$

****

One of the most important results of calculus is that antiderivatives of continuous functions always exist. The above method can always be used to define an antiderivative. What makes this result so important is that many (in some sense "most") functions have antiderivatives that are not expressible in terms of the elementary functions of mathematics. Even functions as elementary as the one in the above example can have antiderivatives that are mysteriously beyond our reach in terms of familiar expressions.

3.3 Evaluating Definite and Indefinite Integrals

As long as Maple can find an antiderivative, evaluating a definite or indefinite integral is very straightforward. Maple **does not supply the arbitrary constant** for an indefinite integral.

****

```
> int(cos(x),x=1..3);
```

$$\sin(3) - \sin(1)$$

```
> int(7*x^5+1/sqrt(x)+5,x);
```

$$\frac{7}{6}\,x^6 + 2\,\sqrt{x} + 5\,x$$

****

Example 3.2 *Evaluate the integral* $\int_2^5 x^2 \sin(x)dx$. *How do you know the answer is correct?*

****

```
> int(x^2*sin(x),x=2..5);
```

#This should be the answer. Is it? #

$$-23\cos(5) + 10\sin(5) + 2\cos(2) - 4\sin(2)$$

```
> F:=int(x^2*sin(x),x);
```

#This should be an antiderivative. Is it? #

$$F := -x^2\cos(x) + 2\cos(x) + 2\,x\sin(x)$$

```
> F:=unapply(F,x);
```

$$F := x \to -x^2\cos(x) + 2\cos(x) + 2\,x\sin(x)$$

```
> diff(F(x),x);
```

$$x^2\sin(x)$$

#So F(x) is an antiderivative.#

```
> F(5)-F(2);
```

$$-23\cos(5) + 10\sin(5) + 2\cos(2) - 4\sin(2)$$

#This verifies the result.#

****

Maple can give exact values to definite integrals only if it can find a formula for the antiderivative. For continuous functions, antiderivatives always exist—that's what formula (3.3) says. "Most" of the time, however, there is no formula, other than (3.3), for expressing an antiderivative in finite terms by using elementary functions. This is the nature of mathematics, not a weakness in Maple. Even when a formula exists and it can be found by Maple, it may look so awkward, or be so computationally expensive to compute, that it is undesirable to attempt an exact value. When an antiderivative for a definite integral is hard or impossible to compute exactly, we should turn instead to a numerical approximation. We have already used the following pair of input statements in Example 3.1.

>int(f(x),x=a ..b); (This allows only for an exact answer.)
>evalf(%); (This gives a decimal approximation after we have an exact answer.)

The combination of these two commands,

>evalf(int(f(x),x=a ..b)); (The command int( ) is not capitalized.)

appears to be nothing more than a combination of the two preceding commands. In normal Maple syntax, that is all it would mean, but in this special case, it means a little more than that. Maple first tries an exact evaluation by looking for an antiderivative, and then it gives a decimal approximation, after it has determined an exact answer. This is what one would expect of this combination of commands. However, if an antiderivative cannot be found, it determines a numerical approximation to the integral.

To instruct Maple to make a **genuine decimal approximation** for a definite integral, **without first looking for an antiderivative**, we use the inert form Int( ) (capital I) of the integration command. The input statement,

> **evalf(Int($f(x), x = a..b$))**; (The command Int( ) is capitalized.)

prevents unwanted symbolic integration. If Maple cannot find an antiderivative, then the time it spends searching unsuccessfully for an antiderivative could be saved by capitalizing the "I" in the integration command.

Should we be concerned about the differences in these command combinations? Much of the time, it will not seem to matter which method is used, but in some cases, significant performance differences can be realized between these two methods of getting a decimal evaluation of an integral. For example, compare the two methods by evaluating $\int_0^1 \sec^{20}(x)\,dx$ as a decimal. The function $\sec^{20}(x)$ was selected here, because it has a complicated antiderivative, but one that Maple can find. To fully appreciate the comparison, it may also help to ask Maple to compute the antiderivative of $\sec^{20}(x)$.

Whenever the combination **evalf(Int( ))** is used, everything inside the integral **must evaluate to a number. No unassigned names are allowed.** With the command **evalf(int( ))**, on the other hand, **unassigned names are sometimes allowed inside the integral.** Maple must be able to find an antiderivative to use this combination with an unassigned name, and the result may not be much different from what would be obtained by using just the int( ) command.

****

```
> f:=x->sqrt(1+cos(x)^3);
```

$$f := x \to \sqrt{1+\cos(x)^3}$$

```
> int(f(x),x=0..9);          #Maple can't evaluate this.#
```

$$\int_0^9 \sqrt{1+\cos(x)^3}\,dx$$

```
> evalf(Int(f(x),x=0..9));
```

$$8.741295568$$

```
> evalf(Int(f(a*x),x=1..4)); # The name ''a'' is unassigned.#

Error, (in evalf/int) function does not evaluate to numeric
```

****

How does Maple perform these integral approximations? We could make them ourselves by approximating the integral with an appropriate Riemann sum. This is another reason why this definition is so important. Actually, Maple uses a much more sophisticated approach to approximating the value of a definite integral. We will return to this topic later in this chapter, but let us first consider some other matters.

Example 3.3 *Find a function $y = y(x)$ that satisfies the following differential equation, and verify that that your answer is correct.*

$$\frac{dy}{dx} = \frac{3x^4 + 7x^3}{x^6}, \textit{ and } y(-7) = 13$$

****

```
> F:=int((3*x^4 + 7*x^3)/x^6,x);
```

$$F := -\frac{3}{x} - \frac{7}{2}\frac{1}{x^2}$$

```
> F:=unapply(F+c,x);
```

#Experiment with this and see why this statement won't work with *c* outside of the unapply command.#

$$F := x \to -\frac{3}{x} - \frac{7}{2}\frac{1}{x^2} + c$$

```
> c:=solve(F(-7)=13,c);
```

$$c := \frac{177}{14}$$

```
> F(x);
```

> #This is the solution.#

$$-\frac{3}{x} - \frac{7}{2}\frac{1}{x^2} + \frac{177}{14}$$

```
> F(-7); diff(F(x),x);
```

#This verifies the result.#

$$13$$

$$\frac{3}{x^2} + \frac{7}{x^3}$$

****

3.4 Integration by Substitution

Maple is able to compute exact values for a large number of integrals without any human intervention (other than the initial command to integrate), but its capacity to integrate is not boundless. It is the nature of mathematics, and not a weakness in Maple, that frequently forces Maple to fail at exact evaluation. When this happens, it is likely that we too will fail if we try to get an exact evaluation through some alternate means. We have, however, one advantage over Maple, namely the ability to be creative and original. Occasionally, we can give Maple a push in the right direction by making a clever change in the variable of integration. The Maple command for changing the variable of integration is **changevar( )**. It resides in the **student** package, and so it can be used only after this package is loaded.

There is another, perhaps more important, purpose for this command. When an integral is evaluated, it is natural to wonder, especially if the result is unexpected, "How in the world did Maple get this answer?" By differentiating the answer you can verify that it is correct, but the air of mystery will persist. One way to take the mystery out of the answer is to use the changevar( ) command and other commands to manipulate the integral until it is in a familiar form. Assuming that Maple can actually evaluate the integral, you will want to prevent premature evaluation of the integral until you are ready to see the answer. To this end, we use the integration command in its **inert** form, **Int( )**.

Let's suppose, for the sake of the next example, that we are going to accept only the seven basic integration rules that follow immediately from the corresponding differentiation rules. These are the rules for the following integrals:

$$\int u^n du (n \neq -1), \int \sin(u) du, \int \cos(u) du, \int \sec^2(u) du$$

$$\int \sec(u)\tan(u) du, \int \csc^2(u) du, \int \csc(u)\cot(u) du$$

Example 3.4 *Use Maple to evaluate the integral $\int (5x-3)\sqrt{2-7x}dx$. Justify the answer by manipulating the integral symbolically until it is in one or more of the seven basic forms above.*

We first compute the integral directly with Maple's integration command.

****

```
> with(student):
```

#We intentionally suppressed the output by using a (:).#

```
> int((5*x-3)*sqrt(2-7*x),x);
```

$$\frac{2}{49}(2-7x)^{5/2} + \frac{22}{147}(2-7x)^{3/2}$$

```
> P:=Int((5*x-3)*sqrt(2-7*x),x);
```

#The name I for "integral" might be more appropriate here, but it is reserved by Maple to represent the imaginary number $I = \sqrt{-1}$. We used P for "Problem" instead.#

$$P := \int (5\,x - 3)\,\sqrt{2 - 7\,x}\,dx$$

```
> P1:=changevar(u=2-7*x,P,u);
```

$$P1 := \int -\frac{1}{7}\left(-\frac{11}{7} - \frac{5}{7}\,u\right)\sqrt{u}\,du$$

```
> P2:=expand(P1);
```

$$P2 := \frac{11}{49}\int \sqrt{u}\,du + \frac{5}{49}\int u^{3/2}\,du$$

```
> P3:=value(P2);
```

#The command value( ) is in the student package.#

$$P3 := \frac{22}{147}\,u^{3/2} + \frac{2}{49}\,u^{5/2}$$

```
> P4:=subs(u=2-7*x,P3);
```

$$P4 := \frac{22}{147}\,(2 - 7\,x)^{3/2} + \frac{2}{49}\,(2 - 7\,x)^{5/2}$$

****

Notice that $P4$ agrees with our original answer. Surely, this work gives us a deeper understanding of why the answer takes this form.

We end this section with an abrupt shift in directions to a Maple matter. As some of these commands in the student package (or any other package) become familiar to you, you will undoubtedly experience, eventually, the frustration of trying to use a command from a package before it is loaded. To prepare for this eventuality, **do it intentionally**, so that you can experience first-hand what happens in this situation. Enter the command "restart," so that the student package is no longer loaded, assign P to an inert integral, and try to use the changevar( ) command without loading the student package. Notice the form of the output and make note of it. When you see this behavior in Maple in the future, it might remind you that you haven't loaded a package.

3.5 Numerical Integration

Maple uses fairly sophisticated techniques for approximating the value of a definite integral. Rather than discuss these techniques, we look instead at *Simpson's Rule*, one of two numerical integration techniques typically introduced in a standard calculus course. An understanding of this rule will give us some insight into the general idea of numerical integration and will, we hope, give us an appreciation of Maple's evalf(Int( )) command in the process.

We begin by using Maple to derive the formula for *Simpson's Rule.* This is a fascinating experience to observe. Recall that, in setting up this rule, we divide the interval $[a, b]$ into an even number, $n = 2m$, of subintervals of equal length

$$h = \frac{b-a}{n}.$$

This determines the partition points

$$a = x_0 < x_1 \ldots < x_n = b, \text{ where } x_k = a + kh, (k = 0, 1, \ldots, n).$$

These subintervals are grouped together in adjacent pairs, and, on each of the pairs of subintervals, the function $f(x)$ is approximated by a quadratic function $q(x)$. Rather than integrate $f(x)$ on a subinterval, we integrate its quadratic approximation over the subinterval instead. *Simpson's Rule* is obtained by adding these m $(m = n/2)$ quadratic integrals.

Maple is an excellent environment in which to carry out all of these details. For each k $(k = 1, 2, ..., n)$, let $y_k = f(x_k)$. We **determine the function $q(x)$ on the first pair of subintervals and then integrate.** It is an easy matter to generalize the result to the other subinterval pairs. Let $q(x)$ be the quadratic function with $q(x_0) = y_0, q(x_1) = y_1$, and $q(x_2) = y_2$.

****

```
> q:=x->a*x^2+b*x+c:
> Q:=int(q(x),x=x0..x0+2*h);
```

$$Q := 2\,a\,x0^2\,h + 4\,a\,x0\,h^2 + \frac{8}{3}\,a\,h^3 + 2\,b\,x0\,h + 2\,b\,h^2 + 2\,c\,h$$

```
> eq1:=q(x0)=y0;    eq2:=q(x0+h)=y1;
  eq3:=q(x0+2*h)=y2;
```

$$eq1 := a\,x0^2 + b\,x0 + c = y0$$

$$eq2 := a\,(x0 + h)^2 + b\,(x0 + h) + c = y1$$

$$eq3 := a\,(x0 + 2\,h)^2 + b\,(x0 + 2\,h) + c = y2$$

```
> solve({eq1,eq2,eq3},{a,b,c});
```

$$\{b=-\frac{1}{2}\frac{2\,x0\,y0-4\,x0\,y1+2\,x0\,y2+3\,y0\,h-4\,y1\,h+y2\,h}{h^2},\ c=\frac{1}{2}(\\ x0^2\,y0-2\,x0^2\,y1+x0^2\,y2+3\,x0\,y0\,h-4\,x0\,y1\,h+x0\,y2\,h+2\,y0\,h^2)\Big/h^2\\ ,\ a=\frac{1}{2}\frac{y0-2\,y1+y2}{h^2}\}$$

```
> assign(%);
```

#This is an important new command which we will discuss later.#

```
> Q;
```

$$\frac{(y0-2\,y1+y2)\,x0^2}{h}+2\,(y0-2\,y1+y2)\,x0+\frac{4}{3}\,(y0-2\,y1+y2)\,h\\ -\frac{(2\,x0\,y0-4\,x0\,y1+2\,x0\,y2+3\,y0\,h-4\,y1\,h+y2\,h)\,x0}{h}-2\,x0\,y0\\ +4\,x0\,y1-2\,x0\,y2-3\,y0\,h+4\,y1\,h-y2\,h+\\ (x0^2\,y0-2\,x0^2\,y1+x0^2\,y2+3\,x0\,y0\,h-4\,x0\,y1\,h+x0\,y2\,h+2\,y0\,h^2)\Big/h$$

```
> simplify(Q);
```

$$\frac{1}{3}\,h\,(y0+4\,y1+y2)$$

****

An expression identical to this except for the subscripts on the symbols y_j will exist for each of the adjacent pairs of intervals. *Simpson's Rule* is easily seen to be the consequence of adding all m such expressions together.

In this last example, we introduced the important new command **assign( )**. Notice that the output line preceding the use of this command had the form

$$\{\ \mathrm{a}=\mathrm{expr}_1,\mathrm{b}=\mathrm{expr}\ _2,\mathrm{c}=\mathrm{expr}_3\}$$

Maple is telling us what the values of a, b, and c should be to solve the system of equations, but notice that Maple is using (=) **and not** (:=). This means that Maple is just telling us what the values should be, but it is **not making an assignment.** The **assign( )** command **turns these equalities into assignments.**

Example 3.5 *Use* Simpson's Rule *with* $n=50$ *to approximate the value of* $\int_1^4 \frac{x^3}{\sqrt{x^2+9}}dx$. *Estimate the size of the error term, using Simpson's 4th derivative error estimate. Find an exact value for the integral, and compare.*

If the 4 *th* derivative $f^{(4)}$ is continuous on the interval $[a,b]$, then the error term

$$E_S=\left|\int_a^b f(x)dx-S_n\right|$$

involved in approximating $\int_a^b f(x)dx$ by its *Simpson's Rule* approximation, S_n, satisfies

$$E_S < \frac{b-a}{180} M h^4$$

where

$$h = \frac{b-a}{n}, \text{ and } M = \max \mid f^{(4)}(x) \mid \text{ for } x \text{ in } [a,b]$$

While the error term can be a chore to compute by hand, notice in the next Maple work session how effectively this term can be estimated.

****

```
> f:=x->x^3/sqrt(x^2+9):
> a:=1:  b:=4:  n:=50:  m:=n/2:  h:=(b-a)/n:
```

#To set up the sum for *Simpson's Rule*, notice that the multipliers in front of the function values appear as $(1),(4),(2),(4),(2),(4),(2),\ldots,(4),(1)$, ending with a (4) and then a (1). They can be grouped so that they appear as

```
[(1), (4)], [(2), (4)], [(2), (4)], . . . ,  [(2), (4)], [(1)].
```

This may help explain why the pairs with $(2),(4)$ multipliers are grouped together as they are in the sum, leaving us with two exceptional terms in the beginning, and one exceptional term at the end. Then notice that the (2) multipliers appear in front of $f(x_j)$ (in other words $f(a+j*h)$) for j **even**, and the multipliers of (4) appear in front of $f(x_j)$ for j **odd**. This is why we used indices of the form $(2k)$ and $(2k+1)$ in the sum instead of simply (j).#

```
> S:=evalf(h/3*(f(a)+4*f(a+h)+f(b)

  +sum(2*f(a+2*k*h)+4*f(a+(2*k+1)*h),k=1..m-1));
```

$$S := 14.58624000$$

#This is the *Simpson's Rule approximation.* The use of the **decimal command evalf( ) is critical** in the above computation. Maple wants to be exact. **Imagine how complicated such a sum would be if we asked for an exact value.#**

\# We approximate the error term next. The word "error" is a natural name for this term, but Maple owns this name.#

```
> MyError:=(b-a)/180*h^4*M;
```

$$MyError := \frac{27}{125000000} M$$

#The value of M is the maximum of $|f^{(4)}(x)|$ on [1, 4]. We find this below.#

```
> fp4:=diff(f(x),x$4);
```

$$fp4 := -60 \frac{x}{(x^2+9)^{3/2}} + 225 \frac{x^3}{(x^2+9)^{5/2}} - 270 \frac{x^5}{(x^2+9)^{7/2}} + 105 \frac{x^7}{(x^2+9)^{9/2}}$$

#Think about what the value of M is going to be used for, and you will realize that it is inappropriate to attempt to find M to any degree of accuracy. All we need is to replace M by any convenient number a little larger than M. To find such a number, we plot $f^{(4)}(x)$ and by eyesight read off a convenient number slightly bigger than $\mid f^{(4)}(x) \mid$ on the interval. The choice is in the eye of the beholder.#

```
> plot(fp4,x=1..4);
```

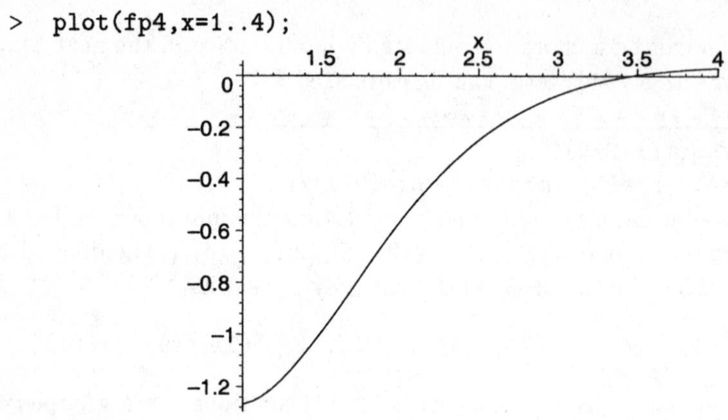


```
> M:=1.5:
> MyError;
```

$$.3240000000\,10^{-6}$$

#So the error term satisfies $MyError < .0000004$.#

```
> int(f(x),x=1..4);
```

$$-\frac{10}{3} + \frac{17}{3}\sqrt{10}$$

```
> evalf(%);
```

$$14.58624008$$

****

The next and last example presents an opportunity to make use of one of the most versatile and important commands in the Maple library. We introduced the procedure command **proc( )** earlier in this manual, but it was used only in a relatively simple way. Now we will use this command to formulate *Simpson's Rule* in a much more convenient way, and in the process we hope to demonstrate the power of this very useful procedure command.

Example 3.6 *Create a procedure command for determining the* Simpson's Rule *approximation of order n for* $\int_a^b f(x)\,dx$. *Use it to determine the* Simpson's Rule *approximation of order* 20 *for* $\int_1^8 \sqrt{3x+1}dx$ *and of order* 50 *for* $\int_0^\pi \sin(x)dx$

Think of f, a, b, n as variables. We create a procedure (a function) called S (for *Simpson's Rule*) such that the value of S at (f, a, b, n), denoted by $S(f, a, b, n)$, is the *Simpson's Rule* approximation of order n for the integral $\int_a^b f(x)dx$. Once the procedure is defined, we can evaluate $S(f, a, b, n)$, for various functions f and values a, b, and n, in much the same way we evaluate any other function. Notice that, in the first input statement below, the symbols f, a, b, and n just represent unassigned names. The main formula is essentially the same as the formula we used for the *Simpson's Rule* approximation in the last example. **The evalf( ) command plays an important role**. An exact sum would be much too complicated. Remember that a **procedure command must end with an "end."**

****

```
> S:=proc(f,a,b,n) h:=(b-a)/n; s1:=f(a)+4*f(a+h)+f(b);

  s:=2*f(a+2*k*h)+4*f(a+(2*k+1)*h); s2:=sum(s,k=1..n/2-1);

  evalf((s1+s2)*h/3); end;
Warning, 'h' is implicitly declared local

Warning, 's1' is implicitly declared local

Warning, 's' is implicitly declared local

Warning, 's2' is implicitly declared local
```

$S :=$ **proc**(f, a, b, n)
local$h$, $s1$, s, $s2$;
$h := (b - a)/n$;
$s1 := f(a) + 4 \times f(a + h) + f(b)$;
$s := 2 \times f(a + 2 \times k \times h) + 4 \times f(a + (2 \times k + 1) \times h)$;
$s2 := \mathrm{sum}(s, k = 1..1/2 \times n - 1)$;
$\mathrm{evalf}(1/3 \times (s1 + s2) \times h)$
end

#The warnings about certain names being **local** are not as ominous as they sound. We'll discuss this issue after we finish the problem.

With this command, look how easy it is to form the approximations. The integrals in this example are easily evaluated by antidifferentiation techniques, so we can compare the approximations with their exact values 26 and 2.#

```
> f:=x->sqrt(3*x+1):
> S(f,1,8,20);
```

$$25.99997554$$

```
> S(sin,0,Pi,50);
```

$$2.000000173$$

****

Now, let us discuss the dire warnings raised above when the proc( ) command was used. The word **local** here means that the **sphere of influence of these assignments begins and ends with the statement itself.** The assigned or unassigned status of these names is unchanged by the definition of the procedure and by its use. This is what we would want. Suppose a name, say, h, were assigned a value through a procedure. When the procedure is used, we would not want the letter h (a letter we no longer even see) to suddenly pick up an assigned value without our knowledge. The short work session below demonstrates this local issue. Look at how h and k are unchanged.

Also of interest, **the arguments of the proc( ) command are always local variables.** This issue was discussed before on page 20. In the following worksession, notice that x was initially assigned a value, but it had no influence on the definition of the procedure or on its use.

Here, then, is a demonstration of these "local" issues. This worksheet has no other purpose.

****

```
> x:=3:  k:=1:
> p:=proc(x,y) h:=x+y;k:=x*y;h+k;end;

Warning, 'h' is implicitly declared local

Warning, 'k' is implicitly declared local
```

$$p := \mathbf{proc}(x,\, y)\ \mathbf{local} h,\, k;\ h := x + y\,;\ k := x \times y\,;\ h + k\,\mathbf{end}$$

```
> p(100,5);
```

$$605$$

```
> x; y; h; k;
```

$$3$$

$$y$$

$$h$$

$$1$$

****

3.6 Exercise Set

New Maple Commands in this Chapter (and a few old commands as a reminder)

assign( )	changevar( )	evalf(Int( ))	evalf(int( ))	expand( )
fsolve( )	int(,x)	int(,x=a..b)	int( )	Int( )
leftbox( )	leftsum( )	middlebox( )	middlesum( )	proc( )
rightbox( )	rightsum( )	sum( )	value( )	with( )

Remember to load the *student* package when it is needed.

1. Set up and evaluate (as decimal numbers) the left endpoint, right endpoint, and midpoint Riemann sums for $n = 30$ corresponding to the integral $\int_{-3}^{2} x^3 dx$. Draw the graphs showing the approximating rectangles. Compute the exact value of the integral by taking the limit of the left endpoint Riemann sums as n, the number of rectangles, goes to infinity.

2. Let $f(x) = \frac{x}{x+\sin(x)}$

a) Use Maple to find a function $F(x)$ such that $F'(x) = f(x)$, for x in the interval $[1, \infty)$.If Maple cannot find the antiderivative directly, use the Fundamental Theorem of Calculus to define it.

b) Evaluate $F(18)$ and $F(3)$ as decimal numbers. Compare the value of $F(18) - F(3)$ with a direct evaluation of $\int_3^{18} f(x)dx$ as a decimal. Why are the answers the same?

c) Use one of Maple's differentiation commands to compute $F'(x)$. Verify the answer by using the definition of a derivative as a limit of a difference quotient to compute $F'(x)$.

d) Use Maple to compute the derivatives of the functions

$$g(x) = F(3x^4 + 7x^2 + 9), h(x) = F(\sin(x)).$$

The answers should be foreseen without using Maple. Why?

e) Find the equation of the tangent line to the graph of g at $x = 4$, for the function g in part d).

3. Find exact values for the following integrals. Verify that your answers are correct.

a) $\int (2 + x + x^2)\sqrt{3x - 5}\, dx$ *b)* $\int \sin^3(2y) \cos^4(2y)\, dy$

c) $\int_0^3 (x^3)\sqrt{25 - x^2}\, dx$

4. Evaluate the integral $\int_2^7 \frac{\sin(x)}{\sqrt{x}} dx$ as a decimal.

5. Find y as a function of x if $\frac{dy}{dx} = 3x^3 + 7x^2 - 4x - 8$, and if $y = -17$ when $x = 6$.

6. Find y as a function of x if $\frac{d^2y}{dx^2} = \frac{7x+5}{(3x+8)^3}$, $y(-1) = 2$, and $y\prime(-1) = 4$. Verify that the answer is correct.

7. Show that $\int_a^b f(x)g(x)\, dx = \int_a^b f(x)\, dx \int_a^b g(x)\, dx$ is *false*, by using an appropriate counterexample.

8. Evaluate the integral $\int \sin(3x)\sqrt{4+\cos(3x)}dx$. Verify Maple's answer, by manipulating the original integral algebraically and changing the variable of integration until it looks like one or more of the seven basic integrals.

9. Evaluate $\int_1^3 \sqrt{x^2+1}\,dx$ as a decimal. Use the *Trapezoidal Rule* and *Simpson's Rule* with $n = 20$ to estimate the integral. Approximate the size of the error terms by using formulas for the error terms given in your calculus text.

10. Maple can "prove" (if that's what you should call it) the *Fundamental Theorem of Calculus.* This is an interesting (and easy) exercise. It will make you think about the proof, and it will demonstrate how good Maple is at handling really abstract mathematical expressions. Define

$$F(x) = \int_a^x f(t)dt$$

and use Maple to compute

$$F'(x) = \lim_{h \to 0}\left(\frac{F(x+h) - F(x)}{h}\right).$$

Incidentally, Maple assumes continuity of the function f in this problem.

Project: Cubic Rule

Use the following instructions to derive another numerical integration rule for $\int_a^b f(x)dx$. You might want to call it the Cubic Rule. Divide the interval $[a, b]$ into $n = 3m$ subintervals of equal length

$$h = \frac{b-a}{n}.$$

This establishes the partition points $a = x_0 < x_1 < .. < x_n = b$ in the usual way. Let $y_k = f(x_k)$ for each $k = 1, 2, \ldots, x_n$. Group the subintervals into adjacent triplets of subintervals. (This can be done, because n is a multiple of 3.) In each triplet of subintervals, construct a cubic polynomial that agrees with $f(x)$ at the four partition points in the interval. Approximate the integral of $f(x)$ on this triplet of subintervals by integrating the cubic instead. Add up all $m = n/3$ of these approximations in order to get the sought after Cubic Rule. (Use the derivation of *Simpson's Rule* in this manual for motivation. Do all your computations on the first triplet of subintervals, simplify, generalize to the other triplets, and add up the results.)

Express your rule as a Maple procedure in the variables f, a, b, and n. Remember that Maple wants to be exact, so put evalf( ) around your expression, or it will lead to undesirable complications.

Approximate the integral $\int_0^\pi \sin(x)dx$, using this new rule with $n = 48$ (a multiple of 3). Compare your answer with the answer obtained by using *Simpson's Rule* with $n = 48$, and also with the approximation obtained by using Maple's evalf(int( )) command. Finally, notice that this is an easy integral with an exact value of 2, so you can compare all three of these approximations to the exact value. One would expect the Cubic Rule to be an improvement over *Simpson's Rule*; after all, we're approximating by cubic polynomials rather than quadratics. Cubics should "fit the curve" better. However, you probably will not experience an improvement in approximating this integral.

Chapter 4

Applications of Integration

We began the last chapter with a remark concerning the importance of the definition of a definite integral as the limit of its Riemann sums. As you can now see in your calculus text, science is rich with examples of situations that have an interpretation as a limit of some Riemann sum, and hence as a definite integral. In this chapter, we follow up on the ideas presented in your calculus text and show how Maple can be used to solve some interesting applied problems.

4.1 Sample Problems

Before we consider our first example, a major new Maple command will be introduced. We remarked earlier that **Maple can plot equations, but not with the plot( ) command**. An equation, recall, is simply an expression that involves the equality symbol (=). The command for plotting equations in two variables resides in a Maple package called the **plots package**. We suggest that you enter the input statement **with(plots)** on your computer followed by a (;) rather than (:) so that you can see the list of special plotting commands available in Maple. Only one of these commands will be used at this time, and so we have suppressed displaying the list of commands just to avoid filling up the page. There will be many occasions to use other commands from this package in future chapters.

The command we use to plot equations in two variables is **implicitplot( )**.

Example 4.1 *Find the area of the region between the circle* $x^2 + y^2 = 25$ *and the cup of the parabola* $x = 7y - 3y^2 - 1$.

To begin this problem, a plot is essential. Notice that we have to give Maple both an x-range and a y-range when we use the implicit plot command.

****

```
> with(plots):

Warning, the name changecoords has been redefined
```

```
> eq1:=x^2+y^2=25:              eq2:=x=-1+7*y-3*y^2:
> implicitplot({eq1,eq2},x=-10..10,y=-10..10)
  ;
```

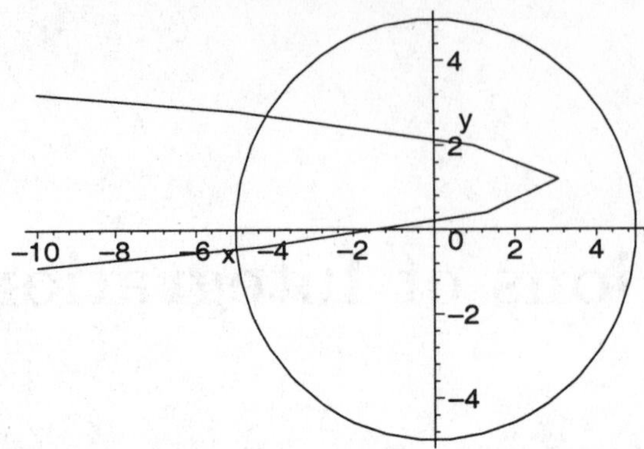

#The command implicitplot( ) will consume a fair amount of memory. The amount of memory that has been used shows in the right-hand corner of the Status Line at the bottom of the computer screen. This is a good time to be reminded of the benefits of quitting Maple and restarting as a means of memory management. From the above plot, it is clear that it would be very difficult to find the area of the region by dividing it up into "thin" vertical rectangles. The formula for the length of a "thin" rectangle located over (or on) the point x on the x-axis appears to change two times (actually three times) depending on what curves the top and bottom sides of the rectangle intersect. (The third change is quite small.)

On the other hand, a "thin" horizontal rectangle always has its right endpoint on the parabola and its left endpoint on the left half of the circle. Think of a "thin" horizontal rectangle passing through the point y on the y-axis. Its width (the small dimension) is dy and its length is $x_2 - x_1$, where (x_2, y) is a point on the right boundary ($eq2$), and (x_1, y) is a point on the left boundary ($eq1$).

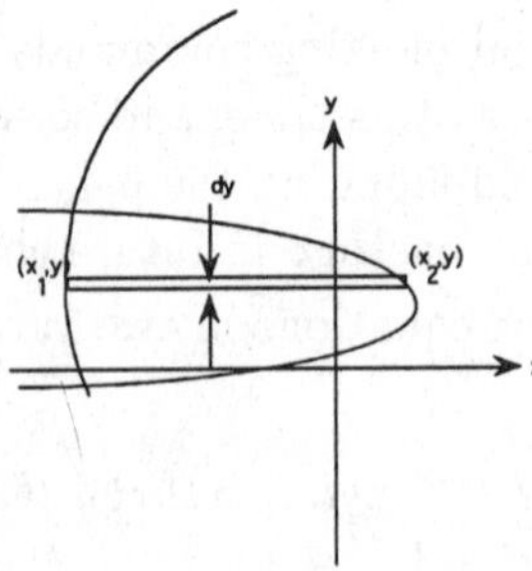

A horizontal distance is always obtained by taking the x-coordinate on the right minus the x-coordinate on the left. We plan to integrate (collect and add rectangles) with respect to y, so these x-coordinates must be expressed in terms of y.#

```
> x1:=solve(eq1,x);   x2:=solve(eq2,x);
```

$$x1 := \sqrt{25 - y^2}, -\sqrt{25 - y^2}$$

$$x2 := -1 + 7\,y - 3\,y^2$$

```
> x1:=x1[2];
```

$$x1 := -\sqrt{25 - y^2}$$

```
> L:=x2-x1;
```

$$L := -1 + 7\,y - 3\,y^2 + \sqrt{25 - y^2}$$

#The area of this "thin" horizontal rectangle is just Ldy. The dy is an assumed part of the integration command. If you write dy inside the integral, Maple will not understand. Collecting and adding, in the sense of integration, all of the "little" rectangular areas, means integrating L with respect to y.#

```
> fsolve(L=0,y);
```

$$-.4725358063$$

```
> a:=%;
```

$$a := -.4725358063$$

```
> b:=fsolve(L=0,y,y=1..3);
```

> **# Notice that Maple needed some help here to find the second point.#**

$$b := 2.724030226$$

```
> area:=evalf(Int(L,y=a..b));
```

$$area := 16.94616710$$

****

When we use the plot( ) command to plot two or more functions, Maple uses a different color for each curve, but it does not tell us which color belongs to which function. With just a little insight about the functions being plotted, we can usually make this identification by ourselves. If it is convenient, Maple can be configured to supply this information. Pull down the Options Menu and click **Display 2-D Legends** so that a check mark appears in front of this menu item. When we use the plot( ) command, Maple will now tell us which color is being used for curve 1, which color for curve 2, and so forth. This is useless information, however, unless we know how the functions are ordered, or in other words, which function is curve 1, which is curve 2, and so forth.

We must digress. **A collection $\{a, b, c, \ldots\}$ in Maple, formed by using curly brackets, is regarded as a set.** Maple's definition of a set is the same as the mathematical definition. The order in which the terms are listed is of no consequence. Maple does not necessarily regard a as the first term. If Maple is asked to list the elements of a set, it will list them in a random fashion.

A collection $[a, b, c, \ldots]$ in Maple, formed by using square brackets, is regarded as a list. A list in Maple **is an ordered set.** The order in which the terms are listed specifies the ordering of the list.

When we plot the curves $f(x), g(x), \ldots, h(x)$ in the same plot, it has been our practice to enclose them inside the plot( ) command with curly brackets to form $\{f(x), g(x), \ldots, h(x)\}$. This has the intended effect of isolating this group to the first argument of the plot command, but the **curves are not ordered, and Maple will use a random method when it labels them curve 1, curve 2, and so forth.**

The functions $f(x), g(x), \ldots, gh(x)$ can also be enclosed inside the plot( ) command with square brackets, to form $[f(x), g(x), \ldots, h(x)]$. **Because this is an ordered list, Maple will definitely label $f(x)$ as curve 1, $g(x)$ as curve 2, and so forth.**

In the next example, we turn on the 2-D Legend and we enclose the functions inside the plot( ) command with square brackets rather than curly brackets so that the functions to be plotted are ordered.

Example 4.2 *Find the volume of the solid of revolution formed by revolving, about the line $x = 10$, the region bounded by the curves $y = x^2 - 8x + 26$ and $y = 4 - 12x - x^2$*

****

```
> f:=x->4-x^2+12*x:                g:=x->x^2-8*x+26:
> plot([f(x),g(x)],x=0..10);
```

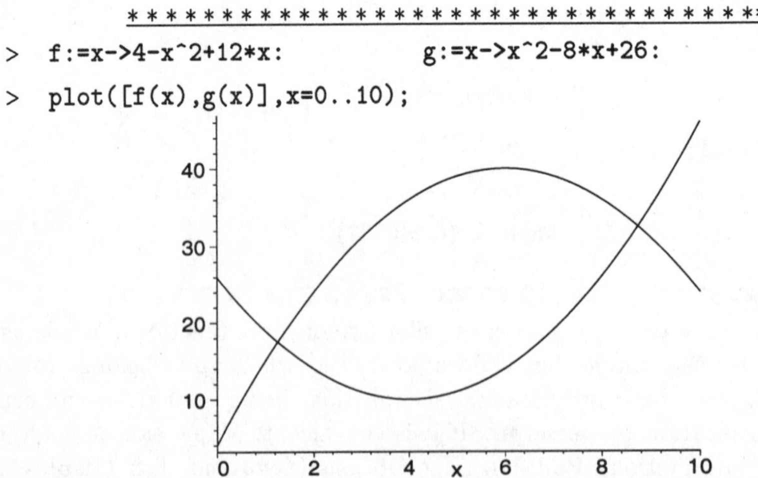

#If this is viewed on a computer screen, the plot legend would appear at the bottom of the plot, indicating which color is assigned to *curve-1* ($f(x)$) and which color is assigned to *curve-2* ($g(x)$). Unfortunately, the legend does not appear in the above figure, and it can be appreciated only if it is viewed in an actual Maple worksheet.

A "thin" vertical rectangle located at the point x on the x-axis inside this region and revolved about the line $x = 10$ would generate a shell. Collecting shells with respect to x, means integrating with respect to x, and so all of the variables must be expressed in terms of x. The volume of a shell is

$$v = 2\pi RLT,$$

where R is the radius, L is the length of the shell, and T is its thickness.

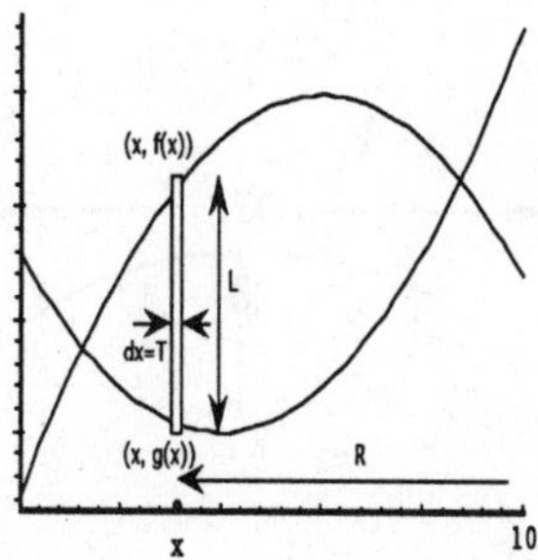

The thickness, $T = dx$, is an assumed part of Maple's integration command, and so it is not a part of our expression for v. Collecting and adding up, in the sense of integration, all of the volumes of the "little" shells means integrating $v = 2\pi RL$ once we have R and L in terms of x.#

```
> L:=f(x)-g(x);
```

#This vertical length is the y-coordinate above minus the y-coordinate below.#

$$L := -22 - 2\,x^2 + 20\,x$$

```
> R:=10-x;
```

#This horizontal length is the x-coordinate on the right minus the x-coordinate on the left.#

$$R := 10 - x$$

```
> v:=2*Pi*R*L;
```

$$v := 2\,\pi\,(10 - x)\,(-22 - 2\,x^2 + 20\,x)$$

```
> s:=fsolve(f(x)=g(x),x);
```

$$s := 1.258342613,\ 8.741657387$$

```
> volume:=evalf(Int(v,x=s[1]..s[2]));
```

$$volume := 4388.444987$$

****

Example 4.3 *A swimming pool has an elliptical shape, as shown in the picture. The water is 3 feet deep at one end, 12 feet deep at the other end, and the depth changes linearly from one end to the other. How much water (in cubic feet) is in the pool, and what is its weight in pounds?*

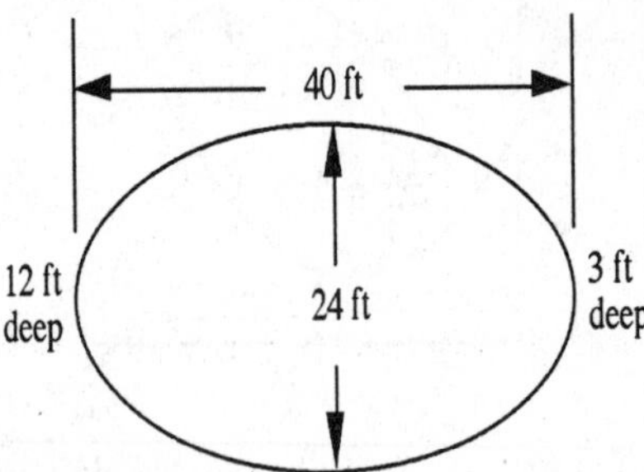

The equation of an ellipse centered at the origin with a horizontal major axis of length 40 and a vertical minor axis of length 24 is

$$\frac{x^2}{20^2} + \frac{y^2}{12^2} = 1$$

It is easy to see that the ellipse crosses the x-axis at $(\pm 20, 0)$ and crosses the y-axis at $(0, \pm 12)$. By using cuts of thickness dx perpendicular to the x-axis, the (3-dimensional) pool can be partitioned into thin, essentially rectangular boxes, or "slabs" of water. The volume of this rectangular "slab" of water is easily expressed as the product of its three dimensions.

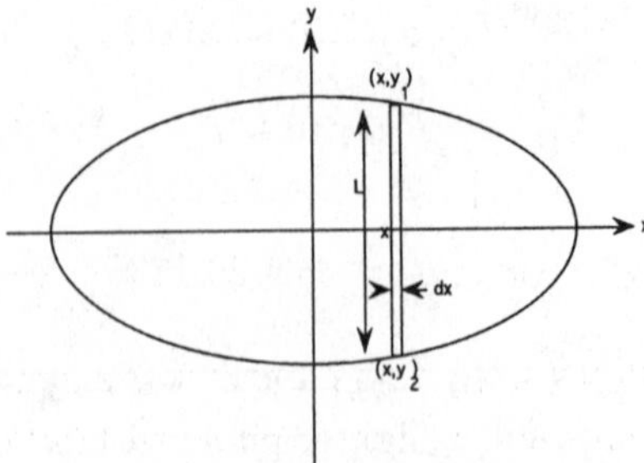

The three dimensions of the slab are L, dx, and the dimension, d (not visible in the diagram), that is perpendicular to the plane of the paper and represents the water depth of the pool at x. The dx is an assumed part of the integration command, and so the volume of the thin slab takes the form $v = Ld$. We add (in the sense of integration) the volumes of all of the thin slabs to form the total volume of the water. The answer, naturally, is expressed in cubic feet, and the total weight of the water is obtained by multiplying its volume by water's density 62.5 pounds per cubic foot.

Both L and d must be expressed in terms of x in order to do the integration. The relationship between d and x is linear. We simply write the equation of the line through the points $(x, d), (20, 3), (-20, 12)$.

****

```
> eq:=x^2/20^2+y^2/12^2=1:
> d:=3+(12-3)/(-20-20)*(x-20);
```

#We used the name d rather than D. The name D is a Maple command.#

$$d := \frac{15}{2} - \frac{9}{40}\,x$$

```
> s:=solve(eq,y);
```

$$s := \frac{3}{5}\,\sqrt{400 - x^2},\ -\frac{3}{5}\,\sqrt{400 - x^2}$$

```
> L:=s[1]-s[2];
```

#This vertical length is the y-coordinate above minus the y-coordinate below.#

$$L := \frac{6}{5}\,\sqrt{400 - x^2}$$

```
> v:=L*d;
```

$$v := \frac{6}{5}\,\sqrt{400 - x^2}\,(\frac{15}{2} - \frac{9}{40}\,x)$$

```
> volume:=evalf(Int(v,x=-20..20));
```

$$volume := 5654.866776$$

```
> weight:=62.5*volume;
```

$$weight := 353429.1735$$

****

Captain Ralph is an outer-space fighter pilot, with whom we will share an occasional adventure in the future. Our first encounter with him is hardly an adventure, but it would be if he fails in his mission.

Example 4.4 *Captain Ralph is piloting a small shuttle between a space station orbiting Mars and the planet's surface. The shuttle (standing vertically up) has the shape of the solid of revolution formed by revolving about the y-axis the region in the* *first quadrant bounded by the coordinate axes and* $y = 36 - (9x^4)/4$ *(The units are all in meters.) As you can see, the ship is quite narrow, being* 36 *meters tall and only* 4 *meters wide at the base. (If you wish to get a "true" picture of the ship by plotting this curve, you should plot with the same scale on both axes.) The ship is also quite nose heavy, with a mass density of* $\delta = 219 + 50\sqrt{y}$ *kilograms per cubic meter at level* y. *Density here should be thought of as the mass of a "thin horizontal slice" at level* y *divided by its volume with the mass of a slice centered on the* y*-axis.*

Captain Ralph has been ordered to transport a 2,000 *pound laser gun down to the planet's surface, and Ralph has decided that the only safe way to do this is to place the gun as close to the ship's center of mass as possible. Please help Captain Ralph avoid an adventure by finding this point.*

To help Captain Ralph, we must find both the total mass M of the ship and its moment M_x with respect to the x-axis. Then the y-coordinate of the center of mass would be $\overline{y} = M_x/M$. The center of mass is on the y-axis, so $\overline{y}$ is all we need. To generate these terms, we partition the ship into "thin" horizontal disk-shaped slices. The thickness of a slice is dy, which is omitted from all of our calculations for the same reason as in previous examples. Except for the omission of dy, we have that the volume v of a slice is $v = \pi r^2$, its mass is $m = vd(d = \delta)$, and its moment with respect to the x-axis is $m_x = my$.

****

```
> f:=x->36-9/4*x^4:
> plot(f(x),x=-18..18,y=0..37):
```

#Output omitted, but notice the scales used to get a "true" picture of the ship. Maple normally takes artistic liberty to fill a plot region with a "best fit" graph. #

```
> r:=solve(f(x)=y,x);
```

$$r := \frac{1}{3}(1296 - 36\,y)^{1/4},\ \frac{1}{3}I\,(1296 - 36\,y)^{1/4},\ -\frac{1}{3}(1296 - 36\,y)^{1/4},\ -\frac{1}{3}I\,(1296 - 36\,y)^{1/4}$$

#Maple, of course, supplies all four solutions. We should be careful which one we select. Is the term $(1296 - 36y)$ positive or negative? The answer determines which one of the expressions is real and positive. In the present case, the choice is the one that "looks" real and positive.#

```
> r:=r[1]:
> v:=Pi*r^2;
```

$$v := \frac{1}{9}\pi\sqrt{-36\,y + 1296}$$

```
> d:=219+50*sqrt(y);
```

$$d := 219 + 50\sqrt{y}$$

```
> m:=d*v;
```

$$m := (219 + 50\sqrt{y})\,\pi\,\sqrt{16 - \frac{4}{9}y}$$

```
> mx:=m*y;
```

$$mx := (219 + 50\sqrt{y})\,\pi\,\sqrt{16 - \frac{4}{9}y}\,y$$

```
> M:=evalf(Int(m,y=0..36));
```

$$M := 119344.7077$$

```
> Mx:=evalf(Int(mx,y=0..36));
```

$$Mx := .1910428901\,10^{7}$$

```
> ybar:=Mx/M;
```

$$ybar := 16.00765495$$

#Before we leave this example, let us return to the solution of the equation $f(x) = y$ for x in terms of y that we discussed above. Luck was with us, and we were able to decide easily which solution was real and positive. The choice might not always be so obvious. For example, if $y > 36$, then $r[4]$ would have been the positive real expression. A new Maple command might help.

If we had solved the equation for x^4 instead of for x, we could have avoided the complications involved in taking roots. The **new Maple command, isolate( )**, is specifically designed to accomplish this kind of manipulation. If we had used the isolate command instead of the solve command in the last example, the problem would have developed as follows.#

```
> isolate(f(x)=y,x^4);
```

$$x^4 = -\frac{4}{9}y + 16$$

```
> r2:=sqrt(rhs(%));
```

#The name r2 is the square of the value we assigned to r in the original solution.#

$$r2 := \frac{2}{3}\sqrt{-y + 36}$$

****

So far, all of the applications we have considered were a consequence of the definition of a definite integral as a limit of its Riemann sums. Many other applications are a consequence of interpreting an integral as an antiderivative.

Example 4.5 *A new mutual fund is growing at the rate of* $.83 - 1/\sqrt{(t+5)}$ *in millions of dollars per year, where t is the age of the fund in years. If the initial value of the fund was $94,000, how long will it take for the fund to reach a value of $7,000,000 ?*

If r denotes the above rate, and if v denotes value at time t in millions of dollars, then $r = \frac{dv}{dt}$.

****

```
> r:=.83-1/sqrt(t+5):
> v:=int(r,t)+c;
```

$$v := .8300000000\, t - 2.\, \sqrt{t+5.} + c$$

```
> c:=fsolve(subs(t=0,v)=.094);
```

#Notice that$94,000 is .094 in millions of dollars.#

$$c := 4.566135956$$

```
> years:=fsolve(v=7,t);
```

$$years := 13.21704965$$

****

4.2 Exercise Set

New Maple Commands in this Chapter (and a few old commands as a reminder)

assume( )	assuming	evalf(int( ))	fsolve( )	implicitplot( )
int( )	isolate( )	readlib( )	solve( )	subs( )

Remember to load the *plots* package when it is needed. In addition, the command isolate( ) must be loaded with the readlib( ) command before it can be used.

To avoid interference from previous Maple activity, start new work with the **Restart command.** In Maple 7, there is a Restart button in the upper right-hand corner of the menu-bar.

1. Find the area of the region in the first quadrant bounded by

$$y = \frac{20}{x^2+1}, \quad x = 2, \text{ and } y = 1.$$

Verify your answer by an alternative calculation.

2. Find the area between the curves

$$y = 3x^4 + 4x^3 - 54x^2 - 108x + 1, \text{ and } y = -75 - 36x.$$

3. Find the area between the curves

$$x + 2y^2 - 28y + 96 = 0, \text{ and } x^2 - 8x - 3y + 7 = 0.$$

4. Find the volume of the solid of revolution formed if the region in the first quadrant bounded by the curves $y = \frac{20}{x^2+1}$, $x = 2$, and $y = 1$ is revolved about
a) the x-axis b) the y-axis
c) the line $x = -4$ d) the line $y = 6$.

5. Verify each of your answers in Problem 4 by an alternative calculation.

6. Use Maple and exact integration to derive a formula for the volume of a cone of height H and radius (of the base) R.

7. Use Maple and exact integration to derive a formula for the volume of a sphere of radius R.

8. Use exact integration to derive a formula for the surface area of a sphere of radius R. With the same variable of integration, this surface area problem can be set up in two substantially different ways. Compute both integrals.

9. Consider the torus generated by revolving about the x-axis the circle of radius r centered at the point $(0, R)$ on the y-axis, where r and R are fixed positive numbers with $R > r$. Use Maple and exact integration to find a formula for the volume of the torus. Verify your answer with an alternative calculation.

To evaluate the integral, Maple must be told that $R > r > 0$. We make an assumption on R and r, and then an additional assumption on r by entering

```
> assume(R>r);additionally(r>0);
```

These commands were introduced in Chapter 2. If you are using Maple 7, try the "assuming" facility. Look it up in the Index or in Maple Help.

10. The region bounded by the curves $y = x^2 - 8x + 8$ and $y = 34x - x^2 - 45$ has the mass density $\delta = 7 + \sqrt{x}$ (in grams per square centimeter) at each of its points on a vertical line passing through x on the x-axis. Find the mass and the center of mass of the region.

11. Use Maple and exact integration to show that the center of mass of a triangle in the (x, y)-plane having uniform (constant) mass density δ lies at the intersection of the medians, one third up from the midpoint of each side towards the opposite vertex.

Surely there is no loss in generality if we place one vertex at the origin and one side along the x-axis. The figure appears on the next page.

12. Find the length of a thin wire defined by the curve $y = \sqrt{x} + x\sin(x)$ for x in the interval $[1, 10]$. Assuming constant mass density δ, find the center of mass.

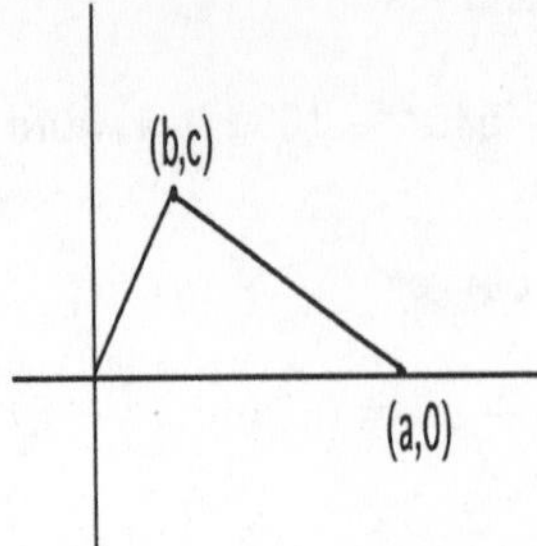

13. A flat circular, glass port hole with a radius of 9 inches has been built into the side of a pool so that underwater activity can be monitored. The top edge of the window is 7 feet below the water surface. What is the total force acting on the glass?

14. A diving bell having a vertical, flat, semicircular glass window with a radius of 4 feet is used to take tourists down into the ocean to view the wonders of nature. The straight side of the semicircle is the horizontal bottom edge, and the window is built to safely withstand a force of 87,000 pounds. How deep (measure from the water surface to the bottom of the window) can the diving bell safely dive? If the diving bell is deep enough, one would expect, intuitively, that pressure is almost constant across the entire window, in which case we could get the force acting on the window simply by multiplying pressure times area. Discuss this matter. Is this a reasonable approximation, at the maximum safe depth?

15. Executives for a logging company are considering the construction of a railroad line from a point where raw lumber is collected to its processing mill, and a map of the region involved is being studied to decide where to build the rail line. A rectangular coordinate system has been placed on the map with the origin at the center of the region. The units are in miles. The raw lumber is collected at the point (0,10), and the processing mill is at the point (20,0). Management must choose between the two possible routes labeled C_1, and C_2 below, which connect these two points. For each of the two rail lines, company engineers have determined a formula for the approximate force that would have to be exerted at the point (x, y) on the road in order to pull a "standard" train at a constant 30 mph pace throughout the course. These approximate force formulas are labeled f_1, and f_2 below. This information will be used to determine the total work energy required to move a standard train over each of the two rail lines under consideration. Management has decided to choose the rail line that requires less total energy. The road C_1 and the force f_1 (in thousands of pounds) at the point (x, y) on this road are defined by

$$C_1 = \{(x, y)|y = \frac{20 - x}{2x + 2}, 0 \leq x \leq 20\}, f_1 = 317 - x^2 + 10x - 0.4y^2 + 0.8y,$$

The road C_2 and the force f_2 (in thousands of pounds) at the point (x, y) on this road are defined by

$$C_2 = \{(x, y)|y = \frac{2000 - x^3 + 30x^2 - 300x}{200}, 0 \leq x \leq 20\},$$

$$f_2 = 132 - 2x^2 + 40x - 3y^2 + 30y.$$

Notice that both curves begin and end at the points (0,10) and (20,0). It is now your job to supply management with these total energy numbers.

16.A model rocket is fired straight up from the ground. The rocket's engine fires for 10 seconds, producing an acceleration $a(t) = 100 + 58\sqrt[3]{t-5}$, $(0 \leq t \leq 10)$. At $t = 10$ seconds, the engine turns off and the motion of the rocket is subject to just gravitational acceleration. How high does the rocket go, and how long does it take to reach the high point? What is its maximum speed? How long does it take to fall back to earth, and at what velocity does it strike the earth? Would this be a good toy for a 10 year old child?

17. A mineral survey of private farm land suggests that the land contains approximately 500,000 barrels of oil. The owner of the land plans to install a low capacity oil well, and the survey crew has estimated that it would be capable of producing oil at the rate of $(\frac{813}{\sqrt{t+7420}} - 1)$ barrels per hour, where t is the age of the well, in hours. The machinery will require very little attention to operate, and so it will be allowed to continue pumping until the production rate drops below 1 barrel per hour, at which time the well will be abandoned. How long, in years, will the well produce oil? How much oil will be recovered, and how much will remain underground when the well is abandoned?

Chapter 5

The Transcendental Functions

According to the dictionary, the word transcendental means something that is beyond common thought or experience, something that is mystical or supernatural. In mathematics, a transcendental expression is an expression that is not algebraic. It is an expression that is somehow beyond algebra. An expression $f(x)$ is algebraic if $y = f(x)$ satisfies an equation of the form

$$a_n(x)y^n + a_{n-1}(x)y^{n-1} + \ldots a_2(x)y^2 + a_1(x)y + a_0(x) = 0,$$

where the expressions $a_0(x)$, $a_1(x), \ldots,$ $a_n(x)$ are all polynomials in x.

The trigonometric functions are all transcendental (beyond algebra), although this is not entirely obvious. In addition to the trigonometric functions, there are several other special transcendental functions that are widely used in mathematics. They will be introduced in this chapter, and their properties relative to calculus will be studied.

Several of these special functions are inverses of other special functions, and so we begin with a section on inverse functions. The chapter ends with a study of the long anticipated *L'Hôpital's Rule.* This important rule gives us "the final word" on how to evaluate limits of expressions when the limits are not obvious. *L'Hôpital's Rule* could have been discussed much earlier in the development of calculus, but it is placed in this chapter because so many really mysterious limit examples involve transcendental functions.

5.1 Inverse Functions

Finding the inverse of a function f involves nothing more than solving the equation $y = f(x)$ for x in terms of y. A symbolic solution is called for here, so the **solve( ) command must be used rather than the fsolve( )** command. Actually, the fsolve( ) command could be used to find the inverse of a function at a particular point, but it cannot be used to find a formula for the inverse. As we have seen, Maple's solve( ) command is a powerful tool for solving equations, but its ability to solve more complicated equations is severely limited by mathematics itself. Mathematicians speak confidently about the existence of solutions, but actually finding solutions is another matter. Frequently, solutions cannot be found, even though they clearly exist, and this happens to be the destiny of mathematics, and not a reflection of the person doing the work.

Surprisingly, even the solution to an equation as straightforward as a polynomial can be impossible to find. A formula for solving a second-degree polynomial equation is well known to every student of algebra, and formulas also exist for solving third-degree and fourth-degree polynomial equations, although they are less well known. The Norwegian mathematician Niels Abel proved in 1824, however, that there is no formula for solving a general fifth degree polynomial equation, and later the French mathematician Evariste Galois proved that there is no formula for solving a general polynomial equation of degree n for any $n \geq 5$. By a formula here, we mean a formula based on root-taking and on the operations $+, -, \times, \div$ that would be applied to the coefficients of the polynomial. Perhaps even more striking, it can be shown, for example, that not even one root of the equation

$$x^5 - 9x + 3 = 0$$

can be expressed by root-taking and by applying the operations of $+, -, \times, \div$ to integers (or rational numbers).

Give Maple a polynomial equation of degree 5 or more, and Maple will not be able to solve it, unless it just happens to be an equation that factors easily. This is unlikely to happen, unless the equation was set up to factor.

We will now use Maple's powerful solve command to find a few inverses, but at the same time let us appreciate how very limited we are in this practice. If you experiment with finding inverses on your own, you are likely to run into rather complicated expressions.

Example 5.1 *Find the intervals where*

$$f(x) = \frac{x}{x^2 - 2x + 4}$$

is one-to-one. In each of these intervals : a) find the corresponding inverse function $f^{-1}(x)$, *b) verify that the identities* $f(f^{-1}(x)) = x$, *and* $f^{-1}(f(x)) = x$ *hold, c) plot* f *and its inverse in a way that shows the symmetry between the graphs of* f *and* f^{-1} *about the line* $y = x$.

A great deal of insight can be gained by first turning our attention to a graph of $f(x)$.

****

```
> f:=x->x/(x^2-2*x+4):

> plot(f(x),x=-10..10);
```

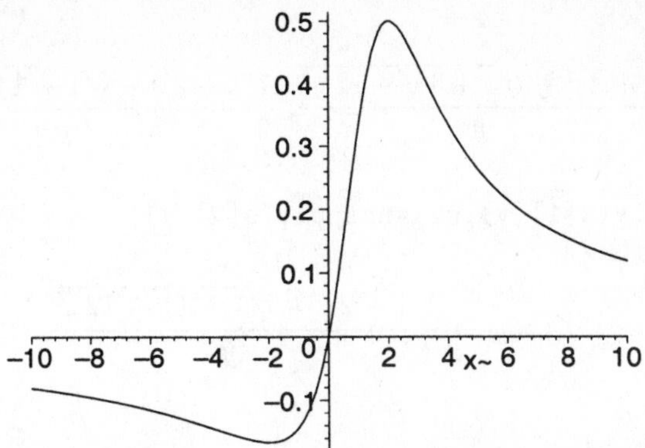

#This plot supplies a wealth of information. The function f appears to be one-to-one on three separate intervals, and so there should be three separate inverses. How do we know, however, that this plot does not have hidden features too small to see (or features beyond this window) which would make these conclusions false? To be on the safe side we first look at the sign of $f'(x)$.#

```
> fp:=diff(f(x),x);
```

$$fp := \frac{1}{x^2 - 2\,x + 4} - \frac{x\,(2\,x - 2)}{(x^2 - 2\,x + 4)^2}$$

```
> fp:=normal(fp);
```

$$fp := -\frac{x^2 - 4}{(x^2 - 2\,x + 4)^2}$$

#Now it is clear that f is a one-to-one increasing function on the interval $[-2, 2]$, because fp is positive on this interval, and f is a one-to-one decreasing function on the intervals $(-\infty, -2]$ and $[2, \infty)$, because fp is negative there.

The intervals for the inverses will be determined by the high and low points on the graph. There is no need to ask Maple to solve $fp = 0$; the formula for fp makes the solution obvious.#

```
> a:=-2:  b:=2:  A:=f(a):  B:=f(b):
```

#From the graph of $f(x)$, we can see the domains and ranges of all three of the inverses, and so we let g_1, g_2, and g_3 be the inverses to f on these various intervals, with inverses and intervals paired as follows.

$$(-\infty, a] \;\underset{\underset{g_1}{\leftarrow}}{\overset{\overset{f}{\rightarrow}}{}}\; [A, 0) \qquad [a, b] \;\underset{\underset{g_2}{\leftarrow}}{\overset{\overset{f}{\rightarrow}}{}}\; [A, B] \qquad [b, \infty) \;\underset{\underset{g_3}{\leftarrow}}{\overset{\overset{f}{\rightarrow}}{}}\; (0, B]$$

The next step is to find formulas for all three of these inverses.#

```
> s:=solve(f(x)=y,x);
```

$$s := \frac{1}{2}\frac{1+2y+\sqrt{1+4y-12y^2}}{y}, \frac{1}{2}\frac{1+2y-\sqrt{1+4y-12y^2}}{y}$$

```
> s1:=unapply(s[1],y);s2:=unapply(s[2],y);
```

$$s1 := y \to \frac{1}{2}\frac{1+2y+\sqrt{1+4y-12y^2}}{y}$$

$$s2 := y \to \frac{1}{2}\frac{1+2y-\sqrt{1+4y-12y^2}}{y}$$

#It is not obvious, quite yet, how to specify the formulas for g_1, g_2, and g_3. Notice that neither $s1$ nor $s2$ is defined at $y = 0$. This is not surprising, considering the sense in which 0 is excluded from the domains of g_1 and g_3. To help settle this issue, we compute the following limits.#

```
> limit(s1(y),y=0); limit(s2(y),y=0);
```

undefined

$$0$$

```
> limit(s1(y),y=0,right); limit(s1(y),y=0,left);
```

$$\infty$$

$$-\infty$$

#It now appears that $g_2(y) = s2(y)$, and that $g_1(y)$ and $g_3(y)$ both have the same formula, $s1(y)$, restricted to the appropriate interval. To establish this beyond any doubt, we show next that, for x in the appropriate interval,

$$f(g_j(x)) = x, g_j(f(x)) = x \text{ for } j = 1, 2, \text{ and } 3.$$

Actually, we will only show this for $j = 1$. The other two cases are similar.

The contrast between these two identities is interesting. The first is fairly easy to prove without any reference to what interval x is in. The second identity, on the other hand, is not so easy to prove. A hard look at the structure of f and g_1 would explain why there is such a contrast between these two identities, but suffice it to say that Maple fails to simplify $g_1(f(x))$. This identity is only meant to hold for $x \leq -2$, and it turns out that this information is needed in order to proceed.

An interesting **new command called assume( )** is used to give Maple this information. If x is an unassigned name (variable) and we place a condition on x by using the assume( ) command, **Maple places a flag on the variable x to indicate that it is no longer entirely "unassigned."** Until x is returned to its unassigned status, **it is represented by the variable $x\tilde{}$**. (This **flag can be removed** if it is regarded as objectionable. Pull down the **Options Menu** and make a choice on **Assumed Variables.**)

With these preliminaries behind us, we begin our work on the above identities by naming the inverse functions.#

```
> g1:=s1:  g2:=s2:  g3:=s1:
> f(g1(x));
```

$$\frac{1}{2}\,\frac{1+2\,x+\sqrt{1+4\,x-12\,x^2}}{x\,(\frac{1}{4}\,\frac{(1+2\,x+\sqrt{1+4\,x-12\,x^2})^2}{x^2}-\frac{1+2\,x+\sqrt{1+4\,x-12\,x^2}}{x}+4)}$$

```
> simplify(%);
```

$$x$$

```
> g1(f(x));
```

$$\frac{1}{2}(x^2-2\,x+4)$$
$$(1+2\,\frac{x}{x^2-2\,x+4}+\sqrt{1+4\,\frac{x}{x^2-2\,x+4}-12\,\frac{x^2}{(x^2-2\,x+4)^2}})/x$$

```
> simplify(%%);
```

#Undesirable output omitted. **Maple needs to know that $x < -2$.**#

```
> assume(x<-2);
```

#With Maple 7, we could instead just write **assuming $x < -2$** at the end of the next simplify command and before the closing semicolon. If we did that, the assumption would not be permanent. Look this up in the Index or in the Help Menu. #

```
> simplify(%%);
```

$$x\tilde{}$$

#This establishes that g_1 is the inverse of f on this interval.

To finish this example, for each $j = 1, 2, 3$, we graph the pair $\{f(x), g_j(x)\}$ along with the line $y = x$, on the same plot so that we can confirm that the graph of a function and its inverse appear to be symmetric images of each other with respect to the line $y = x$. Actually, we will do this for only one of the three (we choose

$j = 2$); the other two are quite similar. In order to create the **appropriate visual image, we must tell Maple to choose the same scale on the x and y axes.** This can be done after the first plot appears. Move the mouse pointer to the plot window and click, so that the plot tool bar is activated. Click the **[1:1] button**, then click the **[R] button** to redraw the graph. We choose an alternative approach: Inserting a second range makes the desired plot appear on the first attempt.

Finally, if we intend to use the plot( ) command, the **equation $y = x$ must be entered as an expression, rather than an equation.** The equality symbol (=) is not allowed in this command. As an alternative, we could use the equality (=) symbol everywhere, and plot all three curves as equations by using the implicitplot( ) command. Keep in mind, however, that the plot( ) command is much faster and uses less memory.#

```
> plot({f(x),g2(x),x},x=-2..2,y=-2..2);
```

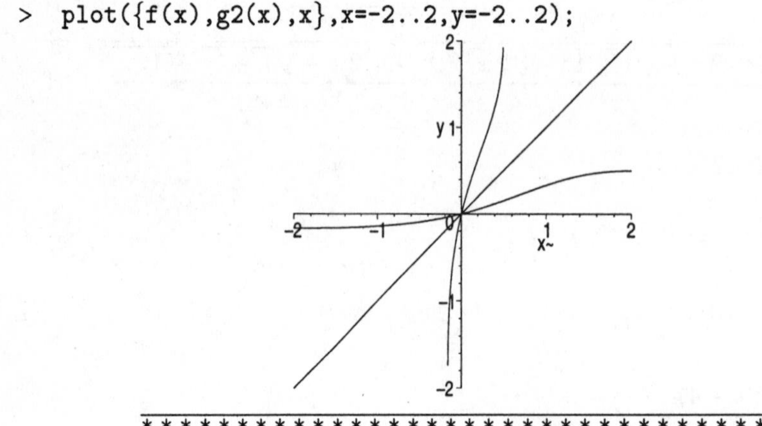

****

In order to save time and space, we will not discuss a second example, but it would be a valuable experience to look for the inverse of a function f that is a cubic polynomial. Such a function either will be globally one-to-one or will be one-to-one on three separate intervals and have three separate inverses. Solving the cubic equation $f(x) = y$ for x in terms of y always leads to three solutions, and **Maple will always give you all three.** If f is globally one-to-one, then two of these solutions will involve the imaginary symbol I, and these two parts can simply be discarded. If f has three separate inverses, then there will be three real solutions to the equation $f(x) = y$. Even with a function as simple as a cubic, the inverse is quite complicated.

As a final topic in this section on inverse functions, we draw your attention to a theorem in your calculus text that allows one to evaluate the derivative of an inverse function at a point, even if a formula for the inverse cannot be found. If $g = f^{-1}$ and if $b = f(a)$, then, according to this result,

$$g'(b) = \frac{1}{f'(a)}.$$

It is usually a straightforward matter to show that an inverse exists, but even with Maple, it is usually difficult or impossible to actually find a formula for the inverse. This result, then, can be very useful.

Example 5.2 *Show that the function $f(x) = 2x + \sin(x)$ is one-to-one on the whole real line. Compute, approximately, the derivative of $f^{-1}(x)$ at the point $x = 17$.*

****

```
> f:=x->2*x+sin(x):
> diff(f(x),x);
```

$$2 + \cos(x)$$

#The derivative is always positive, and so f is an increasing function on the interval $(-\infty, \infty)$. This means that f^{-1} exists. Unfortunately, when we ask Maple to give us a formula for the inverse, it parrots back the same equation. This is Maple's way of responding to an equation it cannot solve. Recall that letters like $\_Z$ are new unassigned names that are invented by Maple. Maple wisely uses symbols that you are not likely to use as your own names. To find the derivative of the inverse at 17, we use the theorem we mentioned above.#

```
> solve(f(x)=y,x);
```

$$\operatorname{RootOf}(2\_Z + \sin(\_Z) - y)$$

```
> a:=fsolve(f(x)=17,x);
```

$$a := 8.005747149$$

```
> answer:=1/subs(x=a,diff(f(x),x));
```

$$answer := \frac{1}{2 + \cos(8.005747149)}$$

```
> answer:=evalf(%);
```

$$answer := .5408865878$$

****

5.2 Logarithmic and Exponential Functions

The search for a function whose derivative is $1/x$ ($x > 0$) leads us immediately to the function $\ln(x)$ defined by

$$\ln(x) = \int_1^x \frac{1}{t} dt.$$

Maple denotes this function by ln or log, and, of course, its value at x is the expression $\ln(x)$ or $\log(x)$. Remember that parentheses are used in basically the same way that they are with pencil and paper. The expression **ln(*x*)^2** in Maple **means** $(\ln(x))^2$**, and not** $\ln(x^2)$. Writing (ln(*x*))^2 for ln(*x*)^2 is acceptable, but the extra set of parentheses is unnecessary. As usual, remember that Maple will give only exact answers unless you allow it to approximate.

****

```
> ln(5);
```

$$\ln(5)$$

```
> evalf(ln(5));
```

$$1.609437912$$

****

This activity, however, hardly gives us an appreciation of the definition of this function as the area under the graph of the function $y = 1/t$ from $t = 1$, to $t = x$. In order to gain some insight into this definition of $\ln(x)$, we will approximate some of its values directly from its area definition. In the process, perhaps we will demonstrate the importance of the Fundamental Theorem of Calculus. This central theorem in mathematics provides a fail safe means of always getting an antiderivative to a continuous function. We use the name $\mathrm{LN}(x)$ for our approximation of $\ln(x)$. The value of $\mathrm{LN}(x)$ is the approximate area under the curve of $y = 1/t$, from $t = 1$ to $t = x$, obtained by computing a Riemann sum corresponding to $n = 50$ and by using right endpoints of subintervals to establish heights of the thin rectangles.

****

```
> with(student):
> f:=t->1/t:
> LN:=proc(x) h:=(x-1)/50;
  evalf(sum(f(1+h*j)*h,j=1..50));end;

Warning, 'h' is implicitly declared local
```

$LN :=$ **proc**(x)

local$h$;

$h := 1/50 \times x - 1/50\,;\ \mathrm{evalf(sum(f(}1 + h \times j) \times h,\ j = 1..50))$

end

```
> LN(5);
```

$$1.577949573$$

#Compare this value with evalf(ln(5)) above.#

```
> rightbox(f(t),t=1..5,50);
```

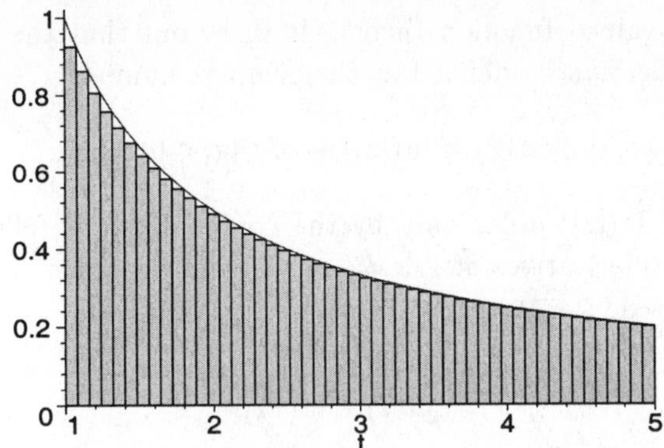


```
> LN(0.2);
```

$$-1.641949573$$

```
> rightbox(f(t),t=0.2..1,50);
```

#Output omitted.#

****

Notice, in this last command, how the range between 1 and $x = 0.2$ had to be reversed so that it went from a smaller to a larger number. Integrating from 1 to the smaller $x = 0.2$ introduces a negative sign in the area interpretation of $\ln(0.2)$, and this is also reflected in our value of LN(0.2). Just how this negative sign entered into this formula can be seen in the definition of LN as a procedure. Immediately, in that definition, we see, from $h := (x-1)/50$, that h, the "width" of a rectangle, will be negative when $0 < x < 1$. More generally, when we integrate $\int_a^b f(x)dx$ from a larger number a to a smaller number b , we can interpret the symbol dx, the "width" of a thin rectangle, as being negative in the same way as h above.

The definition of $\ln(x)$ gives us the important formula

$$\int \frac{1}{x} dx = \ln(x) + C \ (x > 0),$$

and this is easily extended to the negative real axis by the formula

$$\int \frac{1}{x} dx = \ln(|x|) + C \ (x \neq 0).$$

Notice, however, how Maple evaluates this integral.

****

```
> int(1/x,x);
```

$$\ln(x)$$

#Since the expression $1/x$ is well defined on the negative real axis, Maple is certainly not assuming that x is a positive variable. Curiously enough, Maple has not made a mistake, but the way it has chosen to express the antiderivative could cause complications. To fully understand Maple's answer requires a more advanced course in complex-valued function theory. It turns out that the natural logarithm of a negative number can be defined as the complex number

$$\ln(x) = \ln(|x|) + \pi I\ , (x < 0).$$

Because $\ln(x)$ and $\ln(|x|)$ differ only by the constant πI, it follows that they are both legitimate antiderivatives of $1/x$.#

```
> int(1/t,t=-11..-7);
```

$$\ln(7) - \ln(11)$$

#We have the expected answer. Expressions involving $\ln(x)$ will usually be real valued, but we should be prepared for complex-valued complications. Complex-valued logarithms play an important role in mathematics, but they are never appropriate in our real-valued world of calculus. If they are encountered, they should be dealt with accordingly. #

```
> int(1/t,t=-15..a);
```

$$\ln(a) - \ln(3) - \ln(5) - I\,\pi$$

```
> subs(a=-5,%);
```

$$\ln(-5) - \ln(3) - \ln(5) - I\,\pi$$

#In order to express the above output as a real number, we use another form of the eval( ) command. Place a **"c"** (for complex) at the end of **eval( )** and we have the command **evalc( )**, which is the command to **evaluate as a complex number in the form $a + bI$**. In this problem, the term b should turn out to be zero.#

```
> evalc(%);
```

$$-\ln(3)$$

Maple recognizes all of the algebraic properties of $\ln(x)$, but it seems to resist making some obvious logarithmic simplifications in the next expression. There is a reason for this, as we shall see.#

```
> ex:=ln(sqrt((8+2*x)^3)/(a^5*9*b));
```

$$ex := \ln\left(\frac{1}{9}\,\frac{\sqrt{(8+2\,x)^3}}{a^5\,b}\right)$$

```
> ex1:=simplify(ex);
```

$$ex1 := \frac{3}{2}\ln(2) - 2\ln(3) + \ln\left(\frac{\sqrt{(x+4)^3}}{a^5\,b}\right)$$

```
> expand(ex1);
```

$$\frac{3}{2}\ln(2) - 2\ln(3) + \ln(\frac{\sqrt{(x+4)^3}}{a^5\,b})$$

Maple resists making the intended logarithmic simplification, because it is not true unless $x > -4$, $a > 0$, and $b > 0$. Complicating the matter even more, Maple assumes that unassigned variables may be complex-valued. Suppose we wish to make all three assumptions. They are all independent assumptions on independent variables, so we can make all of them with one use of the assume( ) command. In contrast to our last use of this command, there are no additional assumptions on any variable, so the additional( ) command is not needed.#

```
> assume(x>=0,a>=0,b>=0);
```

#In Maple 7, this can be done in a less permanent way with the **"assuming,"** facility. Look it up in the Index or in the Help Menu. #

```
> expand(ex1);
```

$$\frac{3}{2}\ln(2) - 2\ln(3) + \frac{3}{2}\ln(x\tilde{} + 4) - 5\ln(a\tilde{}) - \ln(b\tilde{})$$

****

Look at the consequences of the commands simplify( ) and expand( ) on the above logarithmic expression. Both of these commands have useful options and variations. **They should be looked up in the Help File.** Recall how easy it is to do this. With the cursor on "simplify" or "expand," pull down the **Help Menu** and click on **Help on Context.**

The inverse of the natural logarithm is represented in Maple by the function exp. Its value at x, naturally, is the expression denoted by $\exp(x)$. Ask Maple to solve $\ln(y) = x$ for y in terms of x, and $\exp(x)$ is the expression that will appear as output. In mathematics, this expression also has the form $\exp(x) = e^x$, where e is the exponential constant, whose decimal value is roughly 2.71828. It is important to emphasize that **this constant is not represented in Maple by** e **or** E. The function $\exp(x)$ cannot be expressed in this "power" form, at least not as an input statement. Of course, we could take the initiative, assign the value $exp(1)$ to the name e, and then use the Maple expression $e\hat{}x$ instead of $exp(x)$. There is no real advantage to doing this, and so it is not suggested. Notice (below) that output statements are naturally written in this "power" form anyway.

Before we leave this topic, however, **let us be perfectly clear**! Maple **does not recognize e as the exponential constant.** If you use $e\hat{}x$ instead of $exp(x)$ in an input statement, your output will look perfectly acceptable, but **e will simply be regarded as an unassigned name** by Maple.

****

```
> exp(0),exp(4),exp(-4),evalf(exp(4));
```

#Notice the use of commas in this input statement. The final punctuation is still a semicolon. We shall use this occasionally to save space.#

$$1,\ e^4,\ e^{(-4)},\ 54.59815003$$

```
> solve(ln(y)=x,y);
```

$$e^x$$

#When the exponential function is entered in the form e^x, it is not the exponential function, in spite of its appearance. You can see that something is **not quite right** as soon as it is, for example, differentiated. **Certainly Maple knows that ln(e) = 1, and it would never leave such an obvious simplification unattended.**#

```
> diff(e^x,x);
```

$$e^x \ln(e)$$

****

Just like the names sin, cos, tan, ln, and so on, **exp is the name of a function, not an expression.** Consequently, the input expression **exp^ 4** represents the function whose value at x (also expressed as an input statement) is (exp^ 4)(x) = exp(x)^4. In standard mathematical symbols, (expressed as output) this would evaluate to $(e^x)^4 = e^{4x}$. (We discussed the importance of parentheses in this situation on page 23. Actually, this notation is not particularly convenient anyway; it was mentioned only to explain why **exp^ *x* cannot be used in place of exp(*x*)**, and how Maple would interpret this expression if we made such a mistake.)

Example 5.3 *Find the x-intercept of the line which is tangent to the graph of* $y = \ln(x^3+5)$ *at the point on the curve having the y-coordinate 10.*

****

```
> f:=x->ln(x^3+5):
> a:=solve(f(x)=10,x);
```

$$a := (e^{10}-5)^{1/3},\ -\frac{1}{2}(e^{10}-5)^{1/3}+\frac{1}{2}I\sqrt{3}(e^{10}-5)^{1/3},$$
$$-\frac{1}{2}(e^{10}-5)^{1/3}-\frac{1}{2}I\sqrt{3}(e^{10}-5)^{1/3}$$

```
> a:=a[1];
```

#With this, the point of tangency is (a,10).#

$$a := (e^{10}-5)^{1/3}$$

```
> fp:=diff(f(x),x);
```

$$fp := 3\,\frac{x^2}{x^3+5}$$

```
> m:=subs(x=a,fp);
```

#This is the slope of the tangent line.#

$$m := 3\,\frac{(e^{10}-5)^{2/3}}{e^{10}}$$

```
> eq:=y-10=m*(x-a);
```

$$eq := y - 10 = 3\,\frac{(e^{10}-5)^{2/3}\,(x-(e^{10}-5)^{1/3})}{e^{10}}$$

```
> X:=solve(subs(y=0,eq),x);
```
#This is the x-intercept.#

$$X := -\frac{1}{3}\,\frac{7\,e^{10}+15}{(e^{10}-5)^{2/3}}$$

```
> evalf(X);
```

$$-65.42338903$$

* *

Example 5.4 *Find the area of the region bounded by the graphs of* $y = \exp(2x-3)$, $y = 4$, *and* $x = -1$. *Verify with an alternative calculation.*

* *

```
> f:=x->exp(2*x-3):
> plot({f(x),4},x=-1..2.5,y=0..5);
```

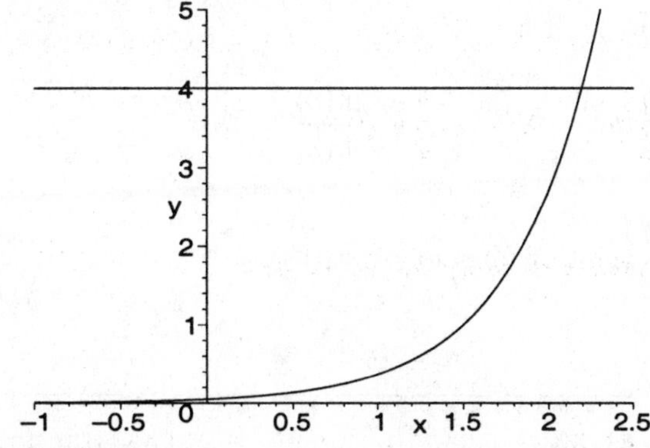


```
> a:=fsolve(f(x)=4,x);
```

$$a := 2.193147181$$

```
> A1:=evalf(Int(4-f(x),x=-1..a));
```

#Area, using "thin" vertical rectangles.#

$$A1 := 10.77595770$$

```
> xr:=solve(f(x)=y,x);
```

$$xr := \frac{3}{2} + \frac{1}{2}\ln(y)$$

```
> b:=evalf(f(-1));
```

#The y-coordinate of the lower left-hand corner is small but not zero.#

$$b := .006737946999$$

```
> A2:=evalf(Int(xr-(-1),y=b..4));
```

#Area, using "thin" horizontal rectangles.#

$$A2 := 10.77595770$$

****

In order to evaluate $\log_b(x)$ on a calculator, we use the well-known formula

$$\log_b(x) = \frac{ln(x)}{ln(b)}.$$

Maple uses the same formula and denotes the base **b logarithm function** by **log[*b*]**. In older versions of Maple, this function is not supplied by Maple but it is an easy matter to define it.

****

```
> log[2](8);
```

$$\frac{\ln(8)}{\ln(2)}$$

```
> simplify(%);
```

$$3$$

****

As you can see, Maple is willing to simplify the above if it is asked to do so, but it prefers to first transform log[2](x) into ln(x) before it does anything else. This can lead to undesirable complications, and frequent calls for simplifications. Sometimes it is easier to give a more direct interpretation to $\log_b(x)$, rather than transform it into $\ln(x)$. One advantage we humans have over computers is the ability of being more selective about how we wish to proceed with a solution.

Maple gets into trouble in the same way in solving both eq1, and eq2 below. Simplifying eq1 is simple enough, but look what happens in eq2.

****

```
> eq1:=log[3](x)=y;
```

$$eq1 := \frac{\ln(x)}{\ln(3)} = y$$

```
> solve(eq1,x);
```

$$e^{(y \ln(3))}$$

```
> simplify(%);
```

$$3^y$$

```
> eq2:=log[3]((x^2+9)^2)=5*(y^2+1);
```

$$eq2 := \frac{\ln((x^2+9)^2)}{\ln(3)} = 5\,y^2 + 5$$

```
> s:=solve(eq2,x);
```

$$s := \{3\sqrt{-1-\sqrt{3}\sqrt{243^{(y^2)}}},\ -3\sqrt{-1-\sqrt{3}\sqrt{243^{(y^2)}}},$$
$$3\sqrt{-1+\sqrt{3}\sqrt{243^{(y^2)}}},\ -3\sqrt{-1+\sqrt{3}\sqrt{243^{(y^2)}}}\}$$

```
> simplify(s);
Error, (in simplify) invalid simplification command
```

#To explain this error, we take a closer look at the simplify command. In its most basic form, the command simplify( ) has only one argument, and even when a more complicated version, simplify(, , ,...) is used, the **object being simplified is entered as the first argument.** Recall that the name s above is a sequence, an open ended list of four objects. Our name s has the form $s := s_1, s_2, s_3, s_4$. When we enter s into simplify(s), Maple tries to simplify just the first term s_1, and it is understandably confused about what to make of the next three arguments s_2, s_3, s_4. The solution to this problem is simple enough. All we have to do is group all four terms together by placing them inside an enclosure. We use set brackets $\{s\}$ in the following, but brackets would do just as well. When Maple looks inside the next simplify command, the first object it sees is $\{s\}$. There are other ways—even better ways—to simplify these terms. They will be introduced in the future.#

```
> s:=simplify({s});
```

$$s := \{3\sqrt{-1-\sqrt{3}\sqrt{243^{(y^2)}}},\ -3\sqrt{-1-\sqrt{3}\sqrt{243^{(y^2)}}},$$
$$3\sqrt{-1+\sqrt{3}\sqrt{243^{(y^2)}}},\ -3\sqrt{-1+\sqrt{3}\sqrt{243^{(y^2)}}}\}$$

#The first two evaluations are not real numbers, so the solutions are as follows:#

```
> s[3],s[4];
```

$$3\sqrt{-1+\sqrt{3}\sqrt{243^{(y^2)}}},\ -3\sqrt{-1+\sqrt{3}\sqrt{243^{(y^2)}}}$$

****

5.3 The Inverse Trigonometric Functions

All six of these functions are known to Maple. Notation, such as $\sin^{-1}(x)$ for the inverse sine function, may well be a familiar sight in print, but, of course, it would not be a suitable way to name these functions in Maple, and so the inverse trigonometric functions are known to Maple only by the names **arcsin, arccos, arctan, arccot, arcsec, arccsc.**

These functions are, of course, not globally one-to-one, and so inverses for them exist only if their domains are suitably restricted. This complicates the identities

$$f(f^{-1}(x)) = x, f^{-1}(f(x)) = x.$$

If $f(x) = \sin(x)$, for example, the first of these identities holds as long as $f(f^{-1}(x))$ is well defined, namely, for all x in the interval $-1 \le x \le 1$. The second identity, on the other hand, holds only for x in the interval $-\pi/2 \le x \le \pi/2$, even though $f^{-1}(f(x))$ is well defined for all values of x. Can you see why these two identities are so different?

In early versions of Maple, this restriction on x was ignored by the simplify( ) command, resulting in a clear error. Fortunately, the mistake was corrected some time ago in Maple VR4. It is mentioned now only as a reminder of why we must remain alert. We will be the source of most of the mistakes in our output, but Maple can be misunderstood, and perhaps it can still make its own mistakes. In spite of Maple's impressive performance, **a questioning attitude should always be maintained**, not just when Maple is in use, but more generally as well.

As we show next, Maple now treats the expressions $f(f^{-1}(x))$, and $f^{-1}(f(x))$ as they should be treated. The graph of $f^{-1}(f(x))$ is shown next. Can you explain why it looks this way? Also notice that a **new Maple command additionally( )** is introduced there, as a means of entering a second assumption made on the variable x.

****

```
> sin(arcsin(x));
```

$$x$$

```
> arcsin(sin(x));
```

$$\arcsin(\sin(x))$$

```
> simplify(%);
```

$$\arcsin(\sin(x))$$

```
> assume(x>=-Pi/2);additionally(x<=Pi/2);
```

#In Maple 7, we can use instead the more convenient (and less permanent) **"assuming"** facility. Look it up in the Index or in the Help Menu.#

```
> arcsin(sin(x));
```

$$\arcsin(\sin(x\tilde{}))$$

```
> simplify(%);
```

$$x\tilde{}$$

```
> plot(arcsin(sin(x)),x=-10..10);
```

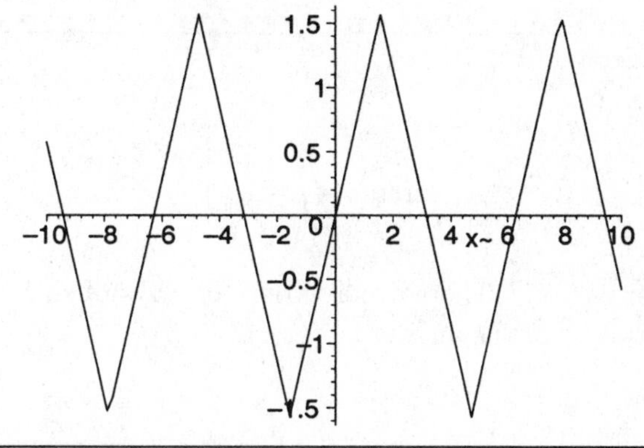

****

The rules for differentiating these functions are well known. In older versions of Maple, the formulas for differentiating the arcsec(x) and arccsc(x) functions were incorrect for $x < 0$. The problem was caused by a more general problem with the manipulation of expressions involving radicals . Fortunately, all of these problems were corrected a long time ago, and there is no need for any concern. The rules for differentiating the arcsec(x) and arccsc(x) may look somewhat unfamiliar, but they are easily seen to be correct.

****

```
> Diff(arcsec(x),x)=diff(arcsec(x),x);
```

$$\frac{\partial}{\partial x}\operatorname{arcsec}(x) = \frac{1}{x^2\sqrt{1-\frac{1}{x^2}}}$$

```
> plot(arcsec);
```

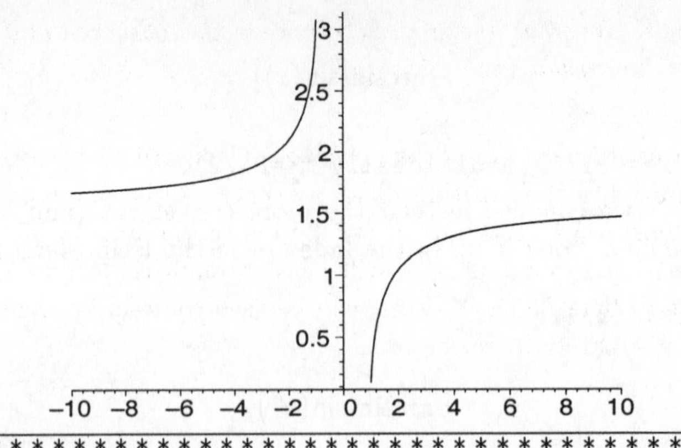

****

As you can see, this arcsec function maps the domain $|x| \geq 1$ onto $[0, \pi]$ (excluding $\pi/2$) which is the most common way of defining the arcsec function. Maple also gives a somewhat unfamiliar answer to the standard integral

$$\int \frac{1}{x\sqrt{x^2-1}}dx = arcsec(|x|) + c.$$

****

```
> f:=x->1/(x*sqrt(x^2-1)):
> int(f(x),x);
```

$$-\arctan(\frac{1}{\sqrt{x^2-1}})$$

#Is this answer correct? There are a variety of ways of showing that it is. We show that the derivative of the answer is $f(x)$.#

```
> diff(%,x);
```

$$\frac{x}{(x^2-1)^{3/2}\,(1+\frac{1}{x^2-1})}$$

```
> simplify(%);
```

$$\frac{1}{\sqrt{x^2-1}\,x}$$

****

Example 5.5 *A 300-foot-long ship whose bow is 57.2 miles east and 37.6 miles north of an observer is steaming due west at 27 miles per hour, while the observer travels due north at 19 miles per hour. When the ship is within visual range, the observer plans to focus a telescope on the ship, adjusting the viewing angle of the lens so that it always fits exactly the 300-foot length of the ship. When is this viewing angle a maximum, and what is the distance between the observer and the bow of the* *ship at that time? When is the distance between the observer and the bow of the ship a minimum, and what is this distance?*

We set up a coordinate system which is easily described by looking at the following figure. Initially, the position of the observer is the point $y = -37.6$ on the y-axis, and the position of the bow of the ship is the point $x = 57.2$ on the x-axis. Clearly the point corresponding to y will pass through the origin before the point corresponding to x. The *Law of Cosines* can be used to describe the angle θ in terms of a and b, and ultimately in terms of t.

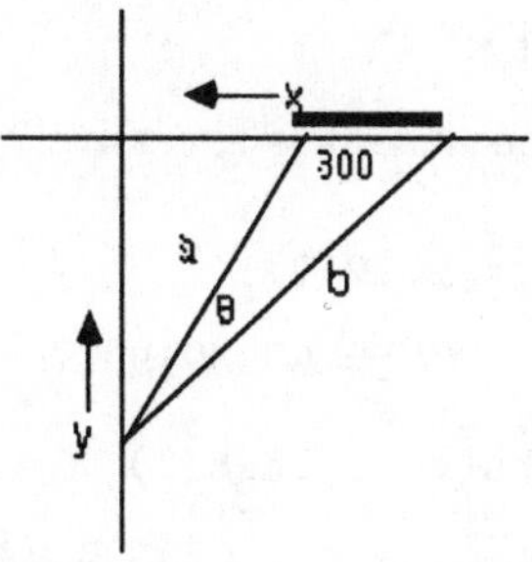

****

```
> y:=19*t-37.6:  x:=57.2-27*t:
```

```
> L:=300/5280:
```

The length of the ship must be expressed in miles to be compatible with the other units.

```
> a:=sqrt(x^2+y^2):b:=sqrt((x+L)^2+y^2):
```

```
> eqn:=expand(L^2=a^2+b^2-2*a*b*cos(theta));
```

$$eqn := \frac{25}{7744} = 9377.703228 - 9038.268182\,t + 2180\,t^2 -$$
$$2\sqrt{4685.60 - 4517.6\,t + 1090\,t^2}\,\sqrt{4692.103228 - 4520.668182\,t + 1090\,t^2}\cos(\theta)$$

```
> Theta:=solve(eqn,theta);
```

$$\Theta := \arccos(.00002066115702($$
$$.1134701700\,10^{13} - .1093630450\,10^{13}t + .2637800000\,10^{12}t^2)/($$
$$\sqrt{117140. - 112940.\,t + 27250.\,t^2}$$
$$\sqrt{.4692103228\,10^{10} - .4520668182\,10^{10}t + .1090000000\,10^{10}t^2}))$$

Both the ship and the observer pass through the origin near time $t = 2$. Knowing this makes it easy to adjust the plotting range to get a good view of the maximum value of Θ.

```
> plot(Theta,t=2..2.5);
```

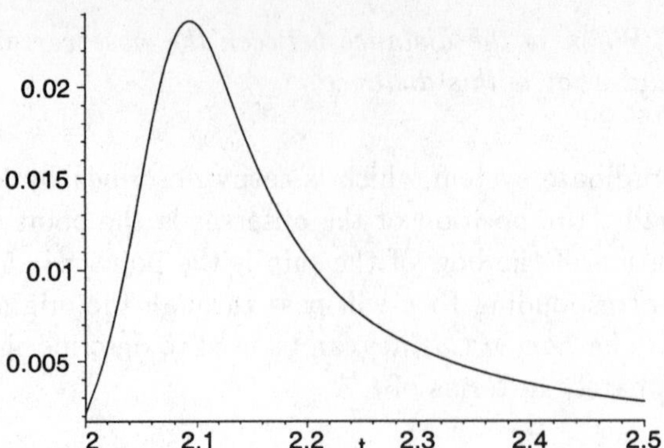


```
> slope:=diff(Theta,t):
```

#The output is omitted because it is excessively complicated.#

```
> t0:=fsolve(slope=0,t,2..2.2);
```

$$t0 := 2.093949077$$

```
> MaxAngle:=evalf(subs(t=t0,Theta));
```

$$MaxAngle := .02359932302$$

```
> DegreeMax:=evalf(180*MaxAngle/Pi);
```

$$DegreeMax := 1.352141608$$

```
> f:=x^2+y^2;
```

This is the square of the distance between the ship and the observer. It will be a minimum at the same time that the distance itself will be a minimum, but it is an easier expression to differentiate.

$$f := (57.2 - 27\,t)^2 + (19\,t - 37.6)^2$$

```
> eq:=diff(f,t)=0;
```

$$eq := -4517.6 + 2180\,t = 0$$

```
> t1:=fsolve(eq,t);
```

$$t1 := 2.072293578$$

```
> MinDistance:=sqrt(subs(t=t1,f));
```

$$MinDistance := 2.168701464$$

```
> AngleDistant:=sqrt(subs(t=t0,f));
```

$$AngleDistant := 2.283513332$$

****

Notice that $t0 > t1$, so the distance is a minimum before the viewing angle is a maximum.

5.4 Exponential Growth and Decay

The differential equation

$$\frac{dy(t)}{dt} = ky(t)$$

is so common in mathematics and its applications that its solution

$$y(t) = Ce^{kt},$$

where C is a constant of integration, quickly becomes a familiar if not memorized formula to most students of mathematics. If the most common initial condition $y(0) = y_0$ is given, then C takes the form $C = y_0$. As much as any other problem, this classic differential equation justifies the introduction of the exponential function into the central core of mathematics.

Example 5.6 *A radioactive sample with a half life of* 117.537 *years must remain in protective storage until less than* 0.005 *grams of the substance are left. If 50 grams of the substance were put into a storage locker, how long will this take?*

****

```
> A:=t->50*exp(k*t):
> k:=solve(A(117.537)=25,k);
```

$$k := -.005897267929$$

```
> A(t);
```

$$50\, e^{(-.005897267929 t)}$$

```
> Time:=solve(A(t)=.005);
```

> #The name "Time" here was capitalized, because "time" is the name of a Maple command. Try to use a predefined Maple name as a user-defined name and notice how Maple responds. #

$$\mathit{Time} := 1561.797850$$

#The answer is expressed in years.#

```
> Check:=A(Time);
```

$$\mathit{Check} := .005000000000$$

****

Newton's Law of Cooling is a principle that leads to a differential equation very similar to the classic equation $\frac{dy}{dt} = ky$ just discussed, and it is frequently presented in calculus courses at this time. This principle states that the **rate at which a body cools is proportional to the difference between the temperature of**

the body and the temperature of the surrounding medium. Translated into mathematics, the equation takes the form

$$\frac{dT(t)}{dt} = K(T(t) - A)$$

where $T(t)$ is the temperature of the body at time t, A is the constant temperature of the surrounding medium, and K is the constant of proportionality—a constant which depends on the physical properties of the medium.

This differential equation quickly turns into the well known equation $\frac{dy}{dt} = Ky$, if we just let $y(t) = T(t) - A$. As a consequence, we can easily specify the solution to *Newton's Law of Cooling* as

$$T(t) = A + Ce^{Kt}$$

where C is a constant of integration, whose value is determined by the initial condition.

Example 5.7 *A steel beam is taken out of a furnace and placed in an environment at* 72° *Fahrenheit. After 10 minutes the beam has a temperature of* 1287°, *and after 15 minutes, its temperature is* 986°. *If we must wait until its temperature is* 100° *before we can continue working with it, then how long must we wait? What was the temperature of the beam when it came out of the furnace?*

****

```
> T:=t->72+C*exp(K*t):
> eq1:=T(10)=1287; eq2:=T(15)=986;
```

$$eq1 := 72 + C\,e^{(10\,K)} = 1287$$

$$eq2 := 72 + C\,e^{(15\,K)} = 986$$

```
> fsolve({eq1,eq2},{C,K});
```

$$\{C = 2147.021742,\ K = -.05693375686\}$$

```
> assign(%);
```

Recall that names enclosed in single quotes are unevaluated. Notice the effect on the output appearance.#

```
> 'T(t)'=T(t);
```

$$\mathrm{T}(t) = 72 + 2147.021742\,e^{(-.05693375686t)}$$

```
> InitialTemp:=T(0);
```

#This is the initial temperature in degrees Fahrenheit.#

$$InitialTemp := 2219.021742$$

```
> wait:=solve(T(t)=100,t);
```

#This is how long it takes, in minutes, to cool to 100°. #

$$wait := 76.22248476$$

```
> T(10); #This checks to see whether
the answer is what it should be.#
```

$$1287.000000$$

****

The study of differential equations is a vast and important branch of applied mathematics. We usually do not memorize solutions to differential equations as we did in this section; we use, instead, the theory and techniques presented in a course in differential equations taken soon after completing the calculus sequence.

Maple, not surprisingly, has some powerful commands for solving differential equations. A brief introduction to differential equations is often presented towards the end of a calculus sequence, and eventually it will be included in this manual—although in a different form. In order to keep size and cost to a minimum, it is not included in the paper form of this manual, but an introduction to differential equations—Chapter 14 of this manual— **is being planned for our web site.**

5.5 The Hyperbolic Functions and Their Inverses

The hyperbolic functions have the same names in Maple as they do in traditional mathematics. Their inverses, just like the inverses of the trigonometric functions, are denoted by attaching the prefix "arc" to the function name. There are many interesting manipulations involving these functions that can be done very effectively with Maple.

Example 5.8 *Derive the differentiation formula for* $\sinh^{-1}(x)$ *in terms of formulas for the exponential and logarithmic functions.*

****

```
> diff(arcsinh(x),x);
```

#Here is the rule.#

$$\frac{1}{\sqrt{1+x^2}}$$

We begin our derivation of this formula, by turning the *arcsinh* function into a more familiar *sinh* function.

```
> eq1:=y=arcsinh(x);
```

$$eq1 := y = \operatorname{arcsinh}(x)$$

```
> eq2:=x=solve(eq1,x);
```

$$eq2 := x = \sinh(y)$$

#The function $\sinh(y)$ is defined in terms of exponentials. We let Maple make the transformation to exponentials for us. Here are two new commands, both of which deserve some attention. We will discuss them below.#

```
> eq3:=x=convert(rhs(eq2),exp);
```

$$eq3 := x = \frac{1}{2}e^y - \frac{1}{2}\frac{1}{e^y}$$

Now that we have the equation in terms of exponential functions, we solve this for y in terms of x.#

```
> y:=solve(eq3,y);
```

$$y := \ln(x + \sqrt{x^2+1}), \ln(x - \sqrt{x^2+1})$$

There are two solutions, but the term y[2] involves the logarithm of an expression that clearly is always negative.#

```
> y:=y[1];
```

$$y := \ln(x + \sqrt{x^2+1})$$

```
> yp:=diff(y,x);
```

$$yp := \frac{1 + \dfrac{x}{\sqrt{x^2+1}}}{x + \sqrt{x^2+1}}$$

```
> yp:=simplify(yp);
```

$$yp := \frac{1}{\sqrt{x^2+1}}$$

****

Two new Maple commands were used in the last example. One of these is very straightforward. If *eq* is an equation, then **rhs(*eq*) and lhs(*eq*) represent the right-hand side and left hand side of *eq*** respectively. The **convert( ) command is a very broad-based command** that can be used in a **variety of situations to convert an expression from one form to another.** We used it to convert $\sinh(y)$ to its exponential form. To fully appreciate this command, **you should definitely look it up in the help file.**

5.6 Indeterminate Forms and L'Hôpital's Rule

If f and g are differentiable functions on an open interval containing $x = a$, except possibly at a itself, and if $h = f/g$ is a $(0/0)$ or (∞/∞) indeterminate form at $x = a$, then, according to *L'Hôpital's Rule*

$$\lim_{x\to a} h(x) = \lim_{x\to a} \frac{f(x)}{g(x)} = \lim_{x\to a} \frac{f'(x)}{g'(x)}$$

provided the limit on the right exists in a finite or infinite sense. Variations of this rule can be stated for left- and right-hand limits and for limits with $a = \pm\infty$.

Maple is capable of computing most limits without any help from us. Its response to a limit request, however, can be somewhat unsettling—the only output is a number. In most other problems, even if we do not verify the result with an alternate computation, we are at least involved enough in the computation to judge whether or not an answer seems reasonable. In problems involving indeterminate forms, it is usually hard to tell, on the face of it, whether the value of a limit seems reasonable. To be reassured, an alternative calculation is called for, and *L'Hôpital's Rule* is a useful tool to use in the verification.

At first glance, it might seem that more elementary methods could be used to judge whether the value of a limit seems reasonable. If we want supportive evidence that Maple's answer to $lim_{x\to a}h(x)$ is correct, why not simply evaluate $h(x)$ for a few values of x close to $x = a$? This frequently will work, but if h is of indeterminate form at $x = a$, then $h(x)$ is computationally unstable for values of x close to $x = a$. In the indeterminate form $(0/0)$, for example, when the numerator and denominator become small enough, round-off errors frequently become sizable. Recall what happen in Example 2.4 when we tried this verification approach.

Example 5.9 *Compute the following limit and verify the answer with an alternative calculation.*

$$\lim_{x\to 0} \frac{6\sin(x) - 6x + x^3}{2x^5}$$

This limit is easily evaluated directly by using Maple's limit command. To verify the answer, notice that the expression is clearly a $(0/0)$ indeterminate form, and so *L'Hôpital's Rule* applies. After we apply the rule, we have another $(0/0)$ indeterminate form, and so *L'Hôpital's Rule* can be applied again and again, until we arrive at a fraction, which is no longer a $(0/0)$. At that point, the value of the limit becomes obvious.

****

```
> f:=(6*sin(x)-6*x+x^3)/(2*x^5): limit(f,x=0);
```

$$\frac{1}{40}$$

```
> f1:=diff(numer(f),x)/diff(denom(f),x);
```

$$f1 := \frac{1}{10}\,\frac{6\cos(x) - 6 + 3\,x^2}{x^4}$$

```
> f2:=diff(numer(f1),x)/diff(denom(f1),x);
```

$$f2 := \frac{1}{40}\,\frac{-6\sin(x) + 6\,x}{x^3}$$

```
> f3:=diff(numer(f2),x)/diff(denom(f2),x);
```

$$f3 := \frac{1}{60}\,\frac{-3\cos(x)+3}{x^2}$$

```
> f4:=diff(numer(f3),x)/diff(denom(f3),x);
```

$$f4 := \frac{1}{40}\,\frac{\sin(x)}{x}$$

```
> f5:=diff(numer(f4),x)/diff(denom(f4),x);
```

$$f5 := \frac{1}{40}\cos(x)$$

#At this point, the limit is no longer a 0/0 indeterminate form, and its value is obviously 1/40.#

****

Example 5.10 *Compute the following limit and verify the answer with an alternate calculation.*

$$\lim_{x\to 0+} (1+\sin(x))^{\cot(x)}$$

This limit is a 1^∞ indeterminate form. To use *L'Hôpital's Rule* on a function f that has a 1^∞ indeterminate form, we first write f in the form $f = e^{\ln(f)}$ and then consider the limit of $p = \ln(f)$. By properties of logarithms, p can then be expressed as a $(0/0)$ or (∞/∞) indeterminate form. **Notice the use of the important option "assume=real" inside the simplify( ) command.**

****

```
> f:=(1+sin(x))^cot(x):  limit(f,x=0,right);
```

$$e$$

```
> p:=ln(f);
```

$$p := \ln((1+\sin(x))^{\cot(x)})$$

```
> p1:=simplify(p,assume=real);
```

#Most of the usual properties of logarithms do not hold for complex-valued expressions. It turns out that each term in this expression has a complex-valued interpretation, and, unless it is told otherwise, Maple always makes allowances for complex values.#

$$p1 := \frac{\cos(x)\ln(1+\sin(x))}{\sin(x)}$$

```
> p2:=diff(numer(p1),x)/diff(denom(p1),x);
```

$$p2 := \frac{-\sin(x)\ln(1+\sin(x)) + \dfrac{\cos(x)^2}{1+\sin(x)}}{\cos(x)}$$

#At this point, the limit is no longer a 0/0 indeterminate form. Its value is obviously 1, and this verifies Maple's answer of e for the original limit.#

****

5.7 Exercise Set

New Maple Commands in this Chapter (and a few old commands as a reminder)

assume( )	assuming	convert( )	evalc( )	expand(,*suppress*)
lhs( )	limit( )	proc( )	rhs( )	simplify(,*options*)

1. Show that $f(x) = 1+\log_2(x+2)$ and $g(x) = 2^{x-1}-2$ are inverse functions of each other. Demonstrate with an appropriate graph the symmetry between the curves. Choose a plot range of an appropriate size, and use the same scale on both axes.

2. Show that $f(x) = \frac{1}{300}x^3 - \frac{3}{200}x^2 + \frac{13}{400}x + 7$ is one-to-one for $-\infty < x < \infty$. Find a formula for its inverse $g(x) = f^{-1}(x)$, and demonstrate the symmetry between the curves $y = f(x)$ and $y = g(x)$ with an appropriate graph. Choose a plot range of an appropriate size, and use the same scale on both axes.

3. Show that $f(x) = x^3 - 6x^2 + 15x + 7$ is one-to-one on $(-\infty, \infty)$ and find a formula for its inverse $g(x) = f^{-1}(x)$. By computing derivatives directly and then evaluating them as decimals, demonstrate, for $p = g(3)$ and $q = f(8)$, that

$$g'(3) = \frac{1}{f'(p)}, g'(q) = \frac{1}{f'(8)}.$$

4. Show that $f(x) = x/\sqrt{x^2+4}$ is one-to-one on $(-\infty, \infty)$ and find a formula for $g(x) = f^{-1}(x)$. Show that $f(g(x)) = x$, and $g((f(x)) = x$ for all real x. Demonstrate with an appropriate graph the symmetry between the curves $y = f(x)$ and $y = g(x)$. Show that, for every a,

$$g'(b) = \frac{1}{f'(a)} \text{for } b = f(a).$$

5. Find the largest interval on which $f(x) = x^2 - e^x$ is one-to-one. You may want to look at a graph for guidance, but don't rely on your eyesight. See how effectively you can use the differential calculus to establish this one-to-oneness. If $g(x)$ denotes its inverse, approximate $g'(7)$ as a decimal.

6. Find a decimal expansion for e that is correct to 30 decimal places.

7. Evaluate the following as decimals. Notice their similarities or differences.
a) $e^{3/2}$ b) $\sqrt{e^3}$ c) $\frac{1}{e^{\sqrt{2}}}$ d) $e^{-\sqrt{2}}$
e) $\ln(7^2)$ f) $2\ln(7)$ g) $\ln(8+9)$ h) $\ln(8)+\ln(9)$
i) $\ln(8\times 9)$ j) $\ln(\frac{5}{3})$ k) $\ln(5)-\ln(3)$ l) $\frac{\ln(5)}{\ln(3)}$

8. Differentiate the following. Express your answer first as an expression, by using diff( ), and then as a function, by using D( ). Notice the answers. According to your knowledge of basic differentiation formulas, are the answers correct?
a) $f(x)=e^{x^2+5x+1}$ b) $g(x)=\ln(1+\cos(2x))$

9. Differentiate the function $f(x) = (\ln(1+\ln(1+x^2)))^2$. Does the answer look correct? Answer this question one way or the other by using the definition of a derivative as a limit of a difference quotient.

10. Set up a procedure for estimating the value of $\ln(x)=\int_1^x \frac{1}{t}dt$ with a Riemann sum obtained by dividing the interval $[1,x]$ into 40 subintervals of equal length and always using left-end points of intervals to establish heights of rectangles. Use the procedure to estimate the value of $ln(0.1)$, $ln(10)$, $ln(100)$.

11. Use implicit differentiation and the diff( ) command to find the slope of the line tangent to the graph of $x^2+y^2=e^{x+y}$ at the point on the graph corresponding to $x=-1/2$. Check your value for the slope, using Maple's implicitdiff( ) command. Plot the equation along with the tangent line. Is there more than one point on the curve corresponding to $x=-1/2$?

12. Find the absolute maximum and absolute minimum, and where they occur, for the function f defined by

$$f(x)=\frac{5x^7}{e^x+e^{-x}}.$$

Express your answers as decimal numbers with 10 digits of accuracy. Include some evidence (not just visual evidence) that larger (or more negative) values will not be found outside of your plot window.

13. Evaluate the following exactly. Simplify your answers.
a) $\log_7(16807)$ b) $\log_{16807}(7)$ c) $\tanh(\ln(e))$
d) $\text{sech}(\ln(17))$ e) $\arctan(-\sqrt{3})$ f) $\sin(\arctan(17))$
g) $\arcsin(\sin(\frac{78}{5}\pi))$ h) $\tanh(\text{arcsinh}(13))$

14. Convert arctanh(3/4) to a form involving evaluations of natural logarithms. It's easy, if you ask Maple to convert it for you. A direct approach gives an answer, but little else. Is it correct? You could evaluate both numbers as decimals, but this approach would only show that the two values are approximately the same. Verify with an alternative, more mathematical approach. Begin with the equation $tanh(y)=3/4$, where y represents the answer.

15. Compute a decimal approximation (10 digits) for each of the following.
a) $\log_8(2.45)$ b) $\sqrt{\arctan(26)}$ c) $e^{\cos(5.3)}$ d) $\sqrt{2\sqrt{2}}$

16. Differentiate the following functions, and simplify your answers if appropriate.
a) $f(x)=\arctan(-\sqrt{x})$ b) $g(t)=\frac{t^5 ln(t^2+1)}{(ln(t))^2}$ c) $h(w)=w^{\cos(w)}$

17. Find the points of intersection of $y = x^3 - 8$ and $y = e^{x/600}$. Before you start looking, how do you know that there must be more than one point? A convincing analytical answer to this question is essential.

18. Evaluate the following integrals, simplify the answers, and verify that the answers are correct with an alternate calculation.

a) $\int \frac{\cos(3x)\sin(3x)}{4+\cos^2(3x)}dx$ b) $\int_{-1}^{3} \frac{e^{2x}}{9+e^{4x}}dx$
c) $\int_{-4}^{-2} \frac{x}{1-x^2}dx$ d) $\int \frac{1}{\sqrt{4-x^2}}dx$
e) $\int \frac{x}{\sqrt{4-x^2}}dx$ f) $\int \frac{x^2}{\sqrt{4-x^2}}dx$

19. Verify the identity $\arctan \frac{1+x}{1-x} = \arctan(x) + \frac{\pi}{4}$. (Hint: Use the idea that, if $F(x)$ and $G(x)$ are both antiderivatives of the same function, then they differ by a constant.)

20. Derive the differentiation formula for $\text{arctanh}(x)$ based on formulas for the exponential and logarithmic functions.

21. Find the area (as a decimal number) of the region bounded by the curves

$$y = x^3 \text{ and } y = 5 - e^{-x}.$$

Verify the answer with an alternative calculation.

22. Find the area of the region bounded by the y-axis and the curves $y = \cos(x)$ and $y = -0.9 + \frac{3}{x-1}$ that lies in the second and third quadrants. The two curves here actually bound infinitely many regions. Consider only the one region that is also bounded by the y-axis.

23. A movie theater has a 20-foot-tall screen mounted on its front wall. The bottom of the screen is 20 feet above the floor, and the entire floor is flat and horizontal. How far from the front wall should a person sit in order to maximize the size of the viewing angle of the screen? What is the maximum viewing angle in degree measure?

24. When an environmental group first began to monitor the population of a threatened species of bird, there were an estimated 8,500,000 of the birds in the country. That was 50 years ago. The results of a current study suggest that the population has dwindled to 1,900,000. Emergency laws designed to save the bird from extinction are automatically placed in effect if and when the population drops to 500,000. Assume that, over the lifetime of the study, the population has been subject to the same law of exponential decay. How much time remains before an emergency is declared?

25. The owner of a small pond is introducing bass into the pond for the first time. The pond is big enough and diverse enough to sustain a maximum population of 10,000 adult bass. It is expected that the population of adult bass will grow at a rate proportional to the difference between the maximum sustainable population and the current population. The pond is initially seeded with 50 adult bass. Three years later the population is estimated to be 1400. What will be the population be

after fifteen years? How long will it take to reach a population that is 95% of the maximum population?

26. A certain bacterium in the human body is capable of doubling its population every 45 minutes. If there were 20 bacteria in the initial contaminant, what will the population be 24 hours later?

27. Compute the following limits, if they exist, and use *L'Hôpital's Rule* to verify the answers. Let a, b, and c denote arbitrary constants. Recall our failed attempt to verify the first limit in Example 2.4.

a) $\lim_{x\to 1} \frac{2\cos(x-1)+x^2-2x-1}{(x-1)^4}$ b) $\lim_{x\to\infty}(1+\frac{2}{x})^x$

c) $\lim_{x\to 0^+} \sin(x)\ln(x)$ d) $\lim_{x\to 0^+}(\cos(ax)+\sin(bx))^{(c/x)}$

Project: Best seat in the theater

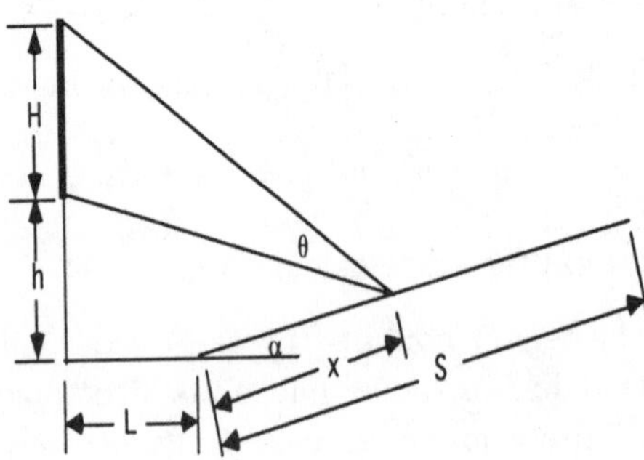

The above diagram describes the basic shape of a movie theater. A screen of height H feet is placed on a wall h feet above a level floor. The floor remains horizontal for the first L feet from that wall and then begins to rise at an incline of angle a as shown. This incline, of course, is where the seats are placed, and its length from the first row of seats to the last row in back is S. How far up this incline should we sit in order to have the best view of the movie? The best view, naturally, is one where θ is as large as possible.

Now suppose we plan to see a movie in a theater where $H = 30$, $h = 10$, $L = 20$, $S = 60$, $\alpha = 20°$, where the lengths are all in feet. Determine the value x that corresponds to a maximum viewing angle and find this angle, θ. If the rows of seats are placed 3 feet apart, with the first row right at the base of the incline, then what row should we sit in?

Suppose management plans to build a new theater where all of the dimensions would be kept the same except for α, which would be changed to $\alpha = 16.5°$. Determine how much this would change the viewing angle by computing the new values of x and θ.

Chapter 6

Techniques of Integration

Neither Maple nor mathematics will ever supply us with enough tools to integrate, in closed form, all of the functions we would like to integrate. This is not for want of trying. We defined the function $\ln(x)$ in order to solve the problem $\int \frac{1}{x}\,dx$, and this same strategy could be used for other integrals just as well. If an antiderivative for the integral $\int f(x)\,dx$ cannot be found, simply pick a number a, define the function $F(x) = \int_a^x f(t)\,dt$, study the properties of F so that it becomes a usable special function, and then add it to a list of known special functions. This would allow us to solve not only $\int f(x)\,dx$, but a host of related integrals as well. Unfortunately, this is an endless task. It turns out that no matter how many special functions we add to this list, we can always break outside of this list by integrating functions within the list. This means that there will always be "standard" functions with "unknown" antiderivatives.

The transcendental functions introduced in the last chapter complete the list of functions commonly referred to as the *elementary special functions* of mathematics. These functions include: the power function $p(x) = x^r$ for a rational number r, the trigonometric functions and their inverses, the various logarithmic and exponential functions, and the hyperbolic functions and their inverses. The functions, which we are able to access or express (in worldly terms) consist of all the combinations of these special functions using finitely many compositions and finitely many algebraic operations. Certainly this is a large list, but, as we stated in the last paragraph, it is not large enough.

Maple has added a few more functions to this list, and so it can evaluate a few more integrals. Maple's integration command is powerful enough that it will usually be able to evaluate an integral if the answer is expressible with these special functions. However, there are exceptions. Integration sometimes requires a creative nudge that only a human being can provide. Occasionally, there are techniques that we can use to help Maple complete an integral it cannot evaluate by itself. Such an example is presented in Section 6.6, and a few other examples will be encountered in the exercise set at the end of this chapter.

In this chapter, we use Maple to practice the same integration techniques that are studied in a traditional calculus text. In the process, we will introduce several new Maple commands and, hopefully learn more about classical integration techniques.

If Maple ever needs help from us in order to evaluate an integral, our work in this chapter should prepare us for such an event.

Maple will usually be able to evaluate, without any help from us, the integrals we consider in this chapter, so we have to turn our work here into somewhat of a "game." To begin with, **we will have to use the inert form Int( ) of the integration command in order to suppress premature evaluation**, otherwise Maple will just compute the answer directly. Our objective here is to manipulate a given integral, just as we do with pencil-and-paper techniques, until it appears as an entry on a simple table of integrals (or as a linear combination of entries on the table). This table of integrals consists of the 15 or so basic antiderivatives that have been developed up through the last chapter. We may as well give a name to this strategy for evaluating integrals, so that we don't have to repeat all of these instructions every time an integral is to be evaluated in this fashion. With this in mind, let us simply call any approach that evaluates integrals in this way an **elementary integral approach**.

We can always differentiate the output of a Maple integral to verify that an answer is correct, but such an approach hardly explains *why* an answer takes the form that it takes. One advantage of using an "elementary integral approach" is that it indeed supplies such an explanation.

Many of the commands we will be using reside in the student package, and this package will have to be loaded routinely throughout the chapter.

6.1 Integration by Substitution Revisited

This topic was studied in Chapter 3, but now our list of special functions is much larger, and so we are able to integrate many more functions.

Suppose we wish to evaluate an integral $\int p(x)\,dx$. Perhaps Maple cannot evaluate the integral, or perhaps we just want to understand how Maple arrived at its answer. If we can identify the integral as being of the form $\int f(g(x))g'(x)\,dx$, then we can let $u = g(x)$, and

$$\int p(x)\,dx = \int f(g(x))g'(x)\,dx = \int f(u)\frac{du}{dx}\,dx = \int f(u)\,du.$$

If the integral $\int f(u)\,du$ happens to be an elementary or simple integral, then it is an easy matter to finish the problem.

Actually, the relationship between the new variable u and the old variable x might not be so explicit. It might instead be defined by some equation in the variables x and u. In either case, **Maple will take care of the details of an integration by substitution with its changevar( ) command, which resides in the student package.**

We introduced the command changevar( ), and used it briefly, back in Chapter 3. In this chapter, we will be making frequent use of this command. Looking ahead at the following example, notice that Maple expects you to **insert the new variable "u" as a third argument** in the change of variable command. This may strike you as being redundant, but it is not. Maple has to be given the name of the new variable, because the first entry in the change of variable command can be **any equation involving the old and new variables. It does not have to be in the explicit form $u = g(x)$**, where u is the new and x is the old variable. So you can see that Maple will pass through the first entry of the change in variable command without knowing which one of the variables is the new variable.

Example 6.1 *Evaluate the integral $\int \frac{e^{3x}}{9+e^{6x}}dx$, using Maple's integration command directly. Verify the answer by using an elementary integral approach.*

****

```
> int(exp(3*x)/(9+exp(6*x)),x);
```

$$\frac{1}{9}\arctan(\frac{1}{3}(e^x)^3)$$

```
> p:=Int(exp(3*x)/(9+exp(6*x)),x):
> with(student):
> p1:=changevar(u=exp(3*x),p,u);
```

$$p1 := \int \frac{1}{27+3u^2}\,du$$

```
> p2:=factor(p1);
```

$$p2 := \int \frac{1}{3}\frac{1}{9+u^2}\,du$$

```
> p3:=expand(p2);
```

$$p3 := \frac{1}{3}\int \frac{1}{9+u^2}\,du$$

```
> p4:=value(p3);
```

$$p4 := \frac{1}{9}\arctan(\frac{1}{3}u)$$

```
> answer:=subs(u=exp(3*x),p4)+C;
```

$$answer := \frac{1}{9}\arctan(\frac{1}{3}e^{(3x)}) + C$$

****

In the next example, we introduce the command **completesquare( )**, which resides in the student package. As you can see in this example, the command is applied to a quadratic expression, but there is no need for the first argument to be exclusively a quadratic expression. Maple's command for completing a square will **search any expression for quadratic terms, and complete the square on those terms.** Occasionally, Maple has difficulty selecting the variable in the quadratic term, and so it is a good idea to **specify the variable as the second argument** in this command.

Example 6.2 *Evaluate the integral* $\int \frac{5}{\sqrt{7-3x-2x^2}}$, *using Maple's integration command directly. Verify the answer by using an elementary integral approach.*

****

```
> int(5/sqrt(7-3*x-2*x^2),x);
```

$$\frac{5}{2}\sqrt{2}\arcsin(\frac{4}{65}\sqrt{65}\,(x+\frac{3}{4}))$$

```
> p:=Int(5/sqrt(7-3*x-2*x^2),x):
> with(student):
> d:=completesquare(7-3*x-2*x^2);
```

$$d := -2\,(x+\frac{3}{4})^2+\frac{65}{8}$$

```
> p1:=changevar(u=sqrt(2)*(x+3/4),p,u);
```

$$p1 := \int \frac{5}{2}\frac{\sqrt{2}}{\sqrt{7-\frac{3}{8}\,(-3\sqrt{2}+4\,u)\,\sqrt{2}-\frac{1}{16}\,(-3\sqrt{2}+4\,u)^2}}\,du$$

```
> p2:=simplify(p1);
```

$$p2 := 10\int \frac{1}{\sqrt{65-8\,u^2}}\,du$$

```
> p2:=expand(p1);
```

$$p2 := \frac{5}{2}\sqrt{2}\int \frac{1}{\sqrt{\frac{65}{8}-u^2}}\,du$$

```
> p3:=value(p2);
```

$$p3 := \frac{5}{2}\sqrt{2}\arcsin(\frac{2}{65}\sqrt{130}\,u)$$

```
> answer:=subs(u=sqrt(2)*(x+3/4),p3)+C;
```

$$answer := \frac{5}{2}\sqrt{2}\arcsin(\frac{2}{65}\sqrt{130}\sqrt{2}(x+\frac{3}{4})) + C$$

```
> answer:=simplify(answer);
```

$$answer := \frac{5}{2}\sqrt{2}\arcsin(\frac{1}{65}\sqrt{65}(4x+3)) + C$$

****

Example 6.3 *Use Maple's integration command to evaluate the integral*

$$\int \sin^4(x)\cos^7(x)dx$$

directly. Verify the answer by using an elementary integral approach.

A substitution of the form $u = \sin(x)$ or $u = \cos(x)$ would seem appropriate. Notice, however, that, if we try $u = \sin(x)$, then one power of the cosine function can be used for the differential, and the remaining $\cos^6(x)$ can easily be turned into $(1-\sin^2(x))^3$, and hence into $(1-u^2)^3$. A substitution of the form $u = \cos(x)$ would not work. One power of the sine function would be needed for the differential, and the remaining 3 powers of $\sin(x)$ could not be turned (without radicals) into an expression involving $u = \cos(x)$.

****

```
> a:=int(sin(x)^4*cos(x)^7,x);
```

$$a := -\frac{1}{11}\sin(x)^3\cos(x)^8 - \frac{1}{33}\sin(x)\cos(x)^8 + \frac{1}{231}\cos(x)^6\sin(x) + \frac{2}{385}\cos(x)^4\sin(x) + \frac{8}{1155}\cos(x)^2\sin(x) + \frac{16}{1155}\sin(x)$$

```
> p:=Int(sin(x)^4*cos(x)^7,x):
> with(student):
> p1:=changevar(u=sin(x),p,u);
```

$$p1 := \int u^4(1-u^2)^3\,du$$

```
> p2:=expand(p1);
```

$$p2 := \int u^4\,du - 3\int u^6\,du + 3\int u^8\,du - \int u^{10}\,du$$

```
> p3:=value(p2);
```

$$p3 := \frac{1}{5}u^5 - \frac{3}{7}u^7 + \frac{1}{3}u^9 - \frac{1}{11}u^{11}$$

```
> p4:=subs(u=sin(x),p3);
```

$$p4 := \frac{1}{5}\sin(x)^5 - \frac{3}{7}\sin(x)^7 + \frac{1}{3}\sin(x)^9 - \frac{1}{11}\sin(x)^{11}$$

#This answer is different from the answer that we obtained when we integrated directly. To show that they are equivalent, we use the identity $\cos^2(x) = 1 - \sin^2(x)$ to eliminate the cosine function from the first expression. **Doing this, however, requires a new Maple command.** When we use the **command subs(pˆ$n = A, expr$)**, Maple replaces p^n by A in $expr$, but it **does not replace powers of p^n by powers of A.** Thus, for example, it does not replace p^{2n} by A^2. **To replace each power of p^n in $expr$ by a power of A, we use the Maple substitution command powsubs(pˆ$n = A, expr$).** This command **lives in the student package**, which we have already loaded. #

```
> a1:=powsubs(cos(x)^2=1-sin(x)^2,a);
```

$$a1 := -\frac{1}{11}\sin(x)^3(-1+\sin(x)^2)^4 - \frac{1}{33}(-1+\sin(x)^2)^4\sin(x) - \frac{1}{231}(-1+\sin(x)^2)^3\sin(x) + \frac{2}{385}(-1+\sin(x)^2)^2\sin(x) - \frac{8}{1155}(-1+\sin(x)^2)\sin(x) + \frac{16}{1155}\sin(x)$$

```
> a2:=expand(a1);
```

$$a2 := \frac{1}{5}\sin(x)^5 - \frac{3}{7}\sin(x)^7 + \frac{1}{3}\sin(x)^9 - \frac{1}{11}\sin(x)^{11}$$

****

6.2 Integration by Parts

Before we work on this topic, we introduce two new basic Maple commands with a broad range of applications. Actually, only one of the commands will be used later in the chapter on an example, but this may well be coincidental. It will seem clear from their introduction, that these commands are bound to be useful.

It can be a frustrating experience to ask Maple to simplify an expression only to find out that it simplifies the desired part of the expression, and makes a mess out of the rest of the expression. Rather than wait for this to happen, we take this opportunity to introduce tools for **isolating parts of a Maple expression**, or more generally, parts of any Maple data type. With these new commands, we will have an opportunity to simplify only one term in an expression. There are many other reasons for wanting to isolate a part of a Maple data type, besides a desire to simplify a subterm. These commands will have a broad base of applications.

The command **op( ) extracts operands from an expression or data type. If *data* is a Maple data type and *j* is a positive integer, then op(*j*, *data*) is the *j*th term in *data*.** This explanation may be somewhat vague, so let us show by way of examples how to use this important command. Our examples are restricted to Maple expressions rather than to a more general Maple data type.

****

```
> p:=a+b+c+d+e+f:          q:=a*b*c*d*e*f:
  r:=a*b+c*d+e*f:
s:=(a+b)*(c+d)*(e+f):
```

#In the expressions p and q, there are 6 terms. In the expressions r and s, there are 3 terms (and each of these has subterms). What do you see when you look at r? It is a sum of 3 terms. What do you see when you look at s? It is a product of 3 terms. We can access individual terms as follows (commas are used to save page space):#

```
> op(2,p),op(5,q),op(2,r),op(3,s);
```

$$b,\ e,\ c\,d,\ e+f$$

#They can be used like other names to create expressions.#

```
> op(3,p)*(op(6,q)+op(2,p))^op(4,q);
```

$$c\,(f+b)^d$$

#Each term within r and s has its own subterms, and these can be accessed by using the op( ) command a second time.#

```
> op(2,op(1,s));
```

$$b$$

#The next example is included because it is easy to make a mistake when subtraction is involved. As you can see, Maple doesn't recognize subtraction as an operation on a par with addition. It prefers to think of this subtraction as $a+(-1)b$. Consequently, Maple sees the second term here as $-b$. To extract the term b itself, just identify it as the second term of $(-1)b$.#

```
> p:=a-b:  op(2,p),op(1,op(2,p));
```

$$-b,\ -1$$

****

The operand command can be used in a wide variety of situations other than on strictly algebraic expressions. **Try it,** for example, on a Maple structure like

Int($f(x), x = a..b$) to find out what the first and second terms are in such a structure.

The left-hand side and the right-hand side of an equation can also be extracted by the op( ) command, but we have better commands for this purpose, commands we have used on only one other occasion, but which will be used more frequently in the future. The commands **lhs( ) and rhs( ) can be used to extract, respectively, the left-hand side and the right-hand side of an equation.**

Now that we understand Maple's op( ) command, we are ready to introduce **applyop( )**, the second of the promised two new Maple commands. **If *comd*( ) is a Maple command (or equivalently any function), if *expr* is a Maple expression, and if *j* is a positive integer, then the input statement**

>applyop(*comd*,*j*,*expr*);

applies the command *comd*( ) to just the *j*th operand, op(*j*,*expr*), of *expr*, and leaves all the other operands of *expr* unaffected. If *comd*( ) has additional arguments, they are included in the applyop( ) command after the third argument *expr*. A short Maple work session will explain how to use this important new Maple command.

* *

```
> f:=(x+2)^2+(y-3)^2+w^5+z^4:
> applyop(expand,2,f);
```

#Notice that there is no left parenthesis at the end of the expand command.#

$$(x+2)^2 + y^2 - 6\,y + 9 + w^5 + z^4$$

```
> applyop(diff,3,f,w);
```

#The diff( ) command has a second argument, w, which specifies the variable of differentiation. Notice where this argument w is placed in the applyop( ) command.#

$$(x+2)^2 + (y-3)^2 + 5\,w^4 + z^4$$

```
> applyop(int,4,f,z=1..2);
```

#The command int( ) has a second argument, $z = 1..2$. Notice where it is placed in the applyop( ) command.#

$$(x+2)^2 + (y-3)^2 + w^5 + \frac{31}{5}$$

* *

These commands might or might not play a critical role in the immediate future, but they are, at least, available, if they are needed.

With these preparations behind us, we are now ready to discuss the main topic in this section, beginning with a new Maple command. Maple's command for integrating by parts resides in the **student package** and is called **intparts( )**. When we integrate by parts (by hand), we start with an integral and choose u (the part that gets differentiated) and dv (the part which gets integrated). Actually, all we really have to choose is u; dv is implied by a choice for u. As expected then, there are only two entries in Maple's integration by parts command, **intparts( )**. The first is the integral, and the second is our choice for u.

If the expression P in intparts(P, u) contains terms that are not associated with an integral, Maple will just pass over these terms and ignore them. This can be very convenient, as we shall see, when integrating by parts more than once. However, if P contains more than one integral, Maple will integrate each of them by parts, even when this is not desirable. Such unintentional integration by parts could be avoided by using the op( ) command to isolate the integral in P that is to be done by parts.

Example 6.4 *Evaluate the integral $\int x^3 e^{5x} dx$ directly, using Maple's integration command. Verify the answer by using an elementary integral approach.*

****

```
> int(x^3*exp(5*x),x);
```

$$\frac{1}{5}x^3 e^{(5x)} - \frac{3}{25}x^2 e^{(5x)} + \frac{6}{125}x\, e^{(5x)} - \frac{6}{625}e^{(5x)}$$

```
> p:=Int(x^3*exp(5*x),x):
> with(student):
> p1:=intparts(p,x^3);
```

$$p1 := \frac{1}{5}x^3 e^{(5x)} - \int \frac{3}{5}x^2 e^{(5x)}\, dx$$

```
> p2:=intparts(p1,x^2);
```

$$p2 := \frac{1}{5}x^3 e^{(5x)} - \frac{3}{25}x^2 e^{(5x)} + \int \frac{6}{25}x\, e^{(5x)}\, dx$$

```
> p3:=intparts(p2,x);
```

$$p3 := \frac{1}{5}x^3 e^{(5x)} - \frac{3}{25}x^2 e^{(5x)} + \frac{6}{125}x\, e^{(5x)} - \int \frac{6}{125}e^{(5x)}\, dx$$

```
> p=value(p3);
```

#Notice the use of (=) here rather than (:=).#

$$\int x^3 e^{(5x)}\, dx = \frac{1}{5}x^3 e^{(5x)} - \frac{3}{25}x^2 e^{(5x)} + \frac{6}{125}x\, e^{(5x)} - \frac{6}{625}e^{(5x)}$$

****

One could argue that the integral appearing in $p3$ is not yet quite on our elementary table of integrals, but it is clear enough how the evaluation should be completed, and so we dropped the last step. This same more relaxed attitude will prevail on future examples. Our work here gives us real mathematical insight about why the original integral, with its short compact form, has an antiderivative that involves so many terms. Even though Maple could have computed this integral directly, we have acquired a deeper understanding of this integral by evaluating it as we did in this example.

Notice that the entries in the Maple work session for the last example were *expressions*, until we expressed our final answer as an equation. Contrast this to the next example, where the entries in the Maple work session are equations.

Example 6.5 *Evaluate the integral $\int e^{5x}\cos(2x)dx$, using Maple's integration command directly. Verify the answer by using an elementary integral approach.*

****

```
> int(exp(5*x)*cos(2*x),x);
```

$$\frac{5}{29}\,e^{(5\,x)}\cos(2\,x)+\frac{2}{29}\,e^{(5\,x)}\sin(2\,x)$$

```
> p:=Int(exp(5*x)*cos(2*x),x):
> with(student):
> eq1:=p=intparts(p,exp(5*x));
```

$$eq1 := \int e^{(5\,x)}\cos(2\,x)\,dx = \frac{1}{2}\,e^{(5\,x)}\sin(2\,x) - \int \frac{5}{2}\,e^{(5\,x)}\sin(2\,x)\,dx$$

```
> eq2:=p=intparts(rhs(eq1),exp(5*x));
```

$$eq2 := \int e^{(5\,x)}\cos(2\,x)\,dx =$$
$$\frac{1}{2}\,e^{(5\,x)}\sin(2\,x)+\frac{5}{4}\,e^{(5\,x)}\cos(2\,x)+\int -\frac{25}{4}\,e^{(5\,x)}\cos(2\,x)\,dx$$

```
> eq3:=simplify(eq2);
```

$$eq3 := \int e^{(5\,x)}\cos(2\,x)\,dx =$$
$$\frac{1}{2}\,e^{(5\,x)}\sin(2\,x)+\frac{5}{4}\,e^{(5\,x)}\cos(2\,x)-\frac{25}{4}\int e^{(5\,x)}\cos(2\,x)\,dx$$

#Notice how the original integral has returned—yet this is far from being a wasted effort; a solution is at hand. The solution, however, is no longer a calculus problem, but rather a purely algebraic one. We **replace the original integral, in both locations by an unassigned name a (a =answer)**, and simply solve the resulting equation for a. **Replacing the integral by an unassigned name is a necessary step.** From Maple's point of view, it does not make mathematical sense to solve an equation for an object (the integral) that already has a value independent of the equation#.

```
> eq4:=subs(p=a,eq3);
```

$$eq4 := a = \frac{1}{2}\,e^{(5\,x)}\sin(2\,x) + \frac{5}{4}\,e^{(5\,x)}\cos(2\,x) - \frac{25}{4}\,a$$

```
> p=solve(eq4,a);
```

$$\int e^{(5\,x)}\cos(2\,x)\,dx = \frac{2}{29}\,e^{(5\,x)}\sin(2\,x) + \frac{5}{29}\,e^{(5\,x)}\cos(2\,x)$$

****

6.3 Reduction Formulas

Most Tables of Integrals include a large number of formulas known as reduction formulas. These are formulas, such as the one in the next example, which express an integral in terms of an "easier form"—usually a lower power— of a similar integral. Maple is a useful tool that can be used to establish many of these formulas.

Example 6.6 *Verify the reduction formula*

$$\int \csc^n(x)dx = -\frac{1}{n-1}\frac{\cos(x)}{\sin^{n-1}(x)} + \frac{n-2}{n-1}\int \csc^{n-2}(x)dx$$

When we integrate by parts, there are several strategies to consider in our choice of u and dv. One of these strategies involves looking for a part of the integral that is easily integrated. Since this is true of $\csc^2(x)$, that would prompt us to choose $dv = \csc^2(x)$. With this choice of dv, $u = \csc^{n-2}(x)$ becomes the rest of the integral by default. Remember that, in Maple's integration-by-parts command, all we have to specify is u.

It is difficult to get Maple to manipulate some trigonometric expressions to our satisfaction, but it seems to do a much better job on sines and cosines. For this reason, we replace $\csc(x)$ by $1/\sin(x)$.

In this example, pay particular attention to some of the **powerful optional enhancements, which are being introduced for the first time, to some of Maple's basic commands**. When the expand( ) command is used in the form **expand(*expr*1, *expr*2, ..., *expr n*), Maple expands *expr*1, but does not expand the subexpressions *expr*1, *expr*2, ..., *expr n*, appearing in *expr*1.**

We have already mentioned some of the optional enhancements to the simplify( ) command. When this command is used in the form **simplify(*expr*, power), only simplification procedures closely related to exponentiation are invoked.**

There are **many other command refinements for specifying exactly what it is that you want Maple to do.** You might well find immediate use for several more of these enhancements, but we shall not introduce any more of them until they are needed. If you are curious, definitely look in the Help File. The commands expand( ) and simplify( ) would be a good place to start.

****

```
> p:=Int(sin(x)^(-n),x):
> with(student):

Warning, new definition for D
> p1:=intparts(p,sin(x)^(-n+2));
```

$$p1 := -\frac{\sin(x)^{(-n+2)}\cos(x)}{\sin(x)} - \int -\frac{\sin(x)^{(-n+2)}(-n+2)\cos(x)^2}{\sin(x)^2}\,dx$$

```
> p2:=subs(cos(x)^2=1-sin(x)^2,p1);
```

$$p2 := -\frac{\sin(x)^{(-n+2)}\cos(x)}{\sin(x)} - \int -\frac{\sin(x)^{(-n+2)}(-n+2)(1-\sin(x)^2)}{\sin(x)^2}\,dx$$

```
> p3:=expand(p2,-n+2);
```

#We don't want the power $-n+2$ expanded.#

$$p3 := -\frac{\sin(x)^{(-n+2)}\cos(x)}{\sin(x)} + (-n+2)\int\frac{\sin(x)^{(-n+2)}}{\sin(x)^2}\,dx$$
$$-(-n+2)\int\sin(x)^{(-n+2)}\,dx$$

```
> p4:=simplify(p3,power);
```

#This option prevents unwanted trigonometric simplifications.#

$$p4 := -\sin(x)^{(-n+1)}\cos(x) + (-n+2)\int\sin(x)^{(-n)}\,dx$$
$$-(-n+2)\int\sin(x)^{(-n+2)}\,dx$$

#Notice that the original integral (originally named *p*) has reappeared on the right-hand side. This situation is similar to one that occurred in Example 6.5. The idea is to set up the equation $p = p4$ ($p4$ is equivalent to the original integral), and solve the equation (algebraically) for the integral. In Example 6.5, we replaced the integral, in both places, by an unassigned name *a* (*a* = answer) and solved the resulting equation for *a*. This was a necessary step. Maple would have been confused if we had tried to solve the equation for the integral, because the integral is not an unknown—it already has a value without reference to the equation. We could use the same procedure in the current situation, but we use another command instead.

Recall that the command **isolate(*eqn*, *expr*)** isolates the expression *expr* occurring within the equation *eqn*. This command is really a more appropriate one to use in this situation. We do not have to first replace the integral by an unassigned letter. Maple is not being asked to solve (read evaluate) the integral, but only to isolate it; Maple is not confused by the instruction to isolate a term that already has a value. #

```
> eq1:=p=p4;
```

$$eq1 := \int \sin(x)^{(-n)}\,dx = -\sin(x)^{(-n+1)}\cos(x) + (-n+2)\int \sin(x)^{(-n)}\,dx - (-n+2)\int \sin(x)^{(-n+2)}\,dx$$

```
> isolate(eq1,p);
```

$$\int \sin(x)^{(-n)}\,dx = \frac{-\sin(x)^{(-n+1)}\cos(x) - (-n+2)\int \sin(x)^{(-n+2)}\,dx}{-1+n}$$

****

6.4 Inverse Trigonometric Substitutions

In this section, we consider integrals involving expressions of the form $\sqrt{x^2+a^2}$, $\sqrt{x^2-a^2}$, $\sqrt{a^2-x^2}$. Frequently, integrals involving other powers of these basic quadratic expressions can be evaluated by using the same techniques. The integrals are, of course, easily evaluated by using Maple's int( ) command directly, but it is not uncommon for answers to bear little resemblance to the original integrand. Why does an answer take a particular form? To answer this question, our work continues in the same spirit as it has in the other sections of this chapter. We use Maple to mimic what we would have to do to evaluate the integral with "pencil-and-paper" techniques.

The germ of the idea is to use the following substitutions and identities:

$$x = a\sin(\theta) \text{ and } 1-\sin^2(\theta) = \cos^2(\theta), \text{ on the expression } a^2 - x^2;$$

$$x = a\tan(\theta) \text{ and } 1+\tan^2(\theta) = \sec^2(\theta), \text{ on the expression } a^2 + x^2;\text{and}$$

$$x = a\sec(\theta) \text{ and } \sec^2(\theta) - 1 = \tan^2(\theta), \text{ on the expression } x^2 - a^2$$

to turn these basic quadratic expressions into perfect squares. This allows us to cancel the radical—a major simplification. The resulting trigonometric integral can then be evaluated by standard techniques.

The approach is really just integration by substitution, and so Maple's changevar( ) command will easily handle the conversion over to a trigonometric integral. Our past substitutions inside the changevar( ) command have been of the form $u = g(x)$ (a new variable as a function of the old variable). With the current technique, we consider substitutions of the form $x = g(\theta)$ (the old variable as a function of a new variable). As we pointed out earlier, this is a legitimate way to enter the change of variable inside the changevar( ) command.

The changevar( ) command automatically converts the differential dx to the new differential $d\theta$. We could ignore these details, but, to make this a good learning experience, we should at least make a mental note of the relationships

$$x = a\sin(\theta), \qquad x = a\tan(\theta), \qquad x = a\sec(\theta),$$
$$dx = a\cos(\theta)\,d\theta, \qquad dx = a\sec^2(\theta)\,d\theta, \qquad dx = a\sec(\theta)\tan(\theta)\,d\theta$$

between the variables and the differentials.

Example 6.7 *Evaluate the following integral, using Maple's integration command directly. Verify the answer by using an elementary integral approach.*

$$\int \frac{x^2+x+1}{\sqrt{x^2+4}}\,dx$$

The solution of this problem is omitted in order to save space. The reader is invited to visit our web site for a complete solution.

6.5 Rational Functions

Maple can compute integrals of rational functions directly. It uses the same approach, partial-fraction decomposition, that would be used to compute the integral of a rational function by hand. As you recall, the first step in this process involves factoring the denominator of the rational function into a product of linear and irreducible quadratic terms.

According to the *Fundamental Theorem of Algebra*, every polynomial of degree one or more can be expressed as a product of linear and quadratic terms, but finding the factors is not always an easy matter. In Section 5.1, we mentioned that there is no formula for finding the roots of a general polynomial of degree n for any $n \geq 5$. For the very same reason, there is no formula for factoring polynomials of degree $n \geq 5$—factoring and finding roots of polynomials are equivalent.

In spite of this mathematical barrier, Maple is quite good at factoring polynomials that have nice factors, and so one can expect Maple to be reasonably successful at integrating quite a few rational functions. Nevertheless, the next example was chosen with some care. The denominator of a randomly chosen rational function of this size may well not have such nice factors.

```
***********************************
> f:=(6*x^6-25*x^5+71*x^4-235*x^3+317*x^2-582*x+468)/
  (2*x^5-7*x^4+16*x^3-56*x^2+32*x-112);
```

$$f := \frac{6\,x^6 - 25\,x^5 + 71\,x^4 - 235\,x^3 + 317\,x^2 - 582\,x + 468}{2\,x^5 - 7\,x^4 + 16\,x^3 - 56\,x^2 + 32\,x - 112}$$

```
> int(f,x);
```

$$\frac{3}{2}x^2 - 2x + \frac{3}{2}\ln(2x-7) + \frac{3}{2}\ln(x^2+4) - \frac{7}{2}\arctan(\frac{1}{2}x) - \frac{3}{x^2+4}$$

#As you can see, there is no need to specifically ask Maple to compute a partial-fraction decomposition. On the other hand, this decomposition is readily available, if it is desired, via Maple's **convert( )** command. Once again, it is worth mentioning that **this command has a broad range of applications**. It is worth the effort to investigate its full potential.

If $r(x)$ is a rational function, **convert($r(x)$,parfrac,x) will convert $r(x)$ to its partial fraction decomposition**. The degree of the numerator of $r(x)$ does not have to be less than the degree of the denominator of $r(x)$, as it does when one is doing the method by hand. Maple first does the necessary long division. This command has an **optional fourth argument consisting of the word "true" or "false."** Use the word **"true"** if you have already **expressed $r(x)$ in its normal form** (using the command normal( )), and if you have already **expressed the denominator of $r(x)$ in its factored form**. If this option is not declared, its **default value is "false"**, and Maple will attempt, among other things, to factor the denominator of $r(x)$.#

```
> f:=numer(f)/factor(denom(f));
```

$$f := \frac{6x^6 - 25x^5 + 71x^4 - 235x^3 + 317x^2 - 582x + 468}{(2x-7)(x^2+4)^2}$$

```
> convert(f,parfrac,x,true);
```

$$3x - 2 + \frac{3}{2x-7} + \frac{-7+3x}{x^2+4} + 6\frac{x}{(x^2+4)^2}$$

****

6.6 When Maple Needs Help

There are many indefinite integrals that Maple is not able to calculate with the command int( ). This is the legacy of integration, not a weakness in Maple. The answer to an integral may not be expressible in "closed form," that is to say, in terms of the familiar functions, of mathematics. Any attempt to evaluate (exactly) such an integral is destined to fail.

On the other hand, Maple might fail to evaluate an indefinite integral even if it *is* expressible in closed form. In this section, we show that, with a little assistance from us, Maple can sometimes be coaxed into computing integrals of this latter type.

Even if Maple can compute an integral, we could find the answer unsatisfactory. Perhaps the answer includes strange functions we have never encountered before. Maple has many exotic special functions stored in its library that are used to evaluate integrals. If this happens, can we express Maple's answer in terms of more "acceptable" familiar functions? Perhaps we cannot do this, but, if we can, it is certainly important to include this step in the evaluation of an integral. Answers can also be unsatisfactory in other ways. Even if an answer is expressed in terms of the familiar elementary special functions of mathematics, the answer might seem inappropriate—unrelated to the expression being integrated. As we become more proficient at integrating in a "pencil-and-paper" sense, we will also get better at recognizing when an answer has that "certain look" that tells us we have the answer we are looking for.

In this section, we take the gloves off Maple and let it work for us. We will take advantage of all of Maple's integration power, but if it doesn't do what we want it to do, then we will have to help it along. While Maple will usually give us a satisfactory answer directly, and without our intervention, we have naturally chosen examples where Maple needs help.

Example 6.8 *Evaluate the integral $\int \cos^3(x)\sin^n(x)\,dx$, where n is any real constant. Verify the answer by differentiation.*

We can see from our initial attempt to evaluate this integral directly that Maple fails to deliver. Actually, this is surprising, because the integral is quite elementary, Letting $u = \sin(x)$ quickly turns this into an elementary integral. This is a good example to demonstrate why it is still important to be able to integrate by hand, even with the aid of Maple's powerful integration engine.

****

```
> assume(n,real);
> f:=cos(x)^3*sin(x)^n;
```

$$f := \cos(x)^3 \sin(x)^n$$

```
> int(f,x);
```

$$\int \cos(x)^3 \sin(x)^n \, dx$$

Maple can't evaluate this integral, so it just returns the unevaluated integral.#

```
> p:=Int(cos(x)^3*sin(x)^n,x);
```

$$p := \int \cos(x)^3 \sin(x)^n \, dx$$

```
> with(student):
> p1:=changevar(u=sin(x),p,u);
```

$$p1 := \int (1 - u^2)\, u^n \, du$$

#This is clearly an elementary integral, so we ask Maple to evaluate the integral directly.#

```
> p2:=value(p1);
```

$$p2 := -\frac{u^3 e^{(n\ln(u))}}{n+3} + \frac{u\, e^{(n\ln(u))}}{n+1}$$

#Well! We have an answer, but it certainly is not the elementary one we were expecting, so we intervene a bit more.#

```
> p2:=expand(p1);
```

$$p2 := \int u^n \, du - \int u^n \, u^2 \, du$$

```
> p3:=simplify(p2);
```

$$p3 := \int u^n \, du - \int u^{(2+n)} \, du$$

```
> p4:=value(p3);
```

$$p4 := \frac{u^{(n+1)}}{n+1} - \frac{u^{(n+3)}}{n+3}$$

```
> p5:=subs(u=sin(x),p4);
```

$$p5 := \frac{\sin(x)^{(n+1)}}{n+1} - \frac{\sin(x)^{(n+3)}}{n+3}$$

#We have a satisfactory answer. Now we verify that the answer is correct by differentiation.#

```
> p6:=diff(p5,x);
```

$$p6 := \frac{\sin(x)^{(n+1)} \cos(x)}{\sin(x)} - \frac{\sin(x)^{(n+3)} \cos(x)}{\sin(x)}$$

```
> p7:=expand(p6);
```

$$p7 := \cos(x) \sin(x)^n - \cos(x) \sin(x)^2 \sin(x)^n$$

```
> p8:=simplify(subs(sin(x)^2=1-cos(x)^2,p7));
```

$$p8 := \cos(x)^3 \sin(x)^n$$

****

Example 6.9 *Evaluate the integral*

$$\int \frac{\left(1 + x^{(1/4)}\right)^n}{\sqrt{x}} \, dx,$$

where n is any real constant. Verify the answer by differentiation.

In this example, Maple gives us an answer, but the answer is not in terms of the elementary special functions of mathematics. It could be that there is no simpler way to write the answer, but let us see what we can do.

****

```
> f:=(1+x^(1/4))^(n)/sqrt(x);
```

$$f := \frac{(1 + x^{(1/4)})^n}{\sqrt{x}}$$

```
> p:=int(f,x);
```

$$p := 2\sqrt{x} \, \mathrm{hypergeom}([2, -n], [3], -x^{(1/4)})$$

```
> p1:=simplify(p);
```

$p1 :=$

$$4 \frac{(1 + x^{(1/4)})^n \, x^{(1/4)} \, n + (1 + x^{(1/4)})^n \sqrt{x} \, n + (1 + x^{(1/4)})^n \sqrt{x} - (1 + x^{(1/4)})^n + 1}{(2+n)\,(n+1)}$$

#We have an acceptable answer, but it looks as if it can be simplified. The expression $\left(1+x^{(1/4)}\right)^n$ appears in every summand in the numerator except for the last term, which is just the constant 1. without this term, the expression would easily factor. **Here is an interesting trick worth remembering. By adding or subtracting any constant from an antiderivative, we get another equivalent antiderivative.** This means that we can discard the unwanted 1. Notice that it is, in fact, also multiplied by 4 and divided by $(2+n)(1+n)$.#

```
> p2:=normal(p1-4/((2+n)*(1+n)));
```

$$p2 := 4\,\frac{(1+x^{(1/4)})^n\,(x^{(1/4)}\,n+\sqrt{x}\,n+\sqrt{x}-1)}{(2+n)\,(n+1)}$$

#We have the answer in a really nice form. Now let us differentiate this expression to see if we can return the original integrand.#

```
> p3:=diff(p2,x);
```

$$p3 := \frac{(1+x^{(1/4)})^n\,n\,(x^{(1/4)}\,n+\sqrt{x}\,n+\sqrt{x}-1)}{x^{(3/4)}\,(1+x^{(1/4)})\,(2+n)\,(n+1)} + \frac{4\,(1+x^{(1/4)})^n\left(\dfrac{1}{4}\,\dfrac{n}{x^{(3/4)}}+\dfrac{\frac{1}{2}\,n}{\sqrt{x}}+\dfrac{\frac{1}{2}}{\sqrt{x}}\right)}{(2+n)\,(n+1)}$$

```
> p4:=simplify(p3);
```

$$p4 := \frac{(1+x^{(1/4)})^{(n-1)}\,(x^{(3/4)}+x)}{x^{(5/4)}}$$

#Notice that, if we could factor out $x^{(3/4)}$ from the second term of the numerator, then it would cancel with the $x^{(5/4)}$ in the denominator. The two terms in the numerator would then combine into just one power. It is clear that the original integrand will return after these simplifications. We should declare this to be sufficient proof that our answer is correct. We continue, however, just to show how the simplifications can be carried out. Maple is quite stubborn about the simplification, and we must tackle the matter aggressively to make Maple do what we want it to do. #

```
> above:=op(1,p4)*expand(op(2,p4)/x^(3/4));
  below:=simplify(denom(p4)/x^(3/4));
```

$$above := (1+x^{(1/4)})^{(n-1)}\,(1+x^{(1/4)})$$

$$below := \sqrt{x}$$

```
> above:=simplify(above);
```

$$above := (1+x^{(1/4)})^n$$

```
> p5:=above/below;
```

$$p5 := \frac{(1+x^{(1/4)})^n}{\sqrt{x}}$$

****

Example 6.10 *Evaluate the integral* $\int e^{2x} \sin(e^x)\,dx$*, where n is any real constant. Verify the answer by differentiation.*

Maple gives us an answer to this integral directly without intervention by us. The answer is in terms of elementary functions, but it seems disconnected from the original integrand. Let us see whether we can determine a more acceptable answer.

****

```
> f:=exp(2*x)*sin(exp(x));
```

$$f := e^{(2\,x)} \sin(e^x)$$

```
> int(f,x);
```

$$\frac{e^x \tan(\frac{1}{2}\,e^x)^2 + 2\tan(\frac{1}{2}\,e^x) - e^x}{1 + \tan(\frac{1}{2}\,e^x)^2}$$

```
> p:=Int(exp(2*x)*sin(exp(x)),x);
```

$$p := \int e^{(2\,x)} \sin(e^x)\,dx$$

```
> p1:=changevar(u=exp(x),p,u);
```

$$p1 := \int u \sin(u)\,du$$

```
> p2:=value(p1);
```

$$p2 := \sin(u) - u\cos(u)$$

```
> p3:=subs(u=exp(x),p2);
```

$$p3 := \sin(e^x) - e^x \cos(e^x)$$

#This answer is certainly more appropriate than Maple's answer. We verify the answer by differentiation.#

```
> p4:=diff(p3,x);
```

$$p4 := (e^x)^2 \sin(e^x)$$

****

6.7 Improper Integrals

The command int$(f(x), x = a..b)$ can be used to compute, directly, improper integrals of the form

$$\int_a^\infty f(x)dx, \int_{-\infty}^b f(x)dx, \int_{-\infty}^\infty f(x)dx,$$

and also those of the form

$$\int_a^b f(x)dx,$$

where f is discontinuous at one or more points in the interval [a,b]. Whenever the command int$(f(x), x = a..b)$ is entered, Maple looks for discontinuities of f in the interval [a,b] and expresses the integral as a sum of integrals of f over subintervals where the discontinuities appear only as endpoints of the subintervals. It then computes each of the resulting improper integrals separately.

Here is a sample of Maple responses to a few different improper integrals.

****

```
> int(10/(x^2+4)^2,x=-infinity..infinity);
```

$$\frac{5}{8}\pi$$

```
> int(x/(x^2+4),x=0..infinity), int(x/(x^2+4),x=-infinity..0);
```

$$\infty, -\infty$$

#Consequently, the next integral does not exist. Look at Maple's response.#

```
> int(x/(x^2+4),x=-infinity..infinity);
```

$$\int_{-\infty}^{\infty} \frac{x}{x^2+4}\,dx$$

```
> Int(sin(x),x=0..infinity)=int(sin(x), x=0..infinity);
```

$$\int_0^{\infty} \sin(x)\,dx = \mathit{undefined}$$

```
> int(1/x^(2/3),x=-1..8 );
```

#This integrand has a discontinuity at x=0.#

$$3\,8^{1/3} - \frac{3}{2} - \frac{3}{2} I\sqrt{3}$$

#This improper integral exists, and it certainly is not complex-valued (not according to our interpretation), but Maple has difficulty with the fractional powers. **Did Maple make a mistake?** Not really! It depends on what interpretation one gives to a negative number raised to a fractional power. Recall that Maple **always evaluates a^r as a complex number, when $a < 0$ and the rational number $r = p/q$ is not an integer.** It is easy to be frustrated by Maple's behavior on matters of this nature, but do not blame Maple—it is a necessary inconvenience. Problems of this sort are dealt with as follows.#

```
> int(1/surd(x,3)^2,x=-1..8);
```

$$3\,8^{(1/3)} + 3$$

```
> simplify(%);
```

$$9$$

#Notice that Maple computes the value of an improper integral, if it exists in either a finite or infinite sense. If the integral diverges in a stronger sense, that is to say, if it does not exist even in an infinite sense, Maple either states this fact or returns the unevaluated integral symbol. We must be careful about how we interpret this latter output. It could mean that the integral does not exist, but, then again, it might only mean that Maple cannot find an antiderivative.#

```
> int(1/sin(x^2),x=0..1);
```

$$\int_0^1 \frac{1}{\sin(x^2)}\, dx$$

#What does this output mean? Does the improper integral converge or diverge?#

```
> int(1/sin(x^2),x);
```

$$\int \frac{1}{\sin(x^2)}\, dx$$

#This means that Maple cannot find an antiderivative.#

```
> evalf(Int(1/sin(x^2),x=-0..1));
```

$$\text{Float}(\infty)$$

#So, what are we to conclude from the above example? Is this a divergent integral?

Frequently, we can tell whether an improper integral converges or diverges by comparing it to an improper integral of known convergence or divergence. We compare $1/\sin(x^2)$ to $1/x$ by looking at the graphs of $sin(x^2)$ and x.#

```
> plot({sin(x^2),x},x=0..1);
```

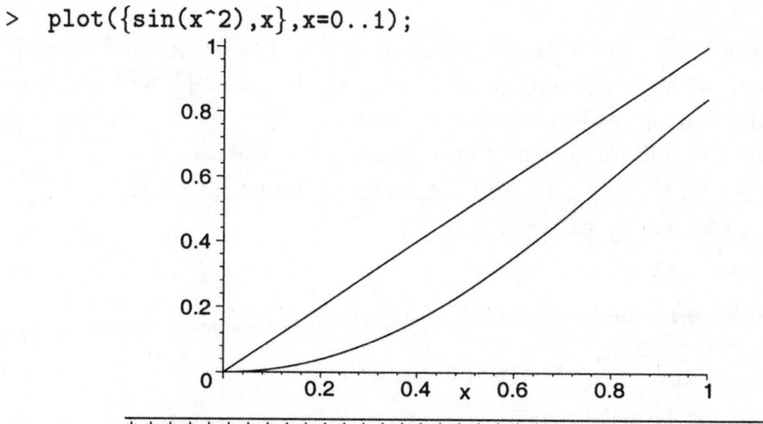

****

This plot reveals that $0 < sin(x^2) < x$ for $0 \le x \le 1$, which means that

$$\frac{1}{sin(x^2)} > \frac{1}{x} \;(0 \le x \le 1),$$

and so

$$\int_0^1 \frac{1}{x} dx = \infty \quad \Rightarrow \quad \int_0^1 \frac{1}{\sin(x^2)} = \infty.$$

Example 6.11 *A large lake is initially seeded with 2,000 smallmouth bass. It is anticipated that the population will grow at the rate* $R(t) = (200 + 10t^3)e^{-0.14t}$ *fish per year t years after the initial seeding. What population does the lake eventually level off to?*

This problem could be handled as a simple antidifferentiation problem, but it could just as well be thought of as the answer to an improper integral. At time t, if dt is a "small" interval of time, then $R(t)dt$ determines the increase in population during this time interval. We simply "add up" all of the little increases over time.

****

```
> R:=t->(200+10*t^3)*exp(-0.14*t):
> pop:=2000+int(R(t),t=0..infinity);
```

$$pop := 159613.4944$$

****

Example 6.12 *Find the arc length of the curve* $f(x) = x * \sin(1/x)$ $(0 < x \leq 1)$ *and the curve* $g(x) = x^2 * \sin(1/x)$ $(0 < x \leq 1)$*, if they exist.*

The formula used to evaluate arc length of the curve $y = h(x)$ $(a \leq x \leq b)$ is

$$L(h) = \int_a^b \sqrt{1 + (h'(x))^2}\, dx.$$

In both of the curves in this example, as $x \to 0^+$, $1/x \to \infty$, and so $\sin(1/x)$ begins to oscillate with greater and greater frequency between $+1$ and -1. Look at the graphs of the two curves in the following work session. Would you ever expect that one of them has infinite length? Our eyes can be deceiving.

****

```
> f:=x*sin(1/x):  g:=x^2*sin(1/x):
> plot(f,x=0..1);
```

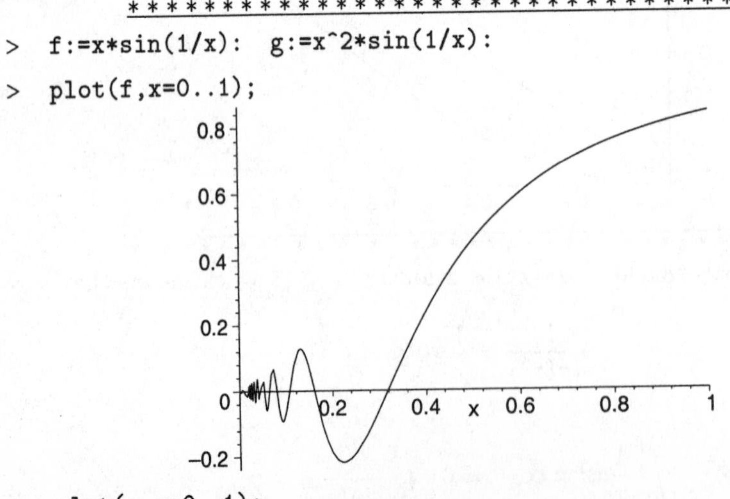


```
> plot(g,x=0..1);
```

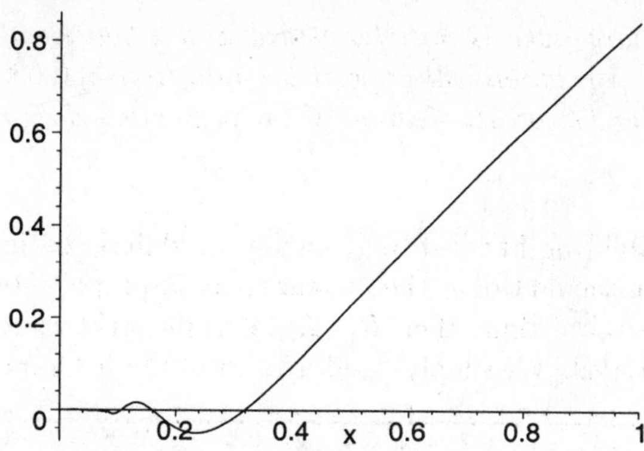


```
> lf:=sqrt(1+(diff(f,x))^2);
```

$$lf := \sqrt{1 + \left(\sin(\frac{1}{x}) - \frac{\cos(\frac{1}{x})}{x}\right)^2}$$

```
> lg:=sqrt(1+(diff(g,x))^2);
```

$$lg := \sqrt{1 + (2\,x\sin(\frac{1}{x}) - \cos(\frac{1}{x}))^2}$$

Both lf, and lg are discontinuous at $x = 0$, and so their integrals are definitely improper integrals. It is interesting to look at the graphs of these expressions before we attempt to compute their integrals.#

```
> plot(lf,x=0..1,0..100);
```

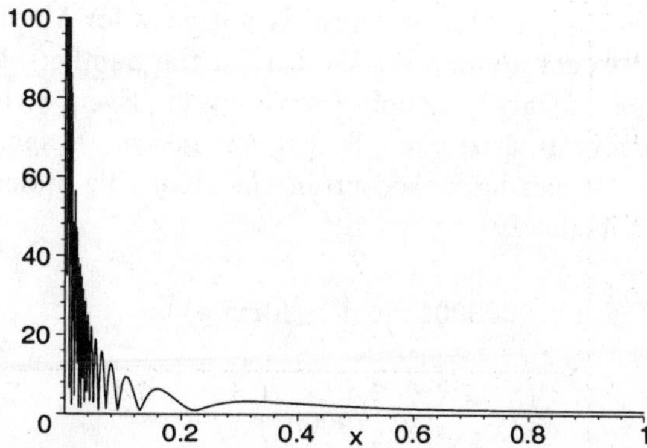


```
> plot(lg,x=0..1);
```

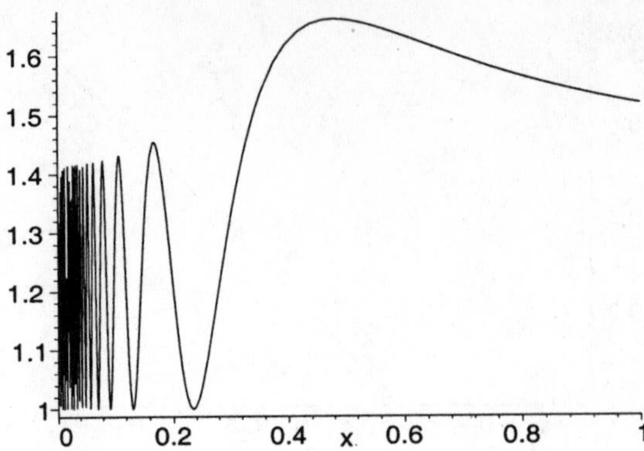


```
> Lf:=evaf(Int(lf,x=0..1));
```

$$Lf := \text{evaf}\left(\int_0^1 \sqrt{1+\left(\sin\left(\frac{1}{x}\right) - \frac{\cos\left(\frac{1}{x}\right)}{x}\right)^2}\,dx\right)$$

#Maple cannot evaluate this, so it returns the integral symbol.#

```
> Lg:=evalf(Int(lg,x=0..1));
```

$$Lg := 1.469378940$$

#The decimal value Lg approximates the arc length of g. The arc length of f happens to be infinite. We are in no position to prove this rigorously, but we can supply some very convincing evidence. The expression lf is continuous on any interval of the form $a \leq x \leq 1$ as long as $0 < a < 1$. Consequently, its integral exists on any such interval. We just can't let $a = 0$.

We let $a = .0000001$, and compute the arc length of f on the interval $a \leq x \leq 1$. Look how big the answer is! This integral is not easy for Maple to do, and some tricks must be used to get an answer. We turned the number of digits of accuracy way down to 2—more accuracy is pointless, anyway. Even with this, Maple spins endlessly until memory is used up. To get an answer, a special approximating technique was used. It can be looked up in the Help File under *evalfint*, but it is not a matter of importance.#

```
> evalf(Int(lf,x=.0000001..1,2,_NCrule));
```

$$79000.$$

****

6.8 Exercise Set

New Maple Commands in this Chapter (and a few old commands as a reminder)

applyop( )	completesquare( )	factor( )	op( )	lhs( )
rhs( )	Int( )	intparts( )	changevar( )	expand(¡*suppress*)
simplify(,*options*)	normal( )	powsubs( )	surd( )	

Remember to load the *student* package for the problems in this exercise set.

In problems 1 through 14, evaluate the integral directly, and then verify your answer by mimicking a pencil-and-paper approach—what we have been calling an elementary integral approach.

1. $\int \sin^3(x)\cos^6(x)\,dx$

2. $\int \sin^2(x)\cos^7(x)\,dx$

3. $\int \frac{x^3}{\sqrt{4-x^2}}\,dx$

4. $\int \frac{p^3}{\sqrt{4-p^4}}\,dp$

5. $\int \sin(5x)\cos(2x)\,dx$ (Integrate by parts twice.)

6. $\int \tan^5(4y)\sec^7(4y)\,dy$

7. $\int \tan^4(4y)\sec^8(4y)\,dy$

8. $\int (5+3x^2)\sqrt{x^2+8}\,dx$

9. $\int \frac{\sqrt{a^2-x^2}}{x^4}\,dx$

10. $\int \frac{1}{(x^2+8)^3}\,dx$

11. $\int (2s^2+7s-8)e^{-5s}\,ds$

12. $\int \frac{\sin(2x)}{9+\cos^4(x)}\,dx$

13. $\int \sin(t)\ln(\cos(t))\,dt$

14. $\int \frac{x^7-12x^3+21x^2+36}{x^5+3x^4+4x^3+12x^2}\,dx$

In the remaining problems, use Maple freely, without restrictions. Answers should, however, be in an appropriate form. They should be free of unfamiliar expressions, expressed in simplest form, and (if possible) related to the original problem.

15. Evaluate the following improper integral directly. Confirm your answer by evaluating it a second time, using the definition of an improper integral as a certain limit of an antiderivative.

$$\int_1^\infty \frac{x^2+25}{x(x^2+4)^2}\,dx$$

16. Evaluate the improper integral $\int_1^\infty (ax^2 + bx + c)e^{-px}dx$, where a, b, c, p are constants, with $p > 0$. Notice that the integral clearly diverges for $p < 0$. (Maple's assume( ) command will be needed in this problem.)

17. The curve $y = x\sin(1/x)$ $(0 < x \leq 1)$ discussed in Example 6.12 is known to have infinite length. Suppose an infinitely long wire has the shape of this curve (with units in centimeters), and suppose that the density of the wire at the point (x, y) on the curve (wire) is $\delta(x) = x$ in grams per centimeter. Compute the total mass of the infinitely long wire (showing, in the process, that its mass is finite).

18. Here is another seemingly paradoxical result concerning the infinitely long curve $y = x\sin(1/x)$ $(0 < x \leq 1)$ discussed in the previous problem and in Example 6.12. Suppose that this curve is the shape of an infinitely long road where the units are in miles. Show, by the following computation, that you can travel the entire length of this road in finite time, if you speed up enough along the way. Suppose that your speed at the point (x, y) on the curve is $s(x) = 1/x$ in miles per hour. Compute the total time required to travel down this infinitely long road (showing, in the process that the time is finite).

19. Use Maple to establish the following reduction formula:

$$\int x^m (\ln(x))^n dx = \frac{x^{m+1}(\ln(x))^n}{m+1} - \frac{n}{m+1}\int x^m (\ln(x))^{n-1} dx (m \neq -1).$$

20. Use Maple to establish the following reduction formula. Consider a careful use of the op( ), expand( ), simplify( ), and subs( ) commands among others. Remember that to prevent Maple from expanding an expression such as (m-1), treating it as two separate numbers, the expression can be inserted as an optional argument in the expand( ) command. Expansion can be suppressed on several subexpressions in the same basic way. See the Help File or Example 6.6 for details. Try to get Maple to do most of the work.

$$\int \cos^m(x)\sin^n(x)dx =$$

$$\frac{\cos^{(m-1)}(x)\sin^{(n+1)}(x)}{m+n} + \frac{m-1}{m+n}\int \cos^{(m-2)}(x)\sin^n(x)dx.$$

21. Evaluate the integral $\int \frac{e^{5x}}{\sqrt{1+e^{10x}}}dx$. Verify that your answer is correct by differentiation.

22. Evaluate the integral $\int x^2 \ln(1 + \sqrt{1+x^2})dx$. Verify that your answer is correct by differentiation.

23. Evaluate the integral $\int \sec^6(x)\tan^n(x)\,dx$. Verify that your answer is correct by differentiation.

24. Evaluate the integral $\int \frac{(1+\sqrt[3]{x})^n}{\sqrt[3]{x}}\,dx$. Verify that your answer is correct by differentiation.

25. Evaluate the integral $\int \frac{(2x^2+1)\ln(x)}{\sqrt{1+x^2}}\,dx$. Verify that your answer is correct by differentiation.

26. Evaluate the integral $\int e^{2x}\arcsin(e^x)\,dx$. Verify that your answer is correct by differentiation.

27. Evaluate the integral $\int x\sin(\ln(x))\,dx$. Verify that your answer is correct by differentiation.

28. Evaluate the integral $\int \sqrt[7]{1+\sqrt[3]{x}}\,dx$. Verify that your answer is correct, using the easiest method. This verification should be effortless. Use fractional powers rather than the surd( ) command to set up the radicals.

29. Evaluate the improper integral $\int_{-a}^{a}\sqrt{\frac{a+x}{a-x}}\,dx$, where a is a positive constant. Is the answer correct? Is the basic antiderivative involved in the problem correct? If Maple doesn't behave to your satisfaction, it might help to express the integrand in a different, equivalent way.

30. The *Horn of Gabriel* is the surface generated by revolving the curve $y = 1/x$ $(1 \leq x < \infty)$ about the x-axis. Show that the surface area of the *Horn of Gabriel* is infinite, but that the volume enclosed by the horn $(1 \leq x < \infty)$ is finite. As the saying goes, *"You can fill the horn with paint, but you can't paint the surface of the horn."*

Chapter 7

Sequences and Series

Imagine the large supply of functions we can create and use with our current knowledge of mathematics. This collection would include the power function, $p(x) = x^r$, where r is a real constant, all of the polynomial and rational functions, the trigonometric, logarithmic, exponential, and hyperbolic functions, and all of their inverses. It would include all of the combinations of these functions using finitely many compositions and finitely many algebraic operations. These are the functions that we have the power to know. We might need Maple's help, but we can evaluate them, graph them, manipulate them, and study their behavior.

The collection is large, and we can certainly use these functions to create a variety of complex mathematical models—yet one of the truly fascinating aspects of mathematics is that this supply of accessible functions occupies only a small corner in a much larger warehouse of "nice," "well-behaved" functions that play a significant role in mathematics and its applications. Most of the functions in this warehouse live in a world that is currently beyond our reach.

Examples come from a variety of mathematical situations. We have already mentioned that antiderivatives of continuous functions are frequently (actually, usually) beyond our reach, even though the *Fundamental Theorem of Calculus* says that they always exist. The same can be said about solutions (of the form $y = f(x)$) to some fairly simple equations in x and y. The same can be said about inverses to some functions, even simple functions like $f(x) = x + \sin(x)$, for example. Solutions to differential equations, which are so important to mathematical applications, are frequently hard to describe.

In all of these situations, and others as well, answers to very practical problems can turn out to exist, yet be beyond our ability to describe in terms of familiar functions. How are we to reach these functions and study their properties? Certainly this is an important issue, if it can be addressed.

In this chapter, we study ideas that will ultimately provide us with a means to get a handle on all of the functions we usually need in mathematics, including those that can be quite inaccessible otherwise. It turns out that functions in this list can be described, surprisingly enough, by expressions that look and behave very much like the simplest functions, namely polynomials

$$p(x) = \sum_{j=0}^{n} a_j x^j, \text{ or } p(x) = \sum_{j=0}^{n} a_j (x-a)^j.$$

The only difference is that we must allow the degree of the polynomial to go to infinity. In the process, "long polynomials"

$$p(x) = \sum_{j=0}^{\infty} a_j x^j, \text{ or } p(x) = \sum_{j=0}^{\infty} a_j (x-a)^j$$

called **power series** and **Taylor series** are created.

Realizing a function as a power series gives us an important means to study the behavior of the function. It turns out that power series behave very much like ordinary polynomials, although this requires a careful analysis. Frequently, the coefficients, a_n $(n = 1, 2, \ldots)$, for a particular power series can all be evaluated, and this gives us a very specific way to formulate the function represented by the power series.

This is not an easy topic, and there are many issues in this chapter that must be discussed along with power series. Power series and Taylor series are introduced in the last section in this chapter, after all the other details are considered. As the most important issue, some text books prefer to discuss power series—especially Taylor series—first, followed by all of the other details. In that case, read a portion of the last section of this chapter first. It could help to postpone the material on power series until later, because this is a bit more abstract. Maple has a command, taylor( ), for producing a Taylor series. You can read the material concerning this command, along with the example, without going through the rest of the chapter first.

7.1 Sequences

While the concept of a sequence is needed for the development of the rest of the chapter, it is also a concept that is interesting in its own right. Intuitively, a sequence is just an ordered list of real numbers, which in this chapter will always be an infinite list. A formal definition, however. is more useful in an environment where technology is used. A sequence, formally, is just a function whose domain is the set of positive integers or a subset of that set. It follows that the standard methods for defining functions in Maple can be used to define sequences just as well. Of course, this means that notation such as $\{a_n\}$ for denoting a sequence will have to be dropped in favor of functional notation when Maple is being used. If a denotes the sequence (function)

$$\left\{\frac{(-1)^n}{n^2}\right\},$$

then the nth term of the sequence and the convergence or divergence of the sequence can be determined in the expected manner.

****

```
> a:=n->(-1)^n/n^2:
> a(7),a(20);
```

$$\frac{-1}{49}, \frac{1}{400}$$

```
> limit(a(n),n=infinity);
```

$$0$$

****

A sequence in Maple means roughly the same thing as it does in mathematics, except that it is more rigidly defined. Computers cannot read and interpret from context, and so computer-driven definitions are necessarily more rigid. It is fair to say that a substantial part of the difficulty we all experience working in with computers is a direct result of our not fully appreciating how rigidly words are interpreted by computers.

In Maple, a **sequence is an open-ended, <u>ordered</u> finite list of objects separated by commas.** By open-ended, we mean that a sequence is not enclosed in parentheses, brackets, or so on. **A list is similar to a sequence, except that it begins with a left bracket ([) and ends with a right bracket (]).** We have encountered these Maple structures on many previous occasions; the output of several Maple commands takes the form of a sequence or list.

In the following Maple work session, the symbol a is a sequence, and the symbol b is a list. As we mentioned, the entries in both a sequence and a list are ordered, and the individual entries can be accessed by using bracketed notation, as shown below. We have used this notation in previous chapters, and so this is only a reminder.

The purpose of our first work session is only to observe the behavior of the seq( ) command and, while we are at it, some of the contrasting elements of the sum( ) command, which will be used in the next section. The input lines for these two commands look similar, but they behave differently.

****

```
> a:=6,-4,p+1,q^2,sqrt(2):
  b:=[6,-4,p+1,q^2,sqrt(2)]:
> a[2],  b[3];
```

$$-4, p+1$$

When a sequence or list has only a few elements, it is easy enough to just list the elements, as we did above. **To define and evaluate a longer sequence or list, it is more convenient to use the new Maple command seq( ).** As you can see below, **this command** must be used only to define or evaluate a **finite sequence**, and the upper number on the index cannot be left unassigned.#

```
> seq(x(n),n=10..20);
```

x(10), x(11), x(12), x(13), x(14), x(15), x(16), x(17), x(18), x(19), x(20)

```
> seq(1/n,n=1..infinity);

Error, unable to execute seq
> seq(1/n,n=1..m);

Error, unable to execute seq
```

The letter n **(the index) is a local variable**, and it is important to understand this in order to avoid confusion. Local issues were discussed on page 20 and again on page 70. The arguments of functions are all local variables, and a sequence is just a function of its index, so it should come as no surprise that Maple treats the index as a local variable. In the following, we give n a value and then compute a sequence, using n as an index. The assigned value of n has no effect on the sequence, because the index is a **local variable.** (The argument of a function behaves in the same way.) Finally, notice that the assigned value of n is unchanged after the sequence is computed.#

```
> n:=250;
```

$$n := 250$$

```
> s:=seq(1/n,n=10..20);
```

$$s := \frac{1}{10}, \frac{1}{11}, \frac{1}{12}, \frac{1}{13}, \frac{1}{14}, \frac{1}{15}, \frac{1}{16}, \frac{1}{17}, \frac{1}{18}, \frac{1}{19}, \frac{1}{20}$$

```
> n;
```

$$250$$

By way of contrast, the **index in Maple's sum( ) command is not a local variable.** This is clearly demonstrated in the next computation.#

```
> sum(1/n,n=10..20);

Error, (in sum) summation variable previously assigned, second
argument evaluates to 250 = 10 .. 20
```

If we plan to use the letter n as an index in the sum( ) command, we must first unassign the letter n. This is not a matter to worry about. If you inadvertently assign a value to n before you use the sum( ) command, **Maple will tell you that you made a mistake**, and it can then be corrected with the following input statement.#

```
> n:='n';
```

$$n := n$$

#**Notice the letter** n **enclosed in single quotes**. We have used this punctuation before. A name enclosed by single quotes is simply left unevaluated. This means that the "letter" n (literally) is assigned to the name n, which leaves the name n unassigned.#

```
> sum(1/n,n=10..20);
```

$$\frac{178964263}{232792560}$$

#Finally, notice that the index n is left unassigned after this sum. We mention this, because, with some operations, as we shall see, the index can unintentionally be assigned its last value.#

```
> n;
```

$$n$$

#Just remember that the arguments of all functions and procedures are local variables, and, because a sequence is a function of its index, its index is also a local variable. A sum, however, is not a function of its index, and so its index is not a local variable.

In the last part of this wandering Maple work session, the letter a represents a sequence (open ended), and b represents a list (enclosed in square brackets).#

```
> a:=seq((2*n+1)/(n+3),n=10..20);
  b:=[seq((2*n+1)/(n+3),n=10..20)];
```

$$a := \frac{21}{13}, \frac{23}{14}, \frac{5}{3}, \frac{27}{16}, \frac{29}{17}, \frac{31}{18}, \frac{33}{19}, \frac{7}{4}, \frac{37}{21}, \frac{39}{22}, \frac{41}{23}$$

$$b := [\frac{21}{13}, \frac{23}{14}, \frac{5}{3}, \frac{27}{16}, \frac{29}{17}, \frac{31}{18}, \frac{33}{19}, \frac{7}{4}, \frac{37}{21}, \frac{39}{22}, \frac{41}{23}]$$

****

While it is customary in mathematics to denote a sequence with notation such as $\{a_n\}$, it should be noted that curly brackets are also used to enclosed the elements of a set, which is mathematically very different from a sequence. **In Maple, curly brackets are only used to define sets.** The elements of a set are not ordered, and elements are entered only once, regardless of how often they are repeated. These two characteristics make sets very different from sequences, and so curly brackets should never be used in a Maple work session to define sequences. If curly brackets are used, Maple will list the elements of the set in some order, but it will not necessarily pay attention to the way you listed the elements.

On the other hand, this curly-bracket notation is a traditional, and widely used, method to denote a sequence in print, and so we will not hesitate to use it outside Maple work sessions.

We will be more interested in **infinite sequences**; they will usually be defined as **Maple functions**, and the seq( ) command will used mainly to evaluate and list some of the terms of sequences.

Example 7.1 *Show that the sequence $\{a_n\}$ defined by $a_n = \frac{n^{100}}{100^n}$ is eventually a decreasing sequence that converges to zero. At what point in the sequence does it become decreasing? Determine an N such that $0 < a_n < 10^{-8}$ for all $n > N$.*

We begin by evaluating, for the sake of interest, the first few terms of the sequence .

****

```
> a:=n->n^100/100^n:
> seq(evalf(a(n),2),n=1..10);
```

$$.010,\ .13\,10^{27},\ .52\,10^{42},\ .16\,10^{53},\ .79\,10^{60},\ .65\,10^{66},\ .32\,10^{71},\ .20\,10^{75},\ .27\,10^{78},\ .10\,10^{81}$$

#Does this sequence diverge to infinity?#

```
> limit(a(n),n=infinity);
```

$$0$$

#So the sequence converges to 0, and **merely listing the first 10 terms, and witnessing the phenomenal growth, did not suggest the long-term pattern of values.** The sequence is defined as a Maple function, so we can study the sequence by plotting the function. We try to do this next, but the numbers involved are too large and we get an overflow message.#

```
> plot(a(n),n=1..40): # Output omitted. Floating point overflow.#
```

#It is easy to overcome these difficulties. We plot the function $\ln(a(n))$ instead of $a(n)$. This brings down the values of the sequence considerably. Since

$$a(n) = e^{\ln(a(n))},$$

it is easy to realize the values of $a(n)$ from its associated logarithmic sequence.#

```
> plot(ln(a(n)),n=1..100);
```

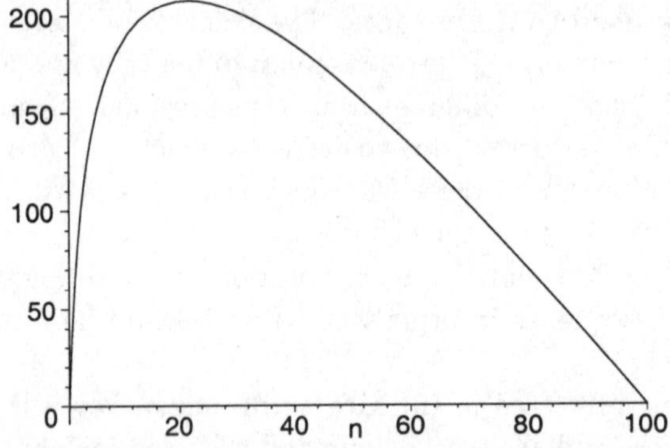

Notice how dramatically large the original sequence gets.It gets larger than e^{200}, but then rapidly decreases to 0. By placing the pointer at the top of the graph and clicking, we can see that the sequence appears to be decreasing starting with the 22nd term. Certainly, this is not a very rigorous approach. Our eyes cannot really establish with any certainty that the graph continues to decrease forever. Thus, to show the decreasing nature of this sequence more rigorously, we follow this plot with a determination of the sign of the first derivative#.

```
> ap:=diff(a(n),n);
```

$$ap := 100\,\frac{n^{99}}{100^n} - \frac{n^{100}\ln(100)}{100^n}$$

```
> ap:=normal(ap);
```

$$ap := -\frac{n^{99}\,(-100 + n\ln(100))}{100^n}$$

```
> solve(-100+n*ln(100)=0,n);
```

$$\frac{100}{\ln(100)}$$

```
> evalf(%);
```

$$21.71472410$$

#It follows that the derivative ap is negative for $n > 22$, and so the sequence decreases for $n > 22$.

For the sake of interest, we mention that we could have computed the next larger integer directly with the following new command.#

```
> ceil(100/ln(100));
```

$$22$$

#The **ceiling and floor commands, ceil( ) and floor( )**, are two of several related integer valued commands, with **ceil(x) returning the smallest integer greater than or equal to x.** These commands are not really essential, but are included so that you may use them if you wish.

Maple has some difficulty with the last part of this example. It cannot solve the inequality, but this is not a major problem. We know that the sequence decreases monotonically, so we need only get a decimal solution to $a_n = 10^{-8}$. Once we have this number, we can say that a_n is smaller yet for all larger values of n. Even here, Maple seems to struggle. Frequently, unsuccessful attempts at using the solve( ) command can be resolved by telling Maple where to look for a solution. In the present case, one can easily tell, by observation, that $a_{100} = 1$. By restricting our search for a solution to values of $n > 100$, we are finally able to get a decimal solution.#

```
> solve(a(n)<10^(-8),n);
> fsolve(a(n)=10^(-8),n);
```

$$.8655891064$$

```
> fsolve(a(n)=10^(-8),n,n=100..infinity);
```

$$105.0749621$$

```
> N:=ceil(%);
```

$$N := 106$$

****

A sequence is just a function whose domain is a set of integers. Frequently, however, the formula used to define a sequence $\{a_n\}$ is well defined, not just for integer values of n, but also for all (or most) real numbers n as well. We took advantage of this when we plotted the sequence in the last example, treating it essentially as a function of a real number. **Some sequences, on the other hand, cannot be extended in such a way.** Mathematical terms like $(-1)^n$ and $n!$ are quite simple when n is an integer, but, when n is not an integer, they represent very advanced mathematical structures that are frequently complex-valued and that have no place in our setting. **Any sequence involving terms such as these cannot be differentiated or plotted as we did the sequence in the last example.** We can, however, still use Maple to plot the sequence over its integer domain. We show the technique by plotting the sequence

$$\{\frac{e^n n!}{\sqrt{n}n^n}\}$$

in the next Maple work session.

****

```
> a:=n->exp(n)*n!/sqrt(n)/n^n:
> s:=seq([n,a(n)],n=1..20):
> plot([s],x=1..20,(2.5)..(2.6));
```

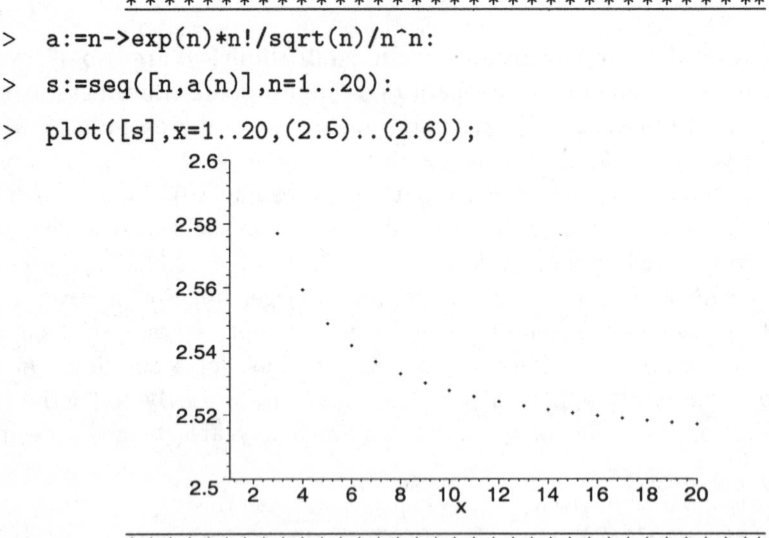

****

Example 7.2 *Show that the sequence $\{a_n\}$ defined by $a_n = (\frac{n+3}{7+n})^{2n}$ converges, and find its limit. Verify its limit by an alternative means.*

****

```
> a:=n->((n+3)/(7+n))^(2*n):
> seq(evalf(a(n),3),n=1..40); n:='n':
```

.250, .0953, .0467, .0269, .0173, .0121, .00900, .00700, .00564, .00468, .00397, .00344, .00302, .00269, .00243, .00221, .00203, .00188, .00175, .00164, .00154, .00146, .00138, .00132, .00126, .00121, .00116, .00112, .00108, .00104, .00101, .000982, .000955, .000930, .000907, .000885, .000865, .000846, .000829, .000812

#Does the sequence converge slowly to zero?#

```
> limit(a(n),n=infinity);
```

$$e^{(-8)}$$

```
> evalf(%);
```

$$.0003354626279$$

#As we can now see, the limit is small, but definitely nonzero.#

****

To gain a deeper understanding of this limit, we could use L'Hôpital's Rule to verify its value. The standard approach calls for us to first rewrite the expression $a(n)$ as $\exp(\ln(a(n)))$, and then to consider the limit of $\ln(a(n))$. This step is omitted, but further insight into this kind of calculation can be found by reviewing our work with *L'Hôpital's Rule* in Example 5.9 and Example 5.10. Using this technique on a sequence $\{a_n\}$ is no different from using it on a function $f(x)$. One simply treats the variable n as a continuous variable similar to the variable x.

Incidentally, if simplification difficulties are encountered when working with a sequence indexed by the variable n, the input statement

```
> assume(n>0);
```

frequently helps overcome the difficulties.

The sequences that we have worked with so far have all been explicitly defined. In other words, they have all been of the form $\{a_n\}$, where a_n is given as a formula in n. A sequence can also be defined by expressing the nth entry in the sequence in terms of some (or all) of the previous $n-1$ entries of the same sequence. Such a sequence is said to be **recursively defined.** It is a more natural way of defining some sequences. Frequently, when a sequence is defined recursively, it is in a form ideally suited for listing values numerically in the most efficient way.

For example, suppose $\{b_n\}$ is the sequence of monthly balances that would result if \$20,000 is borrowed at 9% annual interest compounded monthly, with a monthly payment of \$400. Then, in the most natural way, without any need to look up a financial formula, it is easy to conclude that

$$b_{n+1} = (1 + \frac{.09}{12})b_n - 400, b_0 = 20000.$$

This defines the sequence $\{b_n\}$ recursively. Knowing b_0, we can use this formula to compute b_1, and then, in turn, $b_2, b_3, \ldots$. Because it is, computationally, so easy to go from any one entry to the next, this would be the most efficient way to list, numerically, the whole sequence of balances.

Computations such as these are carried out by using **Maple's programming tools.** Writing programs in Maple is of central importance, and the current problem provides an opportunity to introduce some of the basic ideas. Any problem involving **repetitive calculation** can probably be done **more efficiently** by doing the work inside of a **program.** While programs can sometimes become quite involved, many situations call for really simple programs using the commands **for, from, by, to, while, do, od,** and **break** . In the Maple work session below, notice how simple it is to compute and list the sequence $\{b_n\}$ of balances for the loan discussed above. An unrelated second example is included in this work session just to show some variety and to show how some of the other programming tools are used in a program.

****

```
> b:=20000:  while b>0 do
  b:=(1+.09/12)*b-400;od:
```

#The lengthy output of monthly balances is omitted.#

```
> for n from 2 to 6 do
  sin(n*x)=expand(sin(n*x));od; n:='n':
```

$$\sin(2\,x) = 2\sin(x)\cos(x)$$

$$\sin(3\,x) = 4\sin(x)\cos(x)^2 - \sin(x)$$

$$\sin(4\,x) = 8\sin(x)\cos(x)^3 - 4\sin(x)\cos(x)$$

$$\sin(5\,x) = 16\sin(x)\cos(x)^4 - 12\sin(x)\cos(x)^2 + \sin(x)$$

$$\sin(6\,x) = 32\sin(x)\cos(x)^5 - 32\sin(x)\cos(x)^3 + 6\sin(x)\cos(x)$$

****

Notice that a "do-statement" is always terminated by "od," which is "do" spelled backwards. This is a common practice that we have seen before. Recall that an "if-statement" is always terminated by "fi," which is "if" spelled backwards.

Also notice that the above "do-statement" ends with a directive to unassign the index n. **In any "do-statement" of the form**

> for n from 1 to p do ...;od; (where p is a given constant)

the value of n is $n = p + 1$ after the statement is executed. The extra unassignment directive is included to avoid unintentionally assigning a value to the index n.

Example 7.3 *A very well-known sequence, called the Fibonacci sequence, is defined recursively by*

$$F_{n+2} = F_{n+1} + F_n, \; F_1 = F_2 = 1.$$

List the first 50 terms of the Fibonacci sequence. Determine the 400th term, F_{400}.

This sequence was first formulated 800 years ago to describe the populations of successive generations of breeding rabbits. Since that time, it has found its way into many mathematical problems. In addition, the Fibonacci sequence is found as a number pattern in a surprisingly large number of diverse settings in nature. It is such an interesting sequence that a mathematical journal, The Fibonacci Quarterly, is devoted to its study.

A small matter is mentioned before we start. Maple wants to display the terms created by a program vertically, with one term on each line. In order to avoid this vertical display (and wasted page space), we used full colons everywhere in the program to suppress the output. We then used the seq( ) command to display the results. In order to maintain the relationship between the index j and the value F_j of the jth term, the output is displayed in the form $[j, Fj]$.

```
**********************************
> F[1]:=1:F[2]:=1:for j from 3 to 50 do
  F[j]:=F[j-1]+F[j-2]:od:j:='j':
```

#It is a good idea to always **end a program—indexed by *j*—by unassigning the index *j*.** If we had not included that step in the above program, j would have left the program with a value of $j = 51$.#

```
> seq([j,F[j]],j=1..50);
```

[1, 1], [2, 1], [3, 2], [4, 3], [5, 5], [6, 8], [7, 13], [8, 21], [9, 34], [10, 55], [11, 89],
[12, 144], [13, 233], [14, 377], [15, 610], [16, 987], [17, 1597], [18, 2584],
[19, 4181], [20, 6765], [21, 10946], [22, 17711], [23, 28657], [24, 46368],
[25, 75025], [26, 121393], [27, 196418], [28, 317811], [29, 514229], [30, 832040],
[31, 1346269], [32, 2178309], [33, 3524578], [34, 5702887], [35, 9227465],
[36, 14930352], [37, 24157817], [38, 39088169], [39, 63245986],
[40, 102334155], [41, 165580141], [42, 267914296], [43, 433494437],
[44, 701408733], [45, 1134903170], [46, 1836311903], [47, 2971215073],
[48, 4807526976], [49, 7778742049], [50, 12586269025]

#The speed of modern desk top computers allows us to compute terms like F_{400} in a surprisingly small amount of time. Here is a situation where we definitely want to suppress output.#

```
> F[1]:=1:F[2]:=1:for j from 3 to 400 do
  F[j]:=F[j-1]+F[j-2]:od:j:='j':
> F[400];
```

17602368064501396646822694539241125077038438330449219188672599289\
6575345044216019675

#Interestingly enough, **for some recursively defined sequences $\{x_n\}$**, including the Fibonacci sequence, the term x_n **can be expressed explicitly in**

terms of n. We are not going to pursue this matter in any detail, but we are so close to introducing an interesting and powerful new Maple command, that we will at least mention the topic.

The **rsolve( ) command computes an explicit formula for a sequence that is defined recursively.** Notice how the command is used as we finish the last example. The rsolve command is very similar to the solve and fsolve commands. The first argument is reserved for the equation and its initial values, and the second for the unknown representing the solution to the equation. Three distinct terms must be placed in the argument reserved for the equation, so the terms are grouped together, in the usual way, with set brackets to form one term. Naturally, the sequence itself must be unassigned—it is, after all, an equation in its unknowns.

Oddly enough, the nth term of the sequence must be expressed in the form $x(n)$ with parentheses, rather than in the form $x[n]$ with brackets.#

```
> rsolve({x(n+2)=x(n+1)+x(n),x(1)=1,x(2)=1},x(n));
```

$$\frac{(1-\frac{1}{5}\sqrt{5})\,(\frac{2}{\sqrt{5}-1})^n}{\sqrt{5}-1}+\frac{(-1-\frac{1}{5}\sqrt{5})\,(-\frac{2}{\sqrt{5}+1})^n}{\sqrt{5}+1}$$

```
> x:=unapply(%,n);
```

$$x := n \to \frac{(1-\frac{1}{5}\sqrt{5})\,(\frac{2}{\sqrt{5}-1})^n}{\sqrt{5}-1}+\frac{(-1-\frac{1}{5}\sqrt{5})\,(-\frac{2}{\sqrt{5}+1})^n}{\sqrt{5}+1}$$

```
> evalf(x(400));
```

$$.1760236520\ 10^{84}$$

****

The rsolve( ) command can be used without entering the initial conditions or start-up values of a recursively defined sequence. The terms $x(1) = 1$ and $x(2) = 1$ in the above example could have been omitted. Not surprisingly, the symbols $x(1)$ and $x(2)$ would then appear in the output.

Some discretion should be exercised in using the rsolve command. Problems of this sort are inherently difficult and quickly surpass Maple's ability to deliver.

7.2 Series

There are **two main commands for forming sums**, and we should discuss their similarities and differences before we begin the main topic of this section. Actually, Maple has a package, called "sumtools," for forming sums, but this package of more advanced tools will not be used in this manual.

Along with the sum( ) command that we have used on several occasions, there is a second summation command, called **add( )**. For finite sums, there is little difference between the two commands, until a really large number of terms is involved, but then differences appear. Let's discuss this issue first. Suppose that m and n either are specific integers or are assigned-Maple-names that evaluate to specific integers, and suppose $m < n$. The command **add($b(k)$, $k = m..n$)** always **evaluates to the exact, direct, arithmetical sum of the terms $b(k)$, regardless of the values of m and n.** Imagine how impressive (unwanted?) this would be, if the number $n - m$ is very large. The command **sum($b(k)$, $k = m..n$)**, on the other hand, **evaluates to the exact direct sum of the terms $b(k)$,only if $n-m$ (the number of terms in the sum) is of reasonable size.** If $n - m$ is large enough (on the order of 1000, or more), Maple resorts to non-arithmetical, more theoretical means to evaluate the sum. Most of the sums we consider have a small enough number of terms that we would probably not notice a difference between the two commands.

Here is another way in which the two commands differ. Maple can often **evaluate sum($b(k)$, $k = m$..infinity)**, and it can often **evaluate sum($b(k), k = m..n$) (find a formula), when m and n are unassigned names.** The command **add( ) can never be used in either of these situations.** It can be used only for direct finite sums.

Regardless of which command is used, **remember Maple's penchant for being exact!** If fewer than a thousand terms are involved, Maple will actually add them arithmetically, with either command. If fractions are involved, Maple will, for example, have to compute a common denominator.

Surely, the arithmetic and symbolic complications can quickly get out of hand. For this reason, it is often **advisable to enclose the sum( ) and add( ) commands inside an evalf( ) command.** For large enough sums, the sum( ) command can be used effectively without enclosing it in the evalf( ) command, simply because of the way it works.

Now, let us begin the topic of this section. To say that a series is an infinite sum **misses an important part of the definition** which makes the concept of a series much more comprehensible. If we are to gain real insight into this concept, we need to start with a careful definition. A series is a sequence—**a sequence**—a sequence $\{s_n\}$ of partial sums. Each term in this sequence is much more manageable than the whole series, because each term s_n is a finite sum. We already have a good understanding of finite sums, and the previous section gave us a background in sequences. So now our attention will be focused on this sequence $\{s_n\}$. Does it converge or diverge? If it converges, can we compute its limit? Here are some examples where we use Maple to shift our attention from the series itself to its sequence of partial sums. We start with the example

$$\sum_{j=1}^{\infty} \frac{1}{j^3}.$$

****

```
> ps:=n->evalf(sum(1/j^3,j=1..n),4);
```

$$ps := n \to \text{evalf}\left(\sum_{j=1}^{n} \frac{1}{j^3}, 4\right)$$

```
> seq(ps(n),n=1..40); n:='n':
```

1., 1.125, 1.162, 1.178, 1.186, 1.190, 1.193, 1.195, 1.197, 1.198, 1.198, 1.199, 1.199, 1.200, 1.200, 1.200, 1.200, 1.201, 1.201, 1.201, 1.201, 1.201, 1.201, 1.201, 1.201, 1.201, 1.201, 1.201, 1.201, 1.202, 1.202, 1.202, 1.202, 1.202, 1.202, 1.202, 1.202, 1.202, 1.202, 1.202

****

The above list of terms from the sequence of partial sums would suggest that, roughly speaking,

$$\sum_{j=1}^{\infty} \frac{1}{j^3} \approx 1.202.$$

The above sequence was not computed very efficiently. To compute, for example, the 29th term, psum(29), in the sequence, Maple added up all 29 items that appear in psum(29). Then, to compute psum(30), it duplicated most of its work by adding, essentially all over again, the 30 items that appear in psum(30). It would have been easier for Maple to use the term psum(29) that was just computed in the previous step, and then simply add $1/30^2$ to psum(29) to get psum(30).

Here is a simple program that computes the first 100 terms of this sequence more efficiently. With the speed of modern desk top computers, the program runs very quickly. Try this program yourself, and notice how quickly it is executed.

****

```
> s[0]:=0:for k from 1 to 100 do

  s[k]:=evalf(s[k-1]+1/k^3):od:k:='k':
```

#The vertical display of the results, with one term s_n on each line is omitted to save space.

With such computational power at our finger tips, and a very efficient algorithm for doing the computations, It is an interesting activity to play with this idea. We could, for example, run the program through a thousand terms, suppress the output, and then look at just every 10th term in the sequence of partial sums by using the seq( ) command. In this way we can go quite far out into the sequence without looking at all of the terms. #

```
> s[0]:=0:for k from 1 to 1000 do

  s[k]:=evalf(s[k-1]+1/k^3):od:k:='k':
> seq(s[10*k],k=1..100)k:='k':
```

1.197531986, 1.200867843, 1.201519559, 1.201752119, 1.201860864, 1.201920312, 1.201956311, 1.201979749, 1.201995858, 1.202007401, 1.202015955, 1.202022470, 1.202027544, 1.202031575, 1.202034828, 1.202037493, 1.202039705, 1.202041557, 1.202043126, 1.202044467, 1.202045621, 1.202046621, 1.202047493, 1.202048258, 1.202048935, 1.202049535, 1.202050070, 1.202050550, 1.202050980, 1.202051368, 1.202051720, 1.202052039, 1.202052329, 1.202052594, 1.202052837, 1.202053060, 1.202053265, 1.202053454, 1.202053629, 1.202053791, 1.202053942, 1.202054081, 1.202054211, 1.202054331, 1.202054444, 1.202054550, 1.202054650, 1.202054742, 1.202054831, 1.202054911, 1.202054991, 1.202055061, 1.202055131, 1.202055196, 1.202055256, 1.202055316, 1.202055372, 1.202055422, 1.202055472, 1.202055522, 1.202055567, 1.202055607, 1.202055647, 1.202055687, 1.202055727, 1.202055765, 1.202055795, 1.202055825, 1.202055855, 1.202055885, 1.202055915, 1.202055945, 1.202055975, 1.202056001, 1.202056021, 1.202056041, 1.202056061, 1.202056081, 1.202056101, 1.202056121, 1.202056141, 1.202056161, 1.202056181, 1.202056201, 1.202056221, 1.202056241, 1.202056261, 1.202056274, 1.202056284, 1.202056294, 1.202056304, 1.202056314, 1.202056324, 1.202056334, 1.202056344, 1.202056354, 1.202056364, 1.202056374, 1.202056384, 1.202056394

****

Before we finish our discussion of this series, we show that Maple is able to calculate its value directly. We should not expect such success with just any series. It is easy to ask too much of Maple in the evaluation of series, and, frequently, their values cannot even be expressed in closed form, or, in other words, in terms of familiar mathematical expressions.

****

```
> sum(1/n^3,n=1..infinity);
```

$$\zeta(3)$$

```
> evalf(%);
```

1.202056903

****

Quite often, the output will be, as it is above, in the form of some **unfamiliar special function of mathematics**. This is not the appropriate place to discuss the so-called **Riemann Zeta function**, but notice its decimal value, and how it compares to the values we obtained earlier for its sequence of partial sums. In a calculus course, however, an answer such as this should probably be regarded as **unacceptable**.

We have been discussing the idea that a series is, by definition, just a sequence, the sequence of partial sums. One special example, namely, the series $\sum_{k=1}^{\infty}(-1)^k$, is particularly good for shedding light on this matter. This series may appear to be very puzzling, until it is written as a sequence of partial sums. Take note!

****

```
> psum:=n->sum((-1)^k,k=1..n); s:=seq(psum(n),n=1..30);
```

$$psum := n \to \sum_{k=1}^{n} (-1)^k$$

$$s := -1, 0, -1, 0, -1, 0, -1, 0, -1, 0, -1, 0, -1, 0, -1, 0, -1, 0, -1, 0, -1, 0, -1, 0, -1, 0, -1, 0, -1, 0$$

****

Obviously, this oscillating pattern between -1 and 0 continues forever, so that the sequence, and hence, by definition, the series itself, diverges. The point to be made here is that the practice of using the definition of a series to express it as a sequence of partial sums can, at least in some instances, increase our understanding of a series.

7.3 Positive-Term Series

The sum( ) command has an **inert form Sum( )**, which is formed, as all other inert commands are formed, by **capitalizing the first letter of the command name**. We will use it in most of the forthcoming examples to set up, visually, the series as an infinite sum. In most of our examples, it serves no essential mathematical purpose other than to create a visual display.

The *Comparison Test* (*Inequality Comparison Test*) and the *Limit Comparison Test* are used to determine the convergence or divergence of a given positive-term series, by comparing it to another well-chosen positive-term series of known convergence or divergence. The *Inequality Comparison Test* is not very suitable for Maple involvement. Using this test is more a matter of logical argument than it is a computation. On the other hand, given Maple's ability to compute limits, the *Limit Comparison Test* can be used quite effectively.

To use a comparison test, we need a collection of positive-term series of known convergence or divergence. While any positive-term series, once it is known to converge or diverge, can be put into this collection, it is usually restricted, in a classroom setting, to a certain **"standard set"** consisting of all of the geometric series, together with all of the so-called p-series. The well known convergent series $\sum_{n=0}^{\infty} 1/n!$ will also be placed in our standard set.

Deciding what to compare a given series to involves "educated guess work." Maple's limit command is so easy to use that we can make guesses with less hesitance, knowing that it will not take much work to perform the test.

Example 7.4 *Determine whether the following series converges or diverges.*

$$\sum_{n=1}^{\infty} \frac{\sqrt{n^3+2}}{2^n}$$

The key idea here is that 2^n gets large so fast that it quickly overwhelms the numerator and makes it somewhat insignificant. This would suggest that it behaves somewhat like the series

$$\sum_{n=1}^{\infty} (\frac{1}{2})^n$$

which is a convergent geometric series. If so, then our series probably converges. Unfortunately, this geometric series is smaller than the given series and so it cannot be used in the *Comparison Test.* It is easily seen that the *Limit Comparison Test* fails for pretty much the same reason. What we need is to compare the given series to a somewhat larger convergent series. This is accomplished by using any larger geometric series which still converges. The following limit shows that the series converges.

****

```
> a:=n->sqrt(n^3+2)/2^n:
> b:=n->(2/3)^n:  B:=Sum(b(n),n=1..infinity);
```

$$B := \sum_{n=1}^{\infty} (\frac{2}{3})^n$$

```
> limit(a(n)/b(n),n=infinity);
```

$$0$$

****

Example 7.5 *Determine whether the following series converges or diverges.*

$$\sum_{n=1}^{\infty} \frac{(2n)!}{(n!)^2}$$

The ratio test is always the most promising test to use in series involving factorials.

****

```
> a:=n->(2*n)!/n!^2:
> r:=a(n+1)/a(n);
```

$$r := \frac{(2\,n+2)!\,(n!)^2}{((n+1)!)^2\,(2\,n)!}$$

```
> r:=simplify(r);
```

$$r := 2\,\frac{2\,n+1}{n+1}$$

```
> rho:=limit(r,n=infinity);
```

$$\rho := 4$$

****

Because $\rho > 1$, it follows that the series diverges.

Once it is known that a series converges, a partial sum can be used to approximate the value of the series. This approximation, however, is of little value unless the error between the exact sum and approximate partial sum can be estimated. For example, the series

$$\sum_{j=2}^{\infty}\left(\frac{1}{\ln(\sqrt{j})}\right)^{\ln(\sqrt{j})}$$

can be shown to converge (see Problem 12 in the exercises), but it converges so slowly that the 10,000*th* term is still approximately 0.0009 in order of magnitude. We would obviously have to add tens of thousands of terms to come close to the exact value of the series.

By way of contrast, there are positive-term divergent series whose partial sums get large at an almost imperceptible rate. There are many anecdotes written about how slowly the partial sums of the harmonic series get large.

> Suppose that 10 million years ago one of our ancestors had begun to add up the terms of the harmonic series at the rate of 100 terms per second, and that the task was continued, without fail, each second thereafter until the present time, with the duty to continue the additions passed on smoothly from generation to generation. One can show that, at the present time, the sum of all these terms would be somewhere between 37 and 39, continuing, nevertheless, to increase, in a very slow race, to infinity. (See Problem 13 in the exercises.)

If $\sum_{k=1}^{\infty} b_k$ is a convergent series with sum s and if s_n denotes its nth partial sum, then the nth error term is defined to be

$$e(n) = |\, s - s_n \,| = \left|\sum_{k=1}^{\infty} b_k - \sum_{k=1}^{n} b_k\right| = \left|\sum_{k=n+1}^{\infty} b_k\right|.$$

Deciding how far out one needs to go in the sequence of partial sums of a convergent series in order to get a finite sum that is a good approximation for the whole series is not always an easy matter. Whenever this error term can be measured, it deserves our attention.

One of the important characteristics of the *Integral Test* is that it can be used not only to show convergence, but also to measure the size of the error between the exact sum and a partial sum. Given a series $\sum_{k=1}^{\infty} b_k$, suppose that f is a nonincreasing real-valued continuous function on the interval $[1,\infty]$ such that $f(k) = b_k$ for each k. Basically the same idea used to prove the *Integral Test* can be used to prove that if the series converges according to this test, then

$$s_n + \int_{n+1}^{\infty} f(x)dx \leq s \leq s_n + \int_{n}^{\infty} f(x)dx.$$

If the difference between the two integrals is small, this inequality provides a useful way to estimate the value of s.

In many situations, it is just as effective to measure the error term with the slightly weaker (but simpler) inequality

$$0 \leq e(n) = s - s_n \leq \int_{n}^{\infty} f(x)dx.$$

The nonincreasing nature of the function f in the *Integral Test* is a critical condition, but it can be relaxed. Like most tests for convergence, words like "for all" can usually be replaced by "eventually," and so it is it is only necessary to show that f is eventually nonincreasing. This complicates the error term only slightly. Using the same argument used to prove the *Integral Test*, one can show that, as long as f is nonincreasing for all $x \geq N$, the error term still satisfies

$$e(n) \leq \int_{n}^{\infty} f(x)dx \text{ for } n \geq N.$$

Example 7.6 *Show that the series* $\sum_{k=1}^{\infty} \frac{k^3}{e^k}$ *converges. Approximate the sum with an error not exceeding* 10^{-8}.

Based on growth patterns for the exponential function, it may be clear that the underlining function $f(x) = x^3/e^x$ is eventually decreasing. This may not, however, be clear to everyone, and we cannot use the *Integral Test* until the decreasing nature of this function is established. It is easy enough to see this, if it is not already clear, by plotting $f(x)$. More rigorously, ask Maple to factor $f'(x)$, to see that $f'(x) < 0$ for $x > 3$. This step is not included, and we proceed immediately with the *Integral Test* . The improper integrals $\int_1^{\infty} f(x)dx$ and $\int_n^{\infty} f(x)dx$ $(n \geq 1)$ either all converge or all diverge (for a continuous function f), so there is no need to establish convergence of the series as a separate issue. If the integral involved in any error term converges, then the series converges, and so we turn our attention immediately to this error term.

****

```
> f:=x->x^3/exp(x):
> MyError:=int(f(x),x=n..infinity);
```

$$MyError := e^{(-n)}\,(n^3 + 3\,n^2 + 6\,n + 6)$$

```
> plot({MyError,10^(-8)},n=0..50,0..2*10^(-8));
```

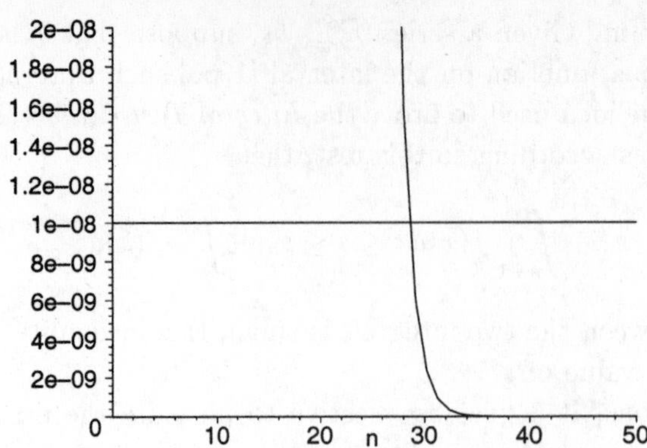

#The error term will clearly be small enough for n=30.#

```
> s30:=evalf(sum(f(k),k=1..30));
```

$$s30 := 6.006512796$$

****

We could have determined a slightly stronger result if we had used the stronger form of the error estimate mentioned in the introduction to this example, but what we have seems strong enough. The partial sum s_n is always smaller than the infinite sum s for any positive-term series, so we have

$$6.00651279 \leq s \leq 6.00651279 + 10^{-8} = 6.00651280.$$

Interestingly enough, our approximate sum for s is correct to this many decimal places, even though we had to sum very few terms in the series to reach this degree of accuracy.

7.4 Alternating Series and Absolute Convergence

The error term can also be estimated quite easily for an alternating series. According to the *Alternating Series Test*, if $\{a_k\}$ is a nonincreasing sequence of positive numbers that converges to 0, then the alternating series $\sum_{k=1}^{\infty}(-1)^{k+1}a_k$ converges, and, for each n, the error term satisfies

$$e(n) \leq a_{n+1}.$$

If the sequence $\{a_k\}$ is nonincreasing only for $k \geq N$, then it is easy to see that the error term still satisfies

$$e(n) \leq a_{n+1} \text{ for } n \geq N.$$

Example 7.7 *Show that the sequence* $\sum_{k=1}^{\infty}(-1)^k\frac{1}{k^2+2^k}$ *converges, and approximate its sum with an error not exceeding* 10^{-8}.

This series clearly converges, by the *Alternating Series Test*, and so we consider the error term immediately.

****

```
> a:=1/(k^2+2^k):
> plot({a,10^(-8)},k=1..50,0..2*10^(-8));
```

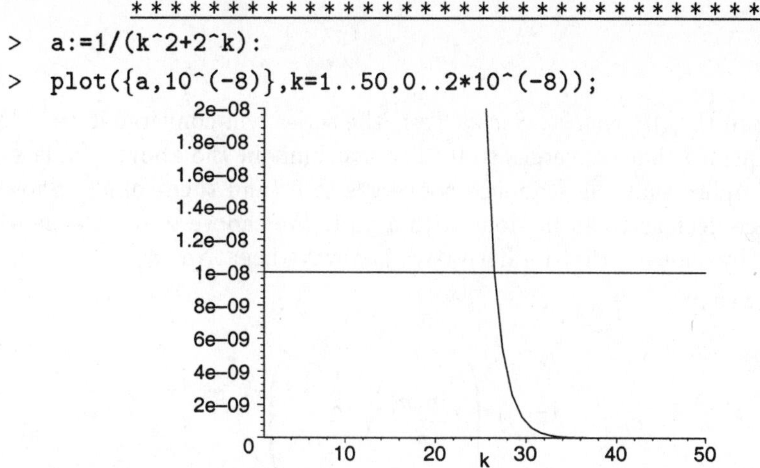

#The error term will clearly be small enough for n=27.#

```
> s:=evalf(sum((-1)^k*a,k=1..27));
```

$$s := -.2470767687$$

****

We should drop the last two digits in this answer, because of the size of the error term, and this gives us an approximate sum of $s = -0.24707676$. This much accuracy is quite striking considering that we only added the first 27 terms of the series to obtain this approximation.

Example 7.8 *Determine whether the series* $\sum_{n=1}^{\infty}(-1)^n\left(n^{-1+\frac{1}{n}}\right)$ *converges absolutely, converges conditionally, or diverges.*

When problems like this are solved by hand, the series under consideration is usually studied with some care before a decision is made regarding which convergence test to use. The labor involved in using these tests is often great enough that there is a reluctance to use a test without a reasonable expectation of a successful outcome. With Maple, on the other hand, the computations are no cause for concern, and so there is no real need to be so cautious before a test is used.

We first consider the absolute-value series and compare it to the $p = 2$ series. This is seen to be inconclusive. The series is then compared to the Harmonic Series, and this test is successful. It follows that the series does not converge absolutely.

****

```
> a:=n^(1/n-1):
> limit(a/(1/n^2),n=infinity);
```

$$\infty$$

```
> limit(a/(1/n),n=infinity);
```

$$1$$

#According to the *Alternating Series Test*, the series will converge if $\{n^{\frac{1}{n}-1}\}$ is a decreasing sequence that converges to 0. The last limit in the above Maple work session clearly implies that the sequence converges to 0 (and then some). Showing that the sequence decreases can be done with a plot. We choose to do it somewhat more rigorously by showing that its derivative is always negative. #

```
> ap:=diff(a,n);
```

$$ap := n^{\left(\frac{1}{n}-1\right)}\left(-\frac{\ln(n)}{n^2}+\frac{\frac{1}{n}-1}{n}\right)$$

#The first term is clearly positive, so we look at the second term.#

```
> normal(op(2,ap));
```

$$-\frac{\ln(n)-1+n}{n^2}$$

****

This last term is clearly always negative, and so the sequence decreases. Consequently, the series converges conditionally.

Example 7.9 *Determine whether the series* $\sum_{n=1}^{\infty}(-1)^n\frac{n^{\ln(n)}}{2^{\sqrt{n}}}$ *converges absolutely, converges conditionally, or diverges.*

Using the *Root Test* on the absolute-value series is seen to be inconclusive. This is followed by a comparison with the $p = 2$ series, which turns out to be successful. It follows that the series converges absolutely.

****

```
> a:=n^ln(n)/2^sqrt(n):
> limit(a^(1/n),n=infinity);
```

$$1$$

```
> limit(a/(1/n^2),n=infinity);
```

$$0$$

****

7.5 Power Series

Example 7.10 *Find the exact interval of convergence of the power series*

$$\sum_{n=1}^{\infty} \frac{n^{\sqrt{n}}}{2^n} x^n.$$

The radius of convergence is found by using the *Ratio* or *Root Test*. These tests can only be used on positive-term series, or, in other words, to test for absolute convergence.

****

```
> a:=n->n^sqrt(n)/2^n:
> P:=Sum(a(n)*x^n,n=1..infinity);
```

$$P := \sum_{n=1}^{\infty} \frac{n^{(\sqrt{n})}\, x^n}{2^n}$$

```
> r:=abs(a(n+1)*x^(n+1)/(a(n)*x^n));
```

$$r := \left| \frac{(n+1)^{(\sqrt{n+1})}\, x^{(n+1)}\, 2^n}{2^{(n+1)}\, n^{(\sqrt{n})}\, x^n} \right|$$

```
> r1:=simplify(r);
```

$$r1 := \frac{1}{2} \left| (n+1)^{(\sqrt{n+1})}\, x\, n^{(-\sqrt{n})} \right|$$

```
> rho:=limit(r1,n=infinity);
```

$$\rho := \frac{1}{2} |x|$$

****

The series converges for $| x |< 2$ (where $\rho < 1$), and diverges for $| x |> 2$ (where $\rho > 1$). It clearly diverges at $x = \pm 2$, and so the exact interval of convergence is the open interval $(-2, 2)$.

Example 7.11 *Find the exact interval of convergence of the power series*

$$\sum_{n=2}^{\infty} \frac{(\ln(n))^2}{n} (x-3)^n.$$

We use the *Ratio Test* to establish the interval of convergence in this example.

****

```
> a:=n->ln(n)^2/n:
> A:=Sum(a(n)*(x-3)^n,n=2..infinity);
```

$$A := \sum_{n=2}^{\infty} \frac{\ln(n)^2\,(x-3)^n}{n}$$

```
> r:=abs(a(n+1)*(x-3)^(n+1)/(a(n)*(x-3)^n));
```

$$r := \left|\frac{\ln(n+1)^2\,(x-3)^{(n+1)}\,n}{(n+1)\ln(n)^2\,(x-3)^n}\right|$$

```
> r1:=simplify(r);
```

$$r1 := \left|\frac{\ln(n+1)^2\,(x-3)\,n}{(n+1)\ln(n)^2}\right|$$

```
> rho:=limit(r1,n=infinity);
```

$$\rho := |x-3|$$

#It follows that the series converges for $\rho = |x-3| < 1$ and diverges for $\rho = |x-3| > 1$. All that remains is to consider the endpoints $x = 2$ and $x = 4$ of the interval of convergence. At $x = 4$, the positive-term series

$$\sum_{n=2}^{\infty} a(n) = \sum_{n=2}^{\infty} \frac{(\ln(n))^2}{n}$$

is clearly larger than the harmonic series, and so it diverges. At $x = 2$, the series is the alternating series

$$\sum_{n=2}^{\infty}(-1)^n a(n) = \sum_{n=2}^{\infty}(-1)^n \frac{(\ln(n))^2}{n}.$$

The *Alternating Series Test* can be used to show that this series converges. We must show that $\{a(n)\}$ is a decreasing sequence that converges to zero.#

```
> limit(a(n),n=infinity);
```

$$0$$

```
> ap:=diff(a(n),n);
```

$$ap := 2\frac{\ln(n)}{n^2} - \frac{\ln(n)^2}{n^2}$$

```
> ap:=normal(ap);
```

$$ap := -\frac{\ln(n)\,(-2+\ln(n))}{n^2}$$

****

The derivative ap is clearly negative for all $n > e^2$, and so the sequence $\{a(n)\}$ eventually decreases. It follows that the series converges, and so the exact interval of convergence is the half open interval $[2, 4)$.

Maple has a command for computing Taylor series. **If f is a function that has derivatives of all orders in some open interval containing $x = a$, and if n is a positive integer, the command**

> t:=taylor(f(x),x=a,n);

computes the Taylor series of f centered at $x = a$ and displays the result as

$$t := f(a) + f'(a)(x-a) + \frac{f''(a)}{2!}(x-a)^2 + \cdots + \frac{f^{(n-1)}(a)}{(n-1)!}(x-a)^{n-1} + O(x^n).$$

The term $O(x^n)$ (read "big Oh of x^n") represents all the terms in the series of order n or larger. To a certain extent, **this display is merely a visual display and cannot be used in subsequent Maple computations until it is altered.** Maple, understandably, has a hard time dealing with the term $O(x^n)$. Unless this display is the end result, most of the time it will be necessary to turn it into a legitimate polynomial before it can be used in any further work. We have discussed the multifaceted **convert( )** command before. It can be used to **convert the taylor series t defined above to a polynomial of degree $(n-1)$** by entering the input statement:

```
> convert(t,polynom);
```

Converting a Taylor series to a polynomial amounts to simply dropping the $O(x^n)$ term. Notice, however, that the end result is the Taylor polynomial, not of order n, but of order $(n-1)$. In other words, to wind up with the Taylor polynomial of order n for f centered at $x = a$, one must start with the command

```
> taylor(f(x),x=a,n+1);
```

If f has derivatives of order $k = 1, 2, \ldots, n$ in some open interval containing $x = a$, then the Taylor polynomial of order m for f at $x = a$ exists for each $m \leq n$. However, if f is not infinitely differentiable in some open interval containing $x = a$, then the Taylor series for f at $x = a$ clearly does not exist, and so it should come as no surprise to discover that Maple returns an error message if an attempt is made to use the Taylor series command in this situation. We mention this all too obvious result, because the Taylor series command can, therefore, not be used to compute the Taylor polynomials of order $m \leq n$, even though they exist.

Example 7.12 *Find the Taylor polynomials of $f(x) = \frac{12x}{x^2+36}$ of order $n = 3, 7, 11$ centered at $x = 0$. Plot all three of them together with the function f on the interval $[-10..10]$. Use Taylor's Theorem to determine the error involved in approximating $f(x)$ by its Taylor polynomial of order 11, on the interval $[-5, 5]$ and on the interval $[-2, 2]$.*

****

```
> f:=12*x/(x^2+36):  t3:=taylor(f,x=0,4);
```

$$t3 := \frac{1}{3}x - \frac{1}{108}x^3 + O(x^5)$$

```
> t3:=convert(t3,polynom);
  t7:=convert(taylor(f,x=0,8),polynom);

  t11:=convert(taylor(f,x=0,12),polynom);
```

$$t3 := \frac{1}{3}x - \frac{1}{108}x^3$$

$$t7 := \frac{1}{3}x - \frac{1}{108}x^3 + \frac{1}{3888}x^5 - \frac{1}{139968}x^7$$

$$t11 := \frac{1}{3}x - \frac{1}{108}x^3 + \frac{1}{3888}x^5 - \frac{1}{139968}x^7 + \frac{1}{5038848}x^9 - \frac{1}{181398528}x^{11}$$

```
> plot({f,t3,t7,t11},x=-10..10,-1..1);
```

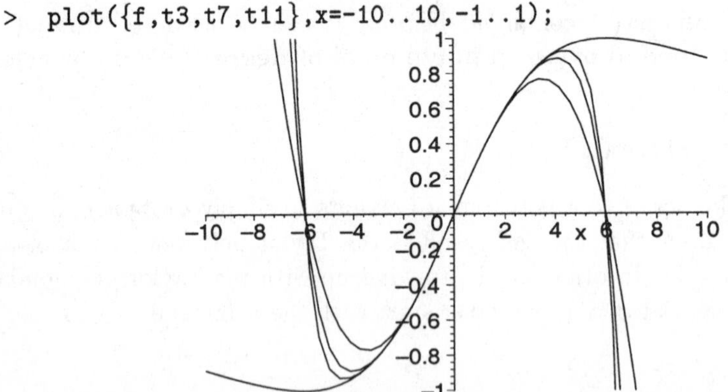

#In the last plot command, we specified a vertical range, rather than allowing Maple to select this range for us. Because polynomials get large rather quickly, Maple would have selected a rather large vertical scale, which would have produced too flat of a picture. Instead, we picked a scale by considering only the values of the original function f.

By looking at this graph, we can see, visually at least, how close the Taylor polynomials are to the graph of the function f that they represent. To get a more reliable estimate of the error term involved, we consider the **remainder term** in

Taylor's Theorem. If $t_n(x)$ is the Taylor polynomial of order n for f based at $x = a$, then the error term is defined by

$$e_n(x) = \mid f(x) - t_n(x) \mid .$$

If f has $n+1$ derivatives on an interval I centered at $x = a$, then *Taylor's Theorem* implies that

$$e_n(x) < \frac{M}{(n+1)!} \mid x - a \mid^{n+1} \quad \text{for all } x \text{ in } I,$$

where M is the maximum value of $\mid f^{(n+1)}(x) \mid$ for x in I. As a practical issue, there is usually no advantage gained in computing M accurately; clearly, any number larger than M can be used in place of M. Using a number larger than M would generate a slightly weaker error term, but the effect is usually so slight that it is of no consequence. Consequently, the usual practice is to use some easily obtained number known to be larger than every value of $\mid f^{(n+1))}(x) \mid$ for x in I. As we finish the above example, notice how a value for M is easily selected, visually, off a graph.#

```
> fp12:=diff(f,x$12):
> plot(fp12,x=-10..10);
```

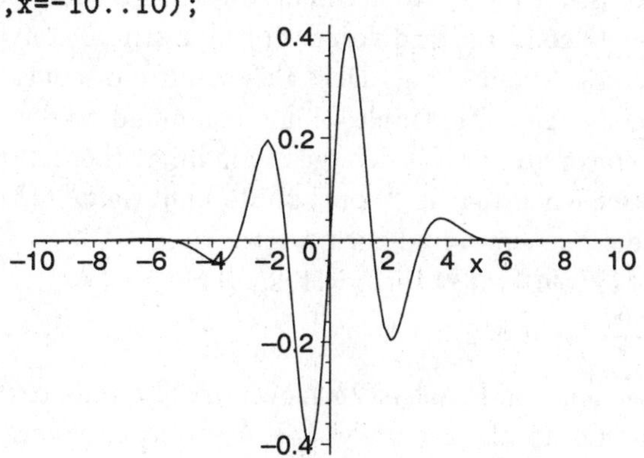


```
> M:=.4;
```

$$M := .4$$

```
> error5:=.4*5^12/12!;
```

$$error5 := .2038745800$$

```
> error2:=.4*2^12/12!;
```

$$error2 := .3420447865\, 10^{-5}$$

The error terms *error*2 and *error*5 are actually upper bounds for the error terms involved in approximating $f(x)$ by its Taylor polynomial $t_{11}(x)$ of order 11 on

the intervals $[-2,2]$ and $[-5,5]$ respectively. Notice how much better the estimate is on $[-2,2]$ than it is on $[-5,5]$.#

****

7.6 Exercise Set

New Maple Commands in this Chapter (and a few old commands as a reminder)

add( )	ceil( )	convert(,*polynom*)	convert( )	denom( )
evalf(,*n*)	floor( )	limit( )	numer( )	print( )
rsolve( )	seq( )	Sum( )	sum( )	surd( )
taylor( )				

1. A sequence $\{a_n\}$ is defined by each of the following formulas. In each case list the first 50 terms of the sequence as decimal numbers. Use Maple's limit command to determine the limit of the sequence, if it converges.
   a) $a_n = \frac{n}{2^n}$ for $n = 1, 2, \ldots$ b) $a_n = \frac{n^2}{10n+100}$ for $n = 1, 2, \ldots$

2. A sequence $\{a_n\}$ is defined by each of the following formulas. In each case list the first 50 terms of the sequence as decimal numbers. By looking at this list, does the sequence appear to converge or diverge? If it appears to converge, try to predict its limit? Compute a_{400} as a decimal. Does this confirm or contradict your previous conclusions. If it converges, use Maple's limit command to determine the limit of the sequence. If it converges, or if it diverges to infinity, then, purely by guess work, find a term in the sequence that is "close" to its limit value. (You may have to go quite far out in the sequence to find such a term.)
   a) $a_n = (n + (-1)^n\sqrt{n})\sin(\frac{1}{n})$ for $n = 1, 2, \ldots$
   b) $a_n = \frac{(\ln(n))^{\ln(n)}}{n^5}$ for $n = 2, 3, \ldots$

3. Show that the sequence in Problem 2b is eventually an increasing sequence. Find the **exact** term (pump up the accuracy) in the sequence where it changes from a decreasing to an increasing sequence.

4. A sequence $\{a_n\}$ is defined by each of the following formulas. In each case, determine the limit of the sequence, if it converges, or state that the sequence diverges. Some of the sequences have obvious limits that can be evaluated without Maple, by a simple observation. Identify such sequences, and determine their limits, before using Maple. If the sequence does not have an obvious limit, identify its indeterminate form before you attempt to compute its limit.
   a) $a_n = \sec(\frac{1}{n})^{5n^2}$ for $n = 1, 2, \ldots$
   b) $a_n = (1 + \tanh(n))^{1/n}$ for $n = 1, 2, \ldots$
   c) $a_n = (1 - \tanh(n))^{1/n}$ for $n = 1, 2, \ldots$
   d) $a_n = \frac{n^{2n+1}}{(n+1)^n}$ for $n = 1, 2, \ldots$

5. The sequence $\{a_n\}$ in problem 4c is quite interesting. Evaluate terms in the sequence as decimals, and try to find a term far enough out in the sequence to be "close" to the sequence's limit value. Prepare, however, to be frustrated in this

endeavor. Is Maple's limit computation correct? To remove any lingering doubt, use *L'Hôpital's Rule* to verify the value of the limit found in Problem 4c.

Why is Maple's decimal evaluator performing in this way? Surprisingly enough, Maple's behavior here is clear and predictable, and in order to understand the limitations of technology, it is important to pursue this question. Devise a way to express a_n as a decimal in such a way that the decimal value of a_n gets close to the limit of the sequence as n gets large. Use Maple's sequence command to list the terms a_n in this manner for n from 10 to 50. Recall that the command evalf(A,k) evaluates A with k digits of accuracy.

6. Let L denote the limit of the sequence $\{a_n\}$ in Problem 4c. Find an N such that $| a_n - L |< 10^{-3}$ for all $n > N$. This turns out to be a nontrivial problem. First, try to find N with a graphical approach. If this fails to be useful, try to find N by experimenting numerically with different values of n. Why do these methods fail so dramatically? How can this failure be avoided? One approach that works involves converting the hyperbolic tangent to its exponential form. This can be done by using its definition or by **using the convert( ) command** to convert an expression to its **exponential form.** Read more about this multifaceted convert command in the help file. Once you find an integer n for which a_n is sufficiently close, how do you know that all subsequent terms in the sequence will also be sufficiently close? It could help to establish that the sequence $\{a_n\}$ (in its converted form) is either increasing or decreasing.

7. Let $\{x_n\}$ be the sequence defined recursively by $x_n = (2 - \frac{4}{n} + \frac{2}{n^2})x_{n-1}$ for $n = 2, 3, \ldots$ and $x_1 = 3$. List the first 100 terms of this sequence. Use Maple's rsolve( ) command to find an explicit formula for x_n Compare the value of x_{100} obtained from the two approaches.

8. Let $\{b_n\}$ be the sequence defined recursively by $b_n = 2b_{n-1}+3b_{n-2}, b_1 = 1, b_2 = 2$. List the first 100 terms of this sequence. Use Maple's rsolve( ) command to find an explicit formula for b_n. Compare the value of b_{100} obtained from the two approaches.

9. Suppose that, one month before the first day of your college education, your parents put a lump sum of cash into an account earning 8% annual interest compounded monthly. You are allowed to draw out $1300 per month from this account for living and educational expenses, with your first draw occurring one month after this lump sum deposit. The account is set up so that the balance is 0 after exactly 48 monthly draws. What is the value of the initial deposit into this account? Let $\{b_n\}$ be the sequence of monthly balances. Define this sequence recursively, and then use the rsolve command to answer the main question. Write a program to list the complete sequence of balances. To avoid excessive time required to display each term in the sequence as it is computed, write a program which suppresses all output, and then, as a final step, displays the complete sequence of balances as one output area. Take note, once this problem is finished, that, by defining the sequence of balances recursively, complicated finance formulas were avoided. No special formulas were needed other than a basic monthly interest and withdrawal statement.

10. It is easy to ask Maple to do too much when using the sum command. Remember that Maple will always be exact unless it is allowed to make a decimal approximation.

In this problem, demonstrate the output, probably undesirable, resulting from an exact calculation of $\sum_{n=1}^{95} \frac{1}{n^2}$. Explain the form of this output, and what Maple did to generate it.

11. Some of the mystery surrounding the notion of a series can be removed by appreciating that a series is just a sequence—the sequence of partial sums. Viewing a series in this way may seem unnecessary in calculations, but to fully understand what a series is, as a concept, it helps to pay some attention to the definition of a series. With this in mind, use Maple to set up the sequence of partial sums of the series $\sum_{n=0}^{\infty} \frac{1}{n!}$. Use the evalf command when you define this sequence. List the first 20 terms of this sequence. (The output should be a sequence of decimal numbers.) Increase the number of digits of accuracy to 15, and repeat these calculations. (Recall that this can be done by entering **15 as an optional second argument in the evalf command.**) You will soon know, if you do not already know, that this series converges to e. Compare the 20 terms in this sequence to a decimal expansion for e.

12. Determine whether the following series converge absolutely, converge conditionally, or diverge (p denotes a fixed real number). Do not use the *Integral Test* on each part, even if it seems to work. Use the *Integral Test* only if the integral looks like a straight-forward computation. Integrals should be evaluated exactly (rather than as decimals) when the test is used.

a) $\sum_{j=1}^{\infty} \frac{j^4}{1.2^j}$ b) $\sum_{j=2}^{\infty} \frac{100^{2j}}{j!}$
c) $\sum_{j=2}^{\infty} (1 - \sqrt[j]{j})^j$ d) $\sum_{j=2}^{\infty} (\sqrt{4+j^2} - j)$
e) $\sum_{j=2}^{\infty} (\frac{1}{\ln(j)})^{\ln(j)}$ f) $\sum_{j=2}^{\infty} (\frac{1}{\ln(\sqrt{j})})^{\ln(\sqrt{j})}$
g) $\sum_{j=3}^{\infty} (\frac{1}{\ln(\ln(j))})^{\ln(\ln(j))}$ h) $\sum_{j=2}^{\infty} (\frac{1}{j\ln(j)})$
i) $\sum_{j=2}^{\infty} (\frac{1}{j(\ln(j))^2}$ j) $\sum_{j=2}^{\infty} (\frac{1}{\ln(j)})^p$
k) $\sum_{j=1}^{\infty} \frac{(j+1)^j}{j^j j!}$ l) $\sum_{j=1}^{\infty} \frac{(j!)^3}{j\sqrt{j}}$

13. The inequality used to prove the *Integral Test* can be used to prove the inequality

$$\ln(n+1) < \sum_{j=1}^{n} \frac{1}{j} < 1 + \ln(n).$$

Use this inequality to establish the claim made in the anecdote about the harmonic series on page 164.

14. Show that the series $\sum_{k=2}^{\infty} \frac{1}{k(\ln(k))^6}$ converges. Determine an n such that the error term $e(n) < 10^{-4}$. Use this to approximate the sum of the series with this much accuracy.

15. Show that the series $\sum_{k=0}^{\infty} \frac{(-4)^k}{k!}$ converges. Determine an n such that the error term $e(n) < 10^{-8}$. Use this to approximate the sum of the series with this much accuracy.

16. Find the exact interval of convergence of each of the power series.

a) $\sum_{n=2}^{\infty} \frac{\ln(n)}{n^2 4^n} x^n$ b) $\sum_{n=2}^{\infty} \ln(n)^{\ln(n)} (x+3)^n$

17. For each of the following, compute the Taylor polynomial of order 8 based at the given point a. Compare each Taylor polynomial with the function it represents, by plotting both of them on the same graph over an appropriate interval.
a) $\cos(x)$, $a = 0$ b) e^x, $a = 0$ c) $\ln(x)$, $a = 1$ d) $\sqrt{x}$, $a = 9$

18. Let $f(x) = (12x^2 - 17x - 5)\sin(3x)$. Determine the Taylor polynomial of order $n = 9$ for $f(x)$ based at $x = 0$. Plot $f(x)$, along with its Taylor polynomial, on the interval $-1 \leq x \leq 1$ Use Taylor's remainder formula to estimate the error involved in approximating $f(x)$ by its Taylor polynomial. Compare this estimate to the actual difference between $f(x)$ and its Taylor polynomial by plotting this difference function on the interval $-1 \leq x \leq 1$.

19. Repeat problem 16, using the Taylor polynomial of order 15 instead.

20. Let $f(x) = x\cos(\sqrt{x})$. Determine the Taylor polynomial of order $n = 3$ for $f(x)$ based at $x = 50$. Plot $f(x)$, along with its Taylor polynomial, on the interval $30 \leq x \leq 70$. Use Taylor's remainder formula to estimate the error involved in estimating $f(x)$ by its Taylor polynomial. Compare this estimate to the actual difference between $f(x)$ and its Taylor polynomial by plotting this difference function on the interval $30 \leq x \leq 70$.

21. Compute the Taylor polynomial of order 8 of $f(x) = (x + b)^8$ centered at $x = 0$, but before you do, forecast what the outcome should be in terms of more elementary structures. Use Maple to verify your claim.

22. Let $f(x) = \frac{1}{1-x}$. Without computing the Maclaurin series for $f(x)$, forecast what you would expect it to be. Verify your hunch by computing the series and displaying all of the terms of order $n \leq 15$. Write the result as an equation named eq. Replace x by $-x^2$ in *eq* and integrate both sides. Verify that the result represents the Maclaurin series for $\arctan(x)$. Replace x by $-x$ in *eq* and integrate both sides. Verify that the result represents the Maclaurin series for $\ln(1 + x)$. A Taylor or Maclaurin series for a function $f(x)$ does not have to converge to $f(x)$. In these two cases, convergence to the corresponding function is immediate. Why? What are intervals of convergence?

23. Express the polynomial $f(x) = 8x^5 - 6x^4 + 27x^2 + 32x + 12$ as a polynomial in powers of $(x + 20)$.

24. The function $f(x)$ defined below appears to be discontinuous at $x = 0$, but the discontinuity turns out to be removable. Starting with a Maclaurin series for $\cos(x)$, find a power series that represents $f(x)$ for all $x \neq 0$. Then define $f(0)$ so that f is continuous at $x = 0$.

$$f(x) = \frac{2\cos(x^2) + x^4 - 2}{x^8}$$

As we pointed out earlier, the output of the command taylor( ) is more a "display" than it is a valid mathematical combination of terms. The "Big-Oh" term appearing in the output frequently prevents the kind of manipulation desired. When this happens, one way to resolve the problem is to convert the Taylor expression into a legitimate polynomial of fairly high degree. If you wish, you can even add a symbol to this polynomial that represents the "Big-Oh" term dropped when you converted to a polynomial. You will have an easier time manipulating the individual terms in this expression.

Problems 25, 26, and 27 are interesting applications of Taylor series and Taylor polynomials. Some important functions of mathematics cannot be expressed in elementary terms, even though they are known to exist. This is an event that happens frequently in mathematics. How do we get a "handle" on such a function? The answer is, of course, through a Taylor series. Hopefully, these problems will demonstrate, in a convincing way, the practical importance of this major topic.

25. The equation $x^3 - y^3 = 64xy$ defines y implicitly as one or more functions of x. Plot the equation. The top half of the inner loop of this curve is the graph of one of the functions, which are defined implicitly by the equation. Let $f(x)$ denote this function of x, and let J denote the interval over which $f(x)$ is defined. Can we find a formula for $f(x)$? We know that Maple will solve a cubic equation exactly, but the solution is easily seen to be too complicated to be of much use.

Let $p_5(x)$ denote the Taylor polynomial of $f(x)$ of order 5 based at a point centrally located on the interval J. Use the implicitdiff( ) command on the original equation to determine the coefficients of the Taylor polynomial $p_5(x)$. Plot the original equation and $y = p_5(x)$ together to see how good an approximation $p_5(x)$ is to the function $f(x)$ on the interval J.

26. The function $f(x) = 2x + \sin(x)$ has a derivative, which is everywhere positive, and so $f(x)$ is globally one-to-one. Its inverse, $f^{-1}(x)$, exists, but it is impossible to find a formula for it in terms of the familiar functions of mathematics. Find instead the Taylor polynomial $p_5(x)$ of order 5 for $f^{-1}(x)$ based at $x = 0$. Use the implicitdiff( ) command to determine the coefficients of $p_5(x)$. Plot $f^{-1}(x)$ and $p_5(x)$ together to see how good an approximation $p_5(x)$ is to $f^{-1}(x)$ on an interval centered about the origin. Choose a large enough interval to show the effective range of the approximation. The inverse function will have to be plotted in the form of the equation $x = f(y)$.

27. Consider the differential equation $\frac{dy}{dx} = (\arctan(x))^2$, with initial condition $y(4) = 13$. There is a unique solution $y = y(x)$ to this equation according to the *Fundamental Theorem of Calculus*, but Maple cannot find its formula. It makes sense, then, to look for a polynomial $y = p(x)$ that could be used as an approximate solution. What would we expect of such a polynomial? Certainly we would want $p(4) = 13$, and for all x in some interval containing $x = 4$, we would want not only $p(x)$ to be close to $y(x)$, but also $p'(x)$ to be close to $y'(x) = (arctan(x))^2$. With this in mind, find a polynomial $p(x)$ such that $p(4) = 13$ and that

$$|y(x) - p(x)| < .0001, \ |y'(x) - p'(x)| < .0001, \text{ for } 3 \leq x \leq 5.$$

Hint: If $p'(x)$, and $y'(x)$ are close on the interval, then $p(x)$ and $y(x)$ will also be close on the interval. (See a proof of this following the hint.) Using plotting tools and Maple's taylor( ) command, find a polynomial that is close enough to $(\arctan(x))^2$. The rest is easy! What would be a good point to use as a base for the Taylor polynomial?

Proof of the inequality: This is a consequence of the *Mean Value Theorem.* Let $G(x) = y(x) - p(x)$. Then $G(4) = 0$ (since $y(4) = 13$, and $p(4) = 13$). By the *Mean Value Theorem*, for each x in $[3, 5]$ there is a c between x and 4 such that

$$\begin{aligned} |y(x) - p(x)| &= |G(x) - G(4)| = |G'(c)|\,|x - 4| \\ &= |y'(c) - p'(c)|\,|x - 4| < .001|x - 4| < .001. \end{aligned}$$

Project: My Bank Job (An Application of Recursively Defined Sequences)

You are a loan officer at a bank, and you have been asked to determine the payment schedule on a new home loan. Your clients have just taken on a mortgage of \$200,000. An annual interest rate of 7.5% compounded monthly is being offered for a 30-year loan (360 payments). Two different payment schemes are offered.

One consists of a series of ballooning house payments which will increase by exactly the same amount each month until the loan is paid off. The first month's payment is only \$1250, which is exactly the interest due during the first month. Determine how much the house payment is to increase each month. List the sequence of payments. List the sequence of balances over the life of the loan.

The other payment scheme is a fixed payment each month over the life of the loan. Determine how much that payment should be. List the sequence of balances over the life of the loan.

The Truth in Lending Law requires that you tell your client the true total cost of the loan for each payment schedule. Determine these amounts for each of the two payment plans.

Chapter 8

Plane Curves and Polar Coordinates

In this chapter, we discuss several different types of curves in the plane and several different methods that can be used to describe planar curves.

8.1 Conic Sections

A conic section is a curve that can be described by intersecting a double cone with a plane. A circle, parabola, ellipse, and hyperbola can all be realized in this way, along with certain so-called "degenerate" cases, such as a line, a pair of lines, and a point. Every conic section is the graph of an equation of the form

$$ax^2 + bxy + cy^2 + dx + ey + f = 0 \tag{8.1}$$

and conversely, any such second-degree equation is a conic section (or an empty set).

A second-degree equation (or any other equation in two variables, for that matter) can easily be graphed by using Maple's **implicitplot( ) command, which resides in the plots package**. A graph will not, however, reveal the exact location of, for example, the foci, or vertices. The geometric information associated with a conic section is found by first putting equation (8.1) into a standard form. If b, the coefficient of the xy term, is 0, the equation can be put into its standard form by first completing a square on the x and/or y terms. If b, the coefficient of the xy term, is not 0, then a rotation is performed to transform the original equation into a new second-degree equation in which the new "b-like" term is 0. Once this coefficient is 0, the new equation can be put into its standard form by completing squares.

Working with these equations is easy enough conceptually, but the computations can be very tedious if they are done by hand. A Maple worksheet is an excellent environment for carrying out the computational details.

With Maple's implicitplot command, it is easy to see the rotational effect on the graph of an equation of the form (8.1) where the xy-term $b \neq 0$. In the next Maple work session, we graph the equation $x^2 + 4xy + 5y^2 + 4x - 6 = 0$.

Graphing the equation is easy enough, but finding the region in the plane where the curve resides is another matter. As we shall see, this **plot region needs to be determined with some care.** It would seem to be a good strategy to start with a large plot range, just to find where the curve resides. This could work, but, **if the range is too big, Maple might not find many points on the curve.** Guesswork is frequently the most efficient strategy. If just one point is found on the curve, it is easy enough to adjust the range by trial and error to see the curve. If the curve is really lost in space, try solving an equation to get a point. Choose a value for x or y, and solve for the other variable.

****

```
> with(plots):
```

```
Warning, the name changecoords has been redefined
```

```
> eq:=x^2+4*x*y+5*y^2+4*x-6=0;
```

$$eq := x^2 + 4\,x\,y + 5\,y^2 + 4\,x - 6 = 0$$

```
> implicitplot(eq,x=-25..3,y=-3..10);
```

#This plot range, determined by trial and error, was selected to just fit the curve.#

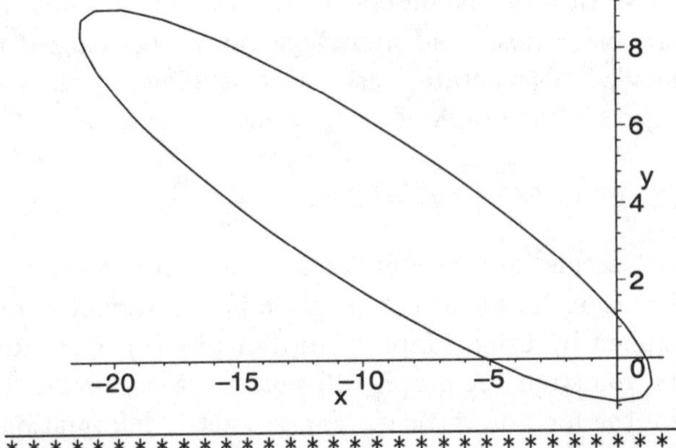

****

A careful adjustment of the x and y ranges inside this plot command will frequently be necessary. If these ranges are much different from what is necessary to just barely see the whole graph, Maple will frequently draw rather strange-looking graphs consisting mostly of line segments. Evidently, Maple's search for points is uniformly distributed over the entire plot window, and, if this window is too big, the critical region near the graph receives insufficient attention, and only a few points on the curve are established. When Maple connects these few points with line segments, it creates a poor representation of the curve.

Most of the work in this section involving translations and rotations of conic sections is omitted to save space, although there are problems of this type at the end of the chapter. The reader is **invited to visit our Web site for a full range of problems and their solutions.** Maple is an excellent environment for these manipulations.

8.2 Parametrically Defined Planar Curves

Suppose that f and g are real-valued continuous functions of a real variable t defined on an interval $D = [a, b]$. The set of all points $(x, y) = (f(t), g(t))$ as t ranges over points in D describes a curve C in the plane. The equations $x = f(t)$, $y = g(t)$ are called parametric equations of the curve C, and C is said to be defined parametrically. The variable t is called the parameter. The point $(f(a), g(a))$ is called the initial point, and $(f(b), g(b))$ is called the terminal point. Intuitively, we think of t as time, so that there is a natural motion along the curve from the initial point to the terminal point.

A parametrically defined curve, such as the one described above, can be plotted with Maple's ordinary plot( ) command. A typical input plot statement would appear in the following way (the second statement includes a control for horizontal and vertical window size):

```
> plot([f(t),g(t),t=a..b]);
```

```
> plot([f(t),g(t),t=a..b],h1..h2,v1..v2);
```

To a certain extent, this is a very natural way to write a parametric plot. After all, the list $[f(t), g(t)]$ is just an **ordered pair of points**, which is exactly what we intend to plot. Notice however, that the **range $t = a..b$ appears inside this list as a third entry**, rather than outside this list as a second argument in the plot command.

This location for the range $t = a..b$ is critical. Recall that the statements plot({f(t),g(t)},t = a..b) and plot([f(t),g(t)],t = a..b) both plot the the two curves $y = f(t)$, and $y = g(t)$ in a **t, y-rectangular coordinate system.** Both the curly and square brackets are used to group the functions together. The square brackets, however, also order the two curves, so that $f(t)$ would definitely be *curve-1* and $g(t)$ would be *curve-2*. We would use this approach, for example, to identify the curves if we wanted to display the 2-D Legend. **Placing the range $t = a..b$ inside the square brackets identifies the plot above as a parametric plot.**

There is another advantage to placing the plot range in this location. Several parametrically defined curves can be plotted together, and, because of this range location, a **different range can be used for each of the curves.** This is shown in the second of the following two examples.

****

```
> f:=3*cos(2*t):  g:=sin(4*t):
```

```
> plot([f,g,t=0..2*Pi]);
```

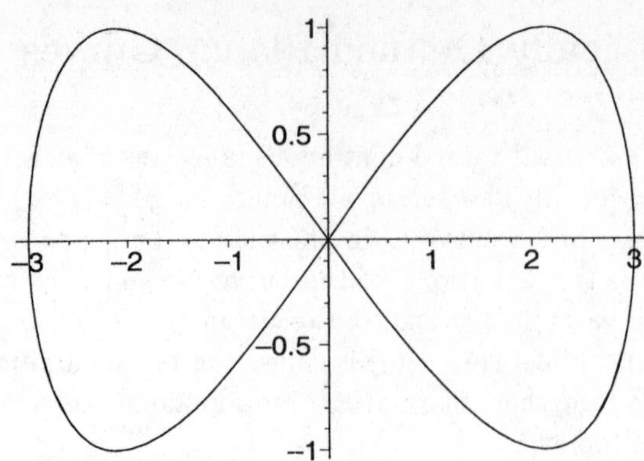


```
> f2:=cos(t)/ln(t):g2:=sin(t)/ln(t):
> C1:=[f,g,t=0..2*Pi]: C2:=[f2,g2,t=3..20]:
> plot({C1,C2});
```

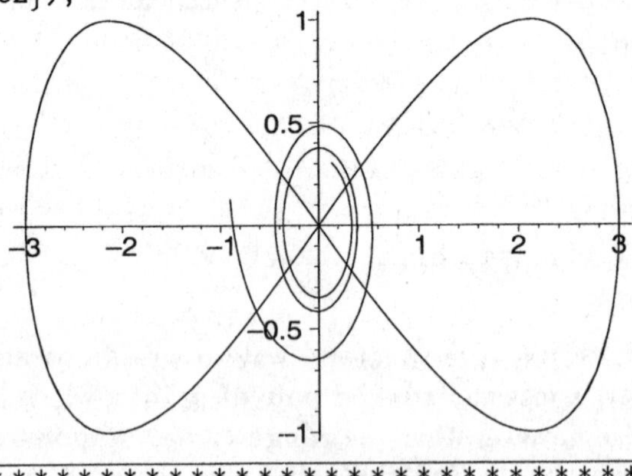

****

The slope of the line tangent to the graph of the curve C defined parametrically by $x = f(t), y = g(t), (a \leq t \leq b)$ at the point $(x, y) = (f(t), g(t))$ is

$$\frac{dy}{dx} = \frac{\frac{dy}{dt}}{\frac{dx}{dt}} = \frac{g'(t)}{f'(t)}.$$

The arc-length differential is

$$ds = \sqrt{dx^2 + dy^2} = \sqrt{\left(\frac{dx}{dt}\right)^2 + \left(\frac{dy}{dt}\right)^2}\, dt = \sqrt{(f'(t))^2 + (g'(t))^2}\, dt$$

so that the length of a curve C is

$$l(C) = \int_a^b \sqrt{\left(\frac{dx}{dt}\right)^2 + \left(\frac{dy}{dt}\right)^2}\, dt = \int_a^b \sqrt{(f'(t))^2 + (g'(t))^2}\, dt.$$

Parametric curves of the form

$$x = a\cos(mt), y = b\sin(nt), 0 \leq t \leq 2\pi$$

are known as Lissajous curves. They are well known to provide a wide variety of interesting, complex shapes. In the next example, the subfamily

$$x = a\cos(3t), y = b\sin(2t), 0 \leq t \leq 2\pi.$$

is considered. We start with a graph of a typical member of this family.

****

```
> plot([2*cos(3*t),7*sin(2*t),t=0..2*Pi]);
```

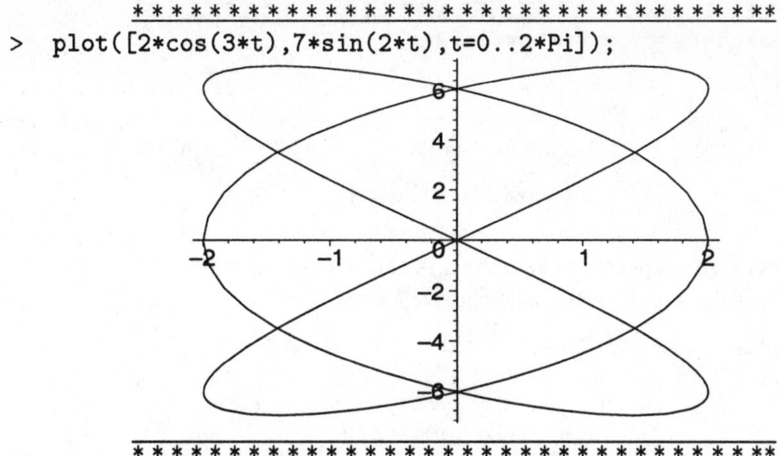

****

All of the members of this family have a similar shape. As you might expect, the value of a controls the amount of horizontal stretch, and the value of b controls the amount of vertical stretch. We can obtain confirmation of this by using a **new and very exciting Maple command called animate( )**, which resides in the plots package. Think of $F(x, t)$ as a function of x for each fixed value of t. Then

> animate(F(x,t),x=c..d,t=p..q);

plots (for t fixed) the graph of $F(x, t)$ on the interval $c \leq x \leq d$ for a specified number of t-values spread out evenly over the interval $p \leq t \leq q$.

The default value for the number of t-values used is 16, but this number can be changed with the option **frames** $= n$, where n is a specified positive integer. Incidentally, **the variable t is called the "frame variable."** With this new Maple command, we can see the horizontal and vertical stretch in the family of curves under consideration with, for example, the commands:

```
> animate([2*cos(3*t),b*sin(2*t),t=0..2*Pi],b=1..20);
> animate([a*cos(3*t),7*sin(2*t),
  t=0..2*Pi],a=4..30);
```

Each member of the family appears to be symmetric to both the x and y axes, and each of them has several self-intersection points. It seems reasonable to suspect that, by choosing a and b appropriately, we can control not only the length of the curve, but the angles at some of the self-intersection points. This is the substance of our next example.

Example 8.1 *Consider the family of curves defined parametrically by*

$$x = a\cos(3t), y = b\sin(2t), (0 \leq t \leq 2\pi), \ a, b > 0$$

i) Show that the curves are symmetric with respect to the x-axis and with respect to the y-axis.
ii)Find a member of this family with an arc length of 20 units, for which the self-intersection points off the coordinate axes are all perpendicular.
iii) Find a point on this curve that is furthest from the origin.

****

```
> x:=t->a*cos(3*t):y:=t->b*sin(2*t):
> P:=[x,y]:
> P(t);
```

$$[a\cos(3\,t),\, b\sin(2\,t)]$$

#Symmetry with respect to the x-axis is fairly easy to show. The relationship between $[x, y]$ and $[x, -y]$ is established as follows.#

```
> P(t);P(-t);
```

$$[a\cos(3\,t),\, b\sin(2\,t)]$$

$$[a\cos(3\,t),\, -b\sin(2\,t)]$$

#Symmetry with respect to the y-axis is less obvious. Plotting one of the curves on different subintervals helps to determine the relationship between $[x, y]$ and $[-x, y]$.#

```
> plot([2*cos(3*t),7*sin(2*t),t=0..Pi]);
```

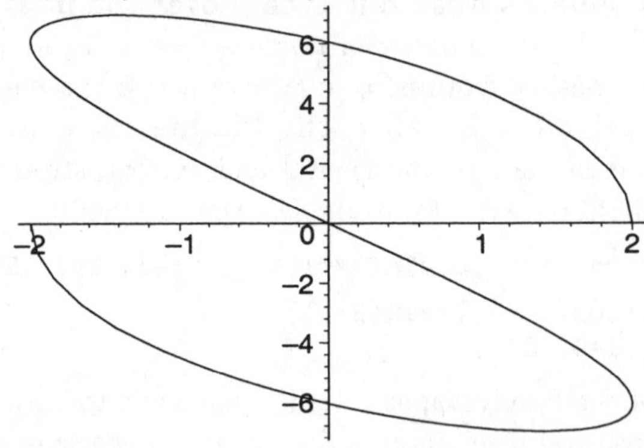


```
> P(t);P(t+Pi);
```

$$[a\cos(3\,t),\, b\sin(2\,t)]$$

$$[-a\cos(3\,t),\, b\sin(2\,t)]$$

#The self-intersection points take the form $P(t1) = P(t2)$ for different t-values $t1, t2$. After equating the x and y coordinates, we get a system of two equations in the two unknowns $t1, t2$. Before we can use Maple's fsolve( ) command on this system, we must rid the equations of all unassigned letters other than $t1, t2$. Fortunately, this is an easy matter. Now, there are many self-intersection points, so we also have to help direct Maple to the one we want to find. There are four off the coordinate axes, but, because of symmetry, we need find only the one in the first quadrant. Notice that the x-coordinate of $P(t)$ is 0 for $t = \pi/6, 3\pi/6, 5\pi/6, 7\pi/6, 9\pi/6, 11\pi/6$. This will help to establish that the portion of the curve for $0 < t < \pi/6$ will cross the portion for $7\pi/6 < t < 9\pi/6$.#

```
> eqx:=x(t1)/a=x(t2)/a; eqy:=y(t1)/b=y(t2)/b;
```

$$eqx := \cos(3\,t1) = \cos(3\,t2)$$

$$eqy := \sin(2\,t1) = \sin(2\,t2)$$

```
> fsolve({eqx,eqy},{t1,t2},t1=0..Pi/6,t2=7*Pi/6..9*Pi/6);
```

$$\{t1 = .2617993878,\ t2 = 4.450589593\}$$

```
> assign(%);
> P(t1);P(t2);
```

$$[.7071067812\,a,\ .5000000000\,b]$$

$$[.7071067796\,a,\ .4999999993\,b]$$

```
> dy:=D(y); dx:=D(x); m:=dy/dx;
```

$$dy := t \to 2\,b\cos(2\,t)$$

$$dx := t \to -3\,a\sin(3\,t)$$

$$m := \frac{dy}{dx}$$

```
> m(t);
```

$$-\frac{2}{3}\,\frac{b\cos(2\,t)}{a\sin(3\,t)}$$

```
> eqm:=m(t1)*m(t2)=-1;
```

$$eqm := -.6666666655\,\frac{b^2}{a^2} = -1$$

```
> B:=solve(eqm,b);
```

$$B := \frac{20000}{1333333331}\sqrt{6666666655}\,a,\ -\frac{20000}{1333333331}\sqrt{6666666655}\,a$$

```
> b:=evalf(B[1]);
```

$$b := 1.224744873\,a$$

```
> P(t);
```

$$[a\cos(3\,t),\ 1.224744873\,a\sin(2\,t)]$$

#Arc length is next on our agenda.#

```
> ds:=sqrt(dx(t)^2+dy(t)^2);
```

$$ds := \sqrt{9\,a^2\sin(3\,t)^2 + 6.000000016\,a^2\cos(2\,t)^2}$$

#Maple cannot evaluate the integral of ds exactly (surely this should be expected), and so numerical approximation techniques must be used. Unfortunately, **we cannot use Maple's numerical integration command, evalf(Int( )), as long as there is an unassigned constant inside the integrand.** Luck is with us, however, because a can be factored out and completely removed from the integral of ds. Maple will resist any attempt to factor a^2 out of this radical. It is easy to see why. We know that $\sqrt{a^2}$ is equivalent to a only for $a > 0$. Consequently, **we must invoke the condition $a > 0$.** It makes sense to try the **factor( ) command,** but "Maple nonsense" prevents its use. Try the simplify( ) command. Maple will do unwanted trigonometric simplification. With the cursor on "simplify," pull down the **Help Menu**, and select **Help on Context**. This will lead you to the many useful options for the simplify command. Choose the **option radical.** We can then simplify ds. All of these details can be overwhelming, but **Maple's Help File is ready and willing to help.**

Alternately, a can be factored out in a less mathematical (but much simpler) way. Substitute $a = 1$ into ds and call the result $ds1$. Use the command evalf(Int( )) to evaluate $\int_0^{2\pi} ds1$ as a decimal, and multiply the answer by the unassigned letter a to get the length of the curve. Given the nature of ds, this will surely lead to the same answer, La, that we determine next with our more mathematical approach.#

```
> assume(a>0);
```

#With Maple 7, the more convenient and less permanent **"assuming"** facility can be used. #

```
> ds1:=simplify(ds,radical);
```

$$ds1 := a\tilde{}\ \sqrt{9\sin(3\,t)^2 + 6.000000016\cos(2\,t)^2}$$

```
> La:=a*evalf(Int(ds1/a,t=0..2*Pi));
```

$$La := 16.50442453\, a\tilde{}$$

#Which curve in this family has an arc length of 20 units?#

```
> a:=solve(La=20,a);
```

$$a := 1.211796265$$

#This establishes the curve we have been looking for. All that is left is to find the point (x, y) on the curve that is furthest from the origin. We take advantage of the simple observation that the distance, $\sqrt{x^2 + y^2}$, will be a maximum at the same point that the square of this distance, $x^2 + y^2$ is a maximum. This helps to alleviate a needless complication with square roots.#

```
> f:=x^2+y^2;
```

#Remember that x and y are functions of t, so f is also a function of t.#

$$f := x^2 + y^2$$

```
> plot(f(x),x=0..Pi/2);
```

#From the plot (omitted to save space), we can see that the maximum occurs near $t = 1$. #

```
> deq:=diff(f(t),t)=0;
```

$$deq := -8.810701128\sin(3\,t)\cos(3\,t) + 8.810701156\cos(2\,t)\sin(2\,t) = 0$$

```
> t3:=fsolve(deq,t,0.8..1.2);
```

$$t3 := .9424777958$$

```
> P(t3);
```

$$[-1.152486734, 1.411502219]$$

```
> MaxDistance:=sqrt(f(t3));
```

$$MaxDistance := 1.822241528$$

****

8.3 Polar Coordinates

Mathematical literature is rich in examples of exotic and beautiful curves that can be drawn in polar coordinates. Some of these appear in the exercise set at the end of the chapter. They should be experienced with delight and enjoyment.

There are several ways to plot a polar coordinate equation of the form $r = f(\theta)$. One of them involves using commands we already have available to us. Using the transformation equations

$$x = r\cos(\theta), y = r\sin(\theta)$$

between the polar and rectangular coordinate systems, the polar equation $r = f(\theta)$ can be written in parametric form

$$x = f(\theta)\cos(\theta), y = f(\theta)\sin(\theta),$$

and the curve can then be plotted by using the techniques of the previous section.

The **command for plotting the curve $r = f(\theta)$ in polar coordinates is polarplot( ), and it resides in the plots package.** Actually, **the ordinary plot( ) command can be used to plot $r = f(\theta)$ directly in polar coordinates if the option "coords=polar"** is inserted in the plot command as shown below. In the commands

```
>polarplot(f(t));
>polarplot(f(t),t=α..β);
>plot(f(t),t=α..β, coords=polar);
```

Maple assumes that the expression $f(t)$, regardless of the letters used, means $r = f(\theta)$, where t plays the role of θ. If the range $t = a..b$ is not specified, it is assumed to be $t = -\pi..\pi$. Naturally, $f(t)$ can be replaced by a collection $\{f(t), g(t), \ldots, h(t)\}$ of polar curves $r = f(\theta), r = g(\theta), \ldots, r = h(\theta)$. The polar expressions must be enclosed, in the usual Maple way, in curly brackets.

Just as with any plot command, when an infinite discontinuity is encountered, Maple tries to fit the whole graph in the window. This produces a window size so large that the graph will be lost. Discontinuities, it should be added, are not altogether uncommon. They occur regularly, for example, when graphing conic sections in polar coordinates. **Window size cannot be controlled in the usual way with this form of the polarplot command.**

Maple has a plot option, which can be used with any plotting tool, for controlling window size. The option **view=[*a..b*, *c..d*]** produces a window with a **horizontal range *a..b***, and a **vertical range *c..d***. A typical input plot statement would look as follows:

>polarplot(f(t),t=$\alpha..\beta$,view=[$a..b, c..d$]);

There is another form of the polarplot( ) command, which permits a greater variety of polar plots other than $r = f(\theta)$. If the polarplot command is used in the following way, **window size can also be controlled in a more familiar way.** The **curve to be plotted is entered as a list.**

>polarplot([f(t),g(t),t=$\alpha..\beta$]);
>polarplot([f(t),g(t),t=$\alpha..\beta$],$a..b, c..d$);

In both of these commands, Maple assumes $\boldsymbol{r = f(t), \theta = g(t)}$. Notice how the **$\boldsymbol{\theta}$-range,$\boldsymbol{t = \alpha..\beta}$, appears inside the bracketed expression.** This is similar to the approach used in the parametric plot command, and the advantage of placing the range here is the same for the polar plot command as it is for the parametric form of the plot command (on page 183). In the second command, the first range, ($a..b$), is the horizontal window size; the second range, ($c..d$), is the vertical window size. The equation $r = f(\theta)$ can be graphed in this form by using the list $[f(t), t]$. As with most Maple commands, there are other variations and options, which are discussed in the Help File.

Maple's **animation command can be used to animate polar plots.** The use of **the option "coords=polar" is critical.**

>animate(f(t,s),t=$\alpha..\beta$,s=$s_0..s_1$,coords=polar);

With the option "coords=polar" inside the animate command, Maple treats $f(t, s)$ as a polar curve of the form $r = f_s(\theta)$, where s (the frame variable) is a constant ranging between s_0 and s_1.

With all of Maple's plot commands, a variety of plot options can be inserted after the necessary arguments are entered.

If the equation $r = f(\theta)$ is written parametrically, then there is no need to depend on any special forms for the slope of the tangent line or for the arc-length differential. We simply use

$$\frac{dy}{dx} = \frac{\frac{dy}{d\theta}}{\frac{dx}{d\theta}}, ds = \sqrt{\left(\frac{dx}{d\theta}\right)^2 + \left(\frac{dy}{d\theta}\right)^2}.$$

However, if desired, these forms readily lead to the formulas

$$\frac{dy}{dx} = \frac{f'(\theta)\sin(\theta) + f(\theta)\cos(\theta)}{f'(\theta)\cos(\theta) - f(\theta)\sin(\theta)},$$

$$ds = \sqrt{r^2 + (\frac{dr}{d\theta})^2 d\theta} = \sqrt{(f(\theta))^2 + (f'(\theta))^2 d\theta}.$$

The arc length of the curve $r = f(\theta)$ on $(\alpha \leq \theta \leq \beta)$ is

$$Length = \int_\alpha^\beta \sqrt{(f(\theta))^2 + (f'(\theta))^2 d\theta}.$$

If R is a region bounded by $r = f(\theta), \theta = \alpha$, and $\theta = \beta$ with $\alpha < \beta$, then the area of R is

$$Area = \int_\alpha^\beta \frac{1}{2}(f(\theta))^2 d\theta.$$

The details of all of these formulas, and others, will be found in any standard calculus text.

Example 8.2 *Find the area of the region that lies inside the four-leaved rose* $r = 7\cos(2\theta + 1)$ *and above the horizontal line* $y = 3$.

The equation $y = 3$ can easily be expressed in its polar form as $r = 3\csc(\theta)$. Unfortunately, discontinuities are involved in its polar form, and this will cause precisely the kind of "large window" problem we referred to above, if we try to plot this pair of equations by using the polarplot( ) command in its simplest form. We mentioned before that there is a way out of this problem. **Look carefully at the exact form of the polarplot command** appearing in the next Maple work session.

The graph of the region shows that there are four points of intersection. It is too much to ask Maple for exact values of the angles at these four points, and so we will use Maple's fsolve( ) command to get decimal approximations for them instead. Maple will need help from us to direct it to the vicinity of each of these four t-angles. By looking at the graph, it appears to be an easy matter to approximate these angles visually. Looks, however, can be deceiving. Remember that negative r-values are involved in both of the curves seen in the graph. A point (r_0, θ_0), which might appear to be a point of intersection, might not satisfy either equation. Instead, the point that satisfies both equations could be $(-r_0, \theta_0 + \pi)$ or some other combination of (r, θ) values. This certainly complicates the matter of guiding Maple to the vicinity of each t-value. To help in this matter, we plot, **as an ordinary Cartesian graph**, the difference in r-values as a function of t. The equation of the horizontal line has infinite discontinuities, so we avoid the large vertical scale that would be caused by these discontinuities by inserting a vertical scale in the plot command. By looking at where this graph crosses the horizontal axis (the t-axis), it is now an easy matter to get a range of t-angles in the vicinity of each intersection point. The rest of the problem is straightforward.

****

```
> r1:=7*cos(2*t+1):  r2:=3*csc(t):
> with(plots):
Warning, the name changecoords has been redefined
> polarplot({[r1,t,t=0..2*Pi],[r2,t,t=0.1..2*Pi-.1]},
  view=[-7..7,-7..7]);
```

```
> plot(r1-r2,t=0..2*Pi,-7..7);
```



```
> a1:=fsolve(r1=r2,t,t=2..2.5);
  a2:=fsolve(r1=r2,t,t=2.5..2.9);
```

$$a1 := 2.119291465$$

$$a2 := 2.695870991$$

```
> a3:=fsolve(r1=r2,t,t=3.6..4);
  a4:=fsolve(r1=r2,t,t=4.5..5);
```

$$a3 := 3.809252544$$

$$a4 := 4.775853085$$

```
> A:=int(r1^2/2,t=alpha..beta)-int(r2^2/2,t=alpha..beta);
```

$$A := \frac{49}{8}\cos(2\beta+1)\sin(2\beta+1)+\frac{49}{4}\beta-\frac{49}{8}\cos(2\alpha+1)\sin(2\alpha+1)-\frac{49}{4}\alpha$$
$$-\frac{9}{2}\frac{-\cos(\beta)\sin(\alpha)+\cos(\alpha)\sin(\beta)}{\sin(\beta)\sin(\alpha)}$$

```
> A1:=evalf(subs(alpha=a1,beta=a2,A));

  A2:=evalf(subs(alpha=a3,beta=a4,A));
```

$$A1 := 3.714198466$$

$$A2 := 11.28274411$$

```
> Area:=A1+A2;
```

$$Area := 14.99694258$$

****

In the next example, another way of dealing with the difficulties caused by the nonuniqueness of polar coordinate points is discussed. The example is noteworthy for several other reasons. **Notice the way that the seq( ) command is used to do much of the computational work.** Also notice, in the computations involving the angle of intersection, the liberal use of functions rather than expressions.

Example 8.3 *Find the polar coordinates of the self-intersection points of the polar coordinate curve*

$$r = 3\sin(7\theta/3).$$

Express the coordinates in a form that satisfies the equation. Find the angle between the intersecting branches of the curve at one of the self-intersecting points (other than the origin).

For the sake of discussion, we let $P(r,\theta)$ denote the point in the plane with polar coordinates (r,θ).

A plot shows an obvious point of intersection at the origin, and so we turn our attention to the seven other points, which appear to be (and surely they are) equally distributed around a circle. Consequently, it should be enough to find just one of the points and then use its coordinates, distributed around a circle in an obvious way, to get the coordinates of the other points.

There are several ways in which the curve could self-intersect. To discuss this matter, let f be the function defined by

$$f(\theta) = 3\sin(7\theta/3),$$

so that we can think of r as a variable. We will not use the letter f in our worksheet, but it will help with this discussion. What if we can find a value θ_0 for which

$$r_0 = f(\theta_0) = f(\theta_0 + 2\pi)?$$

The function f certainly doesn't repeat itself on intervals of length 2π, so the point $P(r_0,\theta_0) = P(r_0,\theta_0+2\pi)$ would have to be one of these self-intersecting points. We might not find a point in this way. Then, perhaps we could find a θ_0 such that

$$r_0 = f(\theta_0) = -f(\theta_0 + \pi)$$

or

$$r_0 = f(\theta_0) = f(\theta_0 + 4\pi)$$

or

$$r_0 = f(\theta_0) = -f(\theta_0 + 3\pi).$$

A solution to any one of these equations would produce a self-intersecting point.

To find the angle between the intersecting branches of the curve, we will use the trigonometric identity

$$\tan(\alpha - \beta) = \frac{\tan(\alpha) - \tan(\beta)}{1 + \tan(\alpha)\tan(\beta)}.$$

The slope of the tangent line to a curve at a point (actually, the slope of any line) is the tangent of the angle of inclination of the line.

****

```
> with(plots):
```

```
Warning, the name changecoords has been redefined
```

```
> polarplot(3*sin(7*t/3),t=0 ..6*Pi);
```

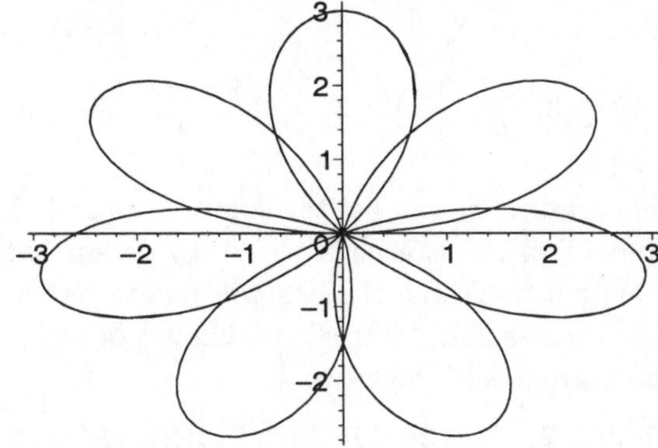


```
> r:=t->3*sin(7*t/3):
```

> #Notice that r represents a <u>function</u>, not an expression.#

```
> t1:=solve(r(t)-r(t+2*Pi),t);
```

$$t1 := \frac{1}{14}\pi$$

```
> t:=seq(t1+j*2*Pi/7,j=0..6);
```

$$t := \frac{1}{14}\pi, \frac{5}{14}\pi, \frac{9}{14}\pi, \frac{13}{14}\pi, \frac{17}{14}\pi, \frac{3}{2}\pi, \frac{25}{14}\pi$$

```
> r1:=r(t1);
```

$$r1 := \frac{3}{2}$$

#From the polar graph of the curve and the basic self-intersecting point we just found, it is clear that the seven points of self-intersection have coordinates (r, θ), where $r = 3/2$, and $\theta = t[j]$, for $j = 1, 2, \ldots, 7$.

There is little doubt that these are, indeed, the coordinates of the points, but only some of them satisfy the equation $r = f(\theta)$. This is a classic polar-coordinate problem caused by the nonunique representation of points in polar space. A point (r_0, θ_0) can be on a curve $r = f(\theta)$, even though $r = r_0$, $\theta = \theta_0$ fails to satisfy the equation of the curve. The points (r_0, θ_0), $(-r_0, \theta_0 + \pi)$, and $(r_0, \theta_0 + 2\pi)$, to list a few, are all polar coordinates of the same point. It is possible for one of these ordered pairs to satisfy a polar equation, and for another ordered pair to fail to satisfy the equation.

With this in mind, let us return to the self-intersection points we determined above. In order to express the coordinates of these points in a form that will satisfy the equation $r = f(\theta)$ of the curve, we compare the value of $r = f(\theta)$ at $\theta = t[j], \theta =$

$t[j]+\pi$, and $\theta = t[j]+2\pi$ for each $j = 1, 2, \ldots, 7$. This comparison will shed light on what coordinates we should use and also on what the "return" coordinates should be. **Notice how easily this is done with the seq( ) command.**#

```
> seq([r(t[j]),r(t[j]+Pi),r(t[j]+2*Pi)],j=1..7)
  ;
```

$$[\frac{3}{2}, 3, \frac{3}{2}], [\frac{3}{2}, \frac{-3}{2}, -3], [-3, \frac{-3}{2}, \frac{3}{2}], [\frac{3}{2}, 3, \frac{3}{2}], [\frac{3}{2}, \frac{-3}{2}, -3], [-3, \frac{-3}{2}, \frac{3}{2}], [\frac{3}{2}, 3, \frac{3}{2}]$$

#Using these comparisons, we can list the coordinates and the "return" coordinates of each self-intersection point, each of which, by design, satisfies the equation of the curve. Later on, our angle calculations will put to rest any lingering doubt about whether the coordinates and "return" coordinates of any point correspond to different or overlapping subarcs of the curve.#

```
> P(3/2,t[1])=P(3/2,t[1]+2*Pi), P(3/2,t[2])=P(-3/2,t[2]+Pi);
```

$$\mathrm{P}(\frac{3}{2}, \frac{1}{14}\pi) = \mathrm{P}(\frac{3}{2}, \frac{29}{14}\pi), \mathrm{P}(\frac{3}{2}, \frac{5}{14}\pi) = \mathrm{P}(\frac{-3}{2}, \frac{19}{14}\pi)$$

```
> P(-3/2,t[3]+Pi)=P(3/2,t[3]+2*Pi), P(3/2,t[4])=P(-3/2,t[4]+Pi);
```

$$\mathrm{P}(\frac{-3}{2}, \frac{23}{14}\pi) = \mathrm{P}(\frac{3}{2}, \frac{37}{14}\pi), \mathrm{P}(\frac{3}{2}, \frac{13}{14}\pi) = \mathrm{P}(\frac{-3}{2}, \frac{27}{14}\pi)$$

```
> P(3/2,t[5])=(-3/2,t[5]+Pi), P(-3/2,t[6]+Pi)=P(3/2,t[6]+2*Pi);
```

$$\mathrm{P}(\frac{3}{2}, \frac{17}{14}\pi) = (\frac{-3}{2}, \frac{31}{14}\pi), \mathrm{P}(\frac{-3}{2}, \frac{5}{2}\pi) = \mathrm{P}(\frac{3}{2}, \frac{7}{2}\pi)$$

```
> P(3/2,t[7])=P(3/2,t[7]+2*Pi);
```

$$\mathrm{P}(\frac{3}{2}, \frac{25}{14}\pi) = \mathrm{P}(\frac{3}{2}, \frac{53}{14}\pi)$$

#Using ideas given in the introduction of this section, we compute next a formula for the slope of the tangent line at a point on the curve. What we really want is the angle of inclination that a tangent line makes with the positive x-axis. This is easy to get from the slope, because slope (rise over run) is just the tangent of the angle of inclination.

Notice the effective use of functions rather than expressions in the following. Look at the Maple assignment for r early in this work session. The letter r actually represents the function of θ that defines the curve. Surely, the terms cos and sin represent functions, not expressions. (Notice that the arguments of the trigonometric functions are missing.)#

```
> x:=r*cos:  y:=r*sin:
  #This means x and y are also functions of θ.#
```

```
> dx:=D(x);dy:=D(y);
```

#Recall that the differential operator D operates on functions, not on expressions, and it returns functions. So, dx and dy represent functions of θ.#

$$dx := (t \to 7\cos(\frac{7}{3}t))\cos - r\sin$$

$$dy := (t \to 7\cos(\frac{7}{3}t))\sin + r\cos$$

```
> m:=dy/dx;
```

#Naturally, m is also a function of θ.#

$$m := \frac{(t \to 7\cos(\frac{7}{3}t))\sin + r\cos}{(t \to 7\cos(\frac{7}{3}t))\cos - r\sin}$$

```
> TanAngle:=(t,s)->abs((m(t)-m(s))/(1+m(t)*m(s)
  ));
```

$$TanAngle := (t, s) \to \left|\frac{\mathrm{m}(t) - \mathrm{m}(s)}{1 + \mathrm{m}(t)\,\mathrm{m}(s)}\right|$$

```
> Angle:=arctan@TanAngle;
```

#Recall that @ is Maple's composition operator (per page 23).#

$$Angle := \arctan@TanAngle$$

```
> RadAngle:=evalf(Angle(t[1],t[1]+2*Pi));
```

$$RadAngle := .4851277480$$

```
> DegAngle:=evalf(RadAngle*180/Pi);
```

$$DegAngle := 27.79577248$$

****

Example 8.4 *Find the arc length of the ellipse defined by the polar-coordinate equation*

$$r = \frac{30}{13 - 12\sin(\theta + \frac{\pi}{5})}.$$

****

```
> with(plots):
Warning, the name changecoords has been redefined
> r:=t->30/(13-12*sin(t+Pi/5)):
> polarplot(r(t),t=0..2*Pi);
```

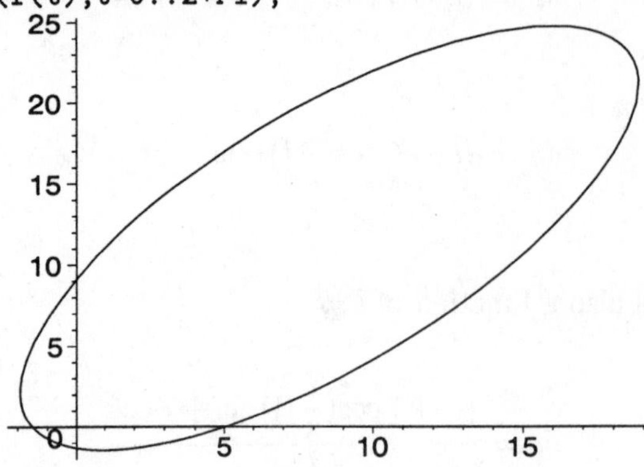

```
> ds:=simplify(sqrt(r(t)^2+diff(r(t),t)^2));
```

$$ds := 30\sqrt{-\frac{-313+312\sin(t+\frac{1}{5}\pi)}{(-13+12\sin(t+\frac{1}{5}\pi))^4}}$$

```
> L:=evalf(Int(ds,t=0..2*Pi));
```

$$L := 71.25302207$$

****

8.4 Exercise Set

New Maple Commands in this Chapter (and a few old commands as a reminder)

animate( )	applyop( )	assume( )	assuming	coeff( )
collect( )	completesquare( )	fsolve( )	implicitplot( )	lhs( )
plot([ ])	polarplot( )	rhs( )	simplify( )	sort( )

Examples about conic sections, along with some commands that could help with the first three problems, can be found at our Web site.

1. Graph the following second-degree equations. If they exist, find the coordinates of the foci and vertices and the slopes of the asymptotes. If the graph happens to be a degenerate conic, is it necessary to go through a formal rotation to confirm this fact?

a) $4x^2 + 15xy + 12y^2 - 20x + 44y + 80 = 0$
b) $9x^2 + 6xy + y^2 + 7x + 3y + 2 = 0$
c) $2x^2 - 5y^2 + 3x + 8y + 9 = 0$
d) $3x^2 - 13xy - 10y^2 + 11x + 13y - 4 = 0$
e) $2x^2 + 5xy + 14y^2 + 3x - 4y - 7 = 0$

2. Use Maple to show that the discriminant of a second-degree equation is unchanged by a rotation, regardless of the angle of the rotation.

3. Use Maple to derive the formula

$$\tan(2\alpha) = \frac{b}{a-c}$$

that is used to analyze the graph of a second-degree equation with a nonzero xy-term. (Hint: Perform an arbitrary rotation and set $B = 0$.)

4. Parametric equations of the form

$$x = x_0 + a\cos(mt), y = y_0 + b\sin(mt), 0 \leq t \leq 2\pi$$

are effective ways of representing a circle or an ellipse having a horizontal or vertical major axis. Ultimately, these parametric equations should become so familiar to a student of mathematics that Maple's plotting tools are not needed. Use pencil-and-paper techniques and Maple's plotting tools to help gain some insight into these forms. Determine the role played by x_0, y_0, a, b, and m. In particular, how can Maple's plot command be used to determine the role played by m?

5.A straight line in the plane is represented parametrically by

$$x = x_0 + pt, y = y_0 + qt, -\infty < t < \infty,$$

but parametric equations of straight lines do not have to be linear. Use Maple to plot

$$x = 3 + 2f(t), y = 5 - 3f(t), t_0 < t < t_1$$

for the following sample of functions f and intervals $[t_0, t_1]$.
a) $f(t) = t$ on $(-\infty, \infty)$ b) $f(t) = t^3$ on $(-\infty, \infty)$
c) $f(t) = t^2$ on $(-\infty, \infty)$ d) $f(t) = \tan(t)$ on $(-\frac{\pi}{2}, \frac{\pi}{2})$
What role does the function f play in these representations? Experiment further, if necessary, to answer this question.

6. Plot the Hypotrochoid

$$x = \cos(t) + 5\cos(3t), y = 6\cos(t) - 5\sin(3t), 0 \leq t \leq 2\pi.$$

7. A circle of radius 1, with a fixed point P marked on the circle, lies inside a circle of radius 10 centered at the origin and is allowed to roll along the circumference of the larger circle. Initially, the point P is at the point (10,0) and the small circle rolls in a counterclockwise direction along the inside of the larger circle. The point P traces out a curve defined parametrically by

$$x = 9\cos(t) + \cos(9t), y = 9\sin(t) - \sin(9t), 0 \leq t \leq 2\pi.$$

Use Maple to plot this curve.

8. Parametric curves of the form

$$x = a\cos(mt), y = b\sin(nt), 0 \leq t \leq 2\pi,$$

are known as Lissajous curves. Use Maple's plot command to graph a sample of them and to determine the role played by each of the constants, a, b, m, n.

a) Fix $n = 2, m = 3$, and plot over a range of a and b values. A nice way to do this is with Maple's animate command. With this command, you could, for example, set $a = s, b = 31 - s$, and $s = 1..30$, where s is the "frame variable."

b) Plot with $a = 2, b = 5, n = 1$ fixed, and $m = 2, 3, 6$.

c) Plot with $a = 2, b = 5, n = 2$ fixed, and $m = 3, 5, 7$.

d) Plot with $a = 2, b = 5$ fixed, and $n = 2s, m \doteq 3s$ for $s = 2, 3, 4$. Notice that a common factor, j, between $n = n_1 j$ and $m = m_1 j$ has no effect on the graph. Why? If we happen to know the basic shapes of these curves for all integers m and n, is there anything to be gained by considering rational numbers p and q (instead of m and n) that are not integers?

9. a) Plot the rose-petal curve defined by the polar coordinate equation $r = 3\cos(2\theta)$. b) Determine the role played by the fixed angle s in $r = 3\cos(2\theta + s)$. Use Maple's animation command with s as the frame variable.

10. Plot the following conic sections in polar coordinates and find the vertices and the second focus in polar and rectangular coordinates. Observe what happens if the simplest form of the polarplot command is used to plot $r = f(\theta)$ when f has an infinite discontinuity. Then use the suggestions given on page 190 to control window size.

$a)\ r = \frac{4}{3+2\cos(\theta-\frac{3\pi}{4})} \quad b)\ r = \frac{8}{3-3\cos(\theta+\frac{\pi}{6})} \quad c)\ r = \frac{21}{5+7\cos(\theta+1)}$

11. Consider the family of curves defined by $r = \sin(m\theta)$. The value of m has a dramatic effect on the graph.

a) When m is an positive integer, the curves are called "rose petal" curves. Plot these curves for $n = 2, 3, 4, 5, 6, 7$. Make a conjecture about value of the integer m and the shape of the curve. Explain, analytically, why the patterns are different for m even or odd.

b) Plot the curves for $m = 5/3$, and then for $m = 7/2, 7/3, 7/4, 7/6, 7/11$. Any conjectures?

c) A larger denominator seems to complicate the curve. What would be the result be if m were irrational? Try to answer this question first, then plot the curve for $m = \sqrt{2}$. Use a large θ-range. Why? Then sit back and relax—this will take some time.

d) What about curves of the form $r = \cos(m\theta)$? Do we have to go through all this work again? Think about Problem 9 and an elementary trigonometric identity.

12. Plot the beautiful **Fay Butterfly**

$$r = e^{\cos(\theta)} - 2\cos(4\theta) + \sin^5(\frac{\theta}{12}).$$

Notice that a θ-interval larger than 2π must be used. How big must the plot range be in order to display the whole curve? This curve appears in "The Butterfly Curve" by Temple H. Fay, Amer.Math.Monthly **96** No.5, May 1989, pp.442-443.

This curve will take some time for Maple to draw, but be patient. The result is worth the wait—enjoy!

13. Plot the curve $x = t^2 - t; y = 2t^2 - t^3 + 7$ in a window that shows its self-intersection point. Find the coordinates of the point of intersection expressed as decimal numbers. Find the angle (acute angle) between the two intersecting subarcs at the intersection point. Express your answer in degree measure. Find the length of the arc (loop) beginning and ending at this intersection point.

14. Consider the torus generated by rotating about the x-axis the circle of radius r centered at $(x_0, y_0) = (0, R)$, where r and R are fixed positive numbers with $R > r$. Find a formula for the surface area of the torus in terms of r and R. Use a parametric form for the equation of the circle.

15. Find the area of the region inside the graph of $r = -4\sin(t)$, yet outside the graph of $r = 6\sin(3t)$.

16. Find the area of the region outside the graph of the cardioid $r = 2+2\cos(\theta)$, yet inside the polar coordinate curve $r^2 = 25\cos(\theta)$. Remember that, when you use the polarplot command in its simplest form, Maple assumes that the input expression $f(\theta)$ means the curve $r = f(\theta)$. Consequently, plotting the second curve will require some special handling. The restricted domain of the second curve will also cause some complications. A full plot of the region involved should be included.

17. Use a parametric form for the equation of the ellipse

$$\frac{x^2}{a^2} + \frac{y^2}{b^2} = 1 \text{ (assume } a > b)$$

to show the following classic reflective property of the ellipse: Let P be a point on the ellipse, let L be the tangent line to the ellipse at P,and let C_1, C_2 be the foci of the ellipse. Then the acute angle between L and the line from C_1 to P, is the same as the acute angle between the L and the line from P to C_2. This result is often phrased as, "The angle of incidence equals the angle of reflection."

How does one show that $expr_1 = expr_2$ when $expr_1$, and $expr_2$ are complicated expressions? One way is to show that $expr_1 - expr_2 = 0$, but there is another way, somewhat similar to this, that might work better. Describe an approach of this sort. It could help in this or some future problem.

Incidentally, in this problem it is enough to show that the tangents of the two angles involved are equal. (Why?)

18. The reflective property for a hyperbola is shown below, where $F1$ and $F2$ denote the foci of the hyperbola

$$\frac{x^2}{a^2} - \frac{y^2}{b^2} = 1.$$

For any point P, the line from P to Q (towards $F1$) and then to $F2$ produces angles α and β of equal measure at the point Q on the hyperbola.

Use the parametric form for the equation of the hyperbola to prove this reflective property. See the comment following the preceding exercise for some advice on showing equality.

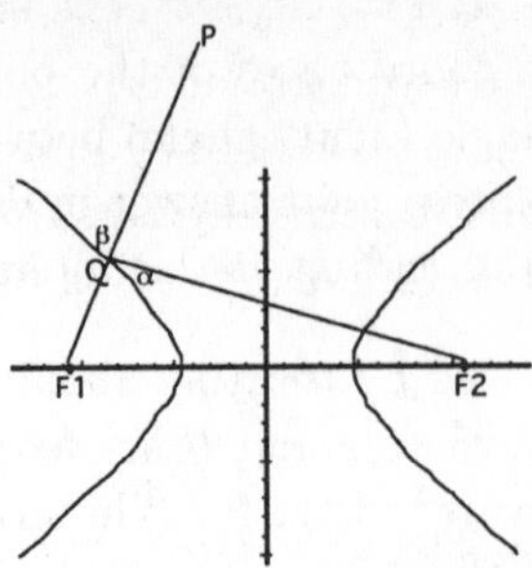

Project:Half of a Rose

Find a horizontal line $y = a$ above the x-axis such that the area of the region both above this line and inside the three-leaved rose $r = 5\sin(3\theta)$ is half the area of the entire rose.

This problem, which resembles Example 8.2, seems simple enough. The area of the entire rose is easy to compute. Then, all one has to do is find the points of intersection of the line and the rose, and the rest is fairly straightforward. An exact solution to the equation involved in this point of intersection, however, is too complicated to be useful, and so it is natural to turn to approximate solutions from Maple's fsolve( ) command. Unfortunately, the value of a that controls the horizontal line is unknown; in order to use the fsolve( ) command in an equation, every name in the equation must evaluate to a numeric except for the unknown that the equation is being solved for.

The easiest way to proceed is by guesswork: Guess at an answer, compute the area involved, use this to adjust your guess. This can be repeated again and again, each step leading to a better answer.

A better way is to write a program that mimics your guess work. Perhaps you can think of an approach that works better than the one discussed below. Notice, however, how "natural" the idea is and how methods of this sort can be created without any background in numerical approximation techniques.

Start with a value for a in $y = a$, call it HI, that is too big and another value for a, call it LO, that is too small. Clearly a =HI= 4 and a =LO= 0 will work. Let MID be the average of LO and HI. One can then check the area corresponding to a=MID to see whether it is larger or smaller than half the area of the rose. Use this to reassign either LO or HI to be the value of MID. This would complete one pass through the program loop, and the steps are then repeated as often as necessary until a =MID is close enough.

Write a program that solves this problem. To avoid integrating each pass through the program loop, **integrate just once, to generate a formula** that can be used instead to evaluate area in each pass through the program loop.

Maple will have to be told to stop the program when the answer is close enough. To do this in a program, the word **"while"** is used in the following way:

> **while** *error*>10^(-8) **do** ... **od**;

The expression *error* is a measure of how close the approximate answer is to the correct answer. There are several ways to measure this, and the choice is left to you. As soon as *error*$\leq 10^{-8}$, the program stops.

One way to speed up the program is to suppress all of the intermediate output by using **a colon (:) instead of semicolon (;) after the delimiter "od" that ends the "do" loop** and then to ask for output only after the program if finished running.

Project:Working on the Railroads

A traffic-control engineer working for a railroad corporation watches the progress of two trains on a computer console. The trains are traveling on different tracks, which intersect at a point ahead of both trains. The coal train is 1.57 miles long and the position of its head (its frontmost point) at time t is

$$x = 17t - 4\cos(2t); y = 2t + 3\sin(5t).$$

The freight train is 1.14 miles long and the position of its head at time t is

$$x = 30 + 11t, y = 117 + 18t - 7t^2.$$

In both cases t represents time, in hours, and x and y are in miles. The engineer makes a quick calculation and decides that the trains will safely pass through the intersection point. Unfortunately, this is not a good decision.

a) Show conclusively that the trains collide, and determine which train crashes into the other.

When the mess is cleaned up, management decides that it better send all of its traffic control engineers to school to learn some calculus. When the course is completed each engineer will be given a test. Failure will mean immediate job dismissal!

Let us tune back in to this drama, then, after some time has passed, and suppose that the moment has arrived. As one of the traffic control engineers, here is the test that you must now take.

b) Given the exact circumstances surrounding the accident, what is the minimal amount of extra time that the coal train must have in order for it to pass the intersection point (crash-free) just ahead of the freight train? What is the minimal amount of extra time that the freight train must have in order for it to pass the intersection point (crash-free) just ahead of the coal train?

If one of the trains had exactly this amount of extra time, then the trains would safely pass through the intersection point of the two railroad tracks, but the trains would come so close to each other, that the experience would clearly scare everyone involved. Consequently, you are now asked to add a margin of safety to your calculations, so that the trains will "safely" pass each other. Think of two distinctly different ways of adding a safety margin, describe them carefully, and recalculate answers with a built-in safety margin.

Part (b) (the test) is not easy. Each answer is the solution to an arc-length equation, but Maple might be unwilling to solve the equation, even with the fsolve( )

command. Nevertheless, setting up the appropriate arc-length equation is a good way to begin the solution.

The easiest way to proceed is by guesswork. Using the results from part a), you know a t-value that is too small (the trains crash). By guesswork, find a t-value that is too big (the trains do not crash). It follows that the correct answer is some t-value in between. Guess again, and decide whether your guess is too big or too small. Now what interval is the correct answer in? Guess again, and again. With each guess, the correct answer is narrowed down to a smaller interval.

A better approach would be to write a program that mimics all of this guess work. Start with a t-value (call it LO) that is too small and a t-value (call it HI) that is too big. Let MID be the average of LO and HI. Decide whether t =MID is too big or too small, and use this to reassign either LO or HI to be the value of MID. This would complete one pass through the program loop, and the steps are then repeated as often as necessary to get MID close enough.

Maple will have to be told to stop the program when the answer is close enough. To do this in a program, the word **"while"** is used in the following way:

```
> while error>10^(-8) do ...od;
```

The expression *error* is a measure of how close the approximate answer is to the correct answer. There are several ways to measure this, and the choice is left to you. As soon as *error*$\leq 10^{-8}$, the program stops.

One way to speed up the program is to suppress all of the intermediate output by using **a colon (:) instead of semicolon (;) after the delimiter "od" that ends the "do" loop** and then to ask for output only after the program is finished running.

Chapter 9

Vectors and Analytic Geometry

Most of the new commands we will need in this chapter reside in the **linalg package**. The name "linalg" is an abbreviation for "linear algebra." Almost any work with vectors requires commands in this package, and so it should be loaded routinely whenever work with vectors is involved. The linear algebra package is quite large; from time to time, it should be loaded with a semicolon (;) in order to see the names of all the commands in the package.

9.1 Vectors in Maple

A vector is usually defined to be a directed line segment, or more precisely, a class of all directed line segments having the same length and direction. Soon after addition and scalar multiplication are defined geometrically, we arrive at the familiar form $\vec{v} = a\vec{i} + b\vec{j}$ for a vector in two-dimensional space and $\vec{w} = a\vec{i} + b\vec{j} + c\vec{k}$ for a vector in three-dimensional space. These forms can be reduced further to the ordered pair $\vec{v} = (a, b)$ and the ordered triplet $\vec{w} = (a, b, c)$.

We take up the subject with this symbolic vector form. Maple's commands for addition and scalar multiplication will be discussed in the next section.

One would think that, as an ordered pair or ordered triplet, a vector would be represented in Maple by using **square brackets** to form a **list**. Much of the time, Maple will treat a simple list such as

```
> w:=[a,b,c];
```

as the vector $\vec{w} = a\vec{i} + b\vec{j} + c\vec{k}$. A list could in fact be used, in some sense, to represent a vector, but there are several other ways to represent a vector as well, and they are all closely related, except for the list. We will adopt the point of view that a **list is not officially a vector**, even though it might sometimes function well in this way, and **we will avoid using a list as a vector**, except for some introductory experimentation.

Let us say that **a vector in Maple is "officially" a matrix—a rectangular array of numbers—with one row**. We will seldom use the more general notion of a matrix, and so that idea will not be introduced until it is needed. It is **usually more convenient, however, to represent a vector as a "row vector" or a**

"column vector"—two data types that are not regarded as matrices by Maple.

The two-dimensional and three-dimensional vectors $\vec{v}$, and $\vec{w}$ mentioned above are represented officially as **matrices** with the input statement:

```
>  v:=vector([a,b]); w:=vector([a,b,c]);
```

Notice that **square brackets must still be used** inside the argument of the vector( ) command. To represent these two vectors as **column vectors** in an input statement, we would write:

```
>  v:=<a,b>; w:=<a,b,c>;
```

Finally, to represented them by **row vectors** in an input statement we would write:

```
>  v:=<a|b>; w:=<a|b|c>;
```

These three forms are different data types in Maple.

In most of mathematics, column vectors are preferred over row vectors, but row vectors are more commonly used in calculus. Consequently, after some initial experimentation, we will no longer use column vectors.

We would use row vectors exclusively, but **Maple occasionally turns row vectors into matrices** without our knowledge. Because of this, it is useful to be familiar with vectors in both their matrix and row-vector form.

In the following introductory experimentation, we use the suffixes "m", "L", "c", and "r" to denote a vector represented as a matrix (an official vector), a list, a column vector, and a row vector, respectively. Notice that the output from a list, a (matrix) vector, and a row vector all look exactly the same. Notice that **a (matrix) vector appears to be quite "closed" to evaluation**, yet all the other forms evaluate automatically. This is characteristic of all matrices. **To "open up" a (matrix) vector, the new command evalm( )—evaluate as a matrix—is used. This command will be used often.**

****

```
>  with(linalg):

Warning, the protected names norm and trace have been redefined and
unprotected

>  wm:=vector([-3,7,2]); wL:=[-3,7,2];wc:=<-3,7,2>;wr:=<-3|7|2>;
```

$$wm := [-3, 7, 2]$$

$$wL := [-3, 7, 2]$$

$$wc := \begin{bmatrix} -3 \\ 7 \\ 2 \end{bmatrix}$$

$$wr := [-3, 7, 2]$$

```
>  wm; wL;wc;wr;
```

$$wm$$

$$[-3, 7, 2]$$

$$\begin{bmatrix} -3 \\ 7 \\ 2 \end{bmatrix}$$

$$[-3, 7, 2]$$

Notice the "closed" nature of the (matrix) vector $\vec{wm}$. The output above is just the name wm.

```
> whattype(wm), whattype(wL),whattype(wc),whattype(wr);
```

$$symbol,\ list,\ Vector_{column},\ Vector_{row}$$

Maple does not see $\vec{wm}$ as a vector—it only sees an unassigned name (**a symbol**)! This is characteristic of all matrices. To open up a Maple vector we use the **evalm( ) command** (evaluate as a matrix).#

```
> evalm(wm);
```

$$[-3, 7, 2]$$

```
> whattype(evalm(wm));
```

$$array$$

```
> wm[1],wL[2],wc[3],wr[2];
```

$$-3, 7, 2, 7$$

```
> vm:=vector([5/3,sqrt(2)]); vr:=<5/3|sqrt(2)>;
```

$$vm := \left[\frac{5}{3}, \sqrt{2}\right]$$

$$vr := \left[\frac{5}{3}, \sqrt{2}\right]$$

```
> um:=evalf(vm); uL:=evalf(vr);
```

$$um := vm$$

$$uL := [1.666666667, 1.414213562]$$

```
> um;
```

$$vm$$

```
> um:=evalm(um);
```

$$um := \left[\frac{5}{3}, \sqrt{2}\right]$$

```
> evalf(evalm(um));
```

$$[1.666666667, 1.414213562]$$

****

As we see above, the ordinary commands of Maple—represented by the typical evalf( ) command—are able to get inside of a "list" to perform their duties in an expected way, but a **"vector" seems to be closed to these commands, until it is opened up with the evalm( ) command.** Also, notice that the **new command whattype( )** was introduced in the above work session. This is a general Maple command that can be used in a variety of situations. It can sometimes shed light on a puzzling response from Maple. **It is often very informative to have Maple tell us how it is reading a symbol.** Notice that the vector $\vec{wm}$ is so "closed" that Maple is unable to read it as anything other than a string (a name).

Finally, **notice in the above that the individual entries of a vector are extracted in the same way as they are extracted from a list, by using square brackets.**

9.2 Addition and Scalar Multiplication

Remember to load the linear algebra package, beause all of the vector commands reside there. We will use Maple's "row-vector" data type for most of our work involving vectors. Occasionally, **Maple will turn our "row vectors" into matrices without our knowledge.** When a **vector-valued espression does not fully evaluate,** it is likely that some part of the expression has turned into a matrix. We should be alert for signs of this happening in our output. **When a vector-valued expression does not fully evaluate, use evalm( ) on the expression, and it should evaluate.**

The familiar symbols (+), (-) and (*) can be used to perform vector addition, subtraction, and scalar multiplication **as long as they are applied to "row vectors"** (or column vectors). These more natural operators **cannot be used on matrices.** (Actually, these operators *can* also be used on matrices, but the calculation must be followed by a use of the evalm( ) command in order to evaluate the expression. Furthermore, this "opening up" of the expression **should be done immediately, or evaluation mistakes can occur** if the expression is further manipulated.)

When vectors are represented as matrices, it is frequently useful to initiate **vector addition, subtraction and scalar multiplication with the commands matadd( ) and scalarmul( ).** If v and w represent the vectors $\vec{v}$ and $\vec{w}$, and a and b are scalars, then

$$\text{matadd}(v, w) = \vec{v} + \vec{w}, \ \ \text{matadd}(v, -w) = \vec{v} - \vec{w},$$

$$\text{scalarmul}(v, a) = a\vec{v}, \ \ \text{matadd}(v, w, a, b) = a\vec{v} + b\vec{w}.$$

Notice the positions of the vectors and scalars in the arguments of these commands. If you make a mistake, Maples's error messages should lead you in the right direction. In the subtraction operation above, the term $-w$ is regarded as the **additive inverse** of w, **not** as a **scalar multiple** of w.

It could be said that Maple wants its vectors to be matrices. It is probably "safer" to work with matrices, and to use these matrix commands to do our work. On the other hand, row vectors and the operator symbols (+), (-), and (*) are a genuine convenience, and so we will use them in most of our work. Occasionally, we may prefer to represent vectors as matrices and to use these matrix commands. Consequently, the matadd( ) and scalarmul( ) commands should not be discarded.

The next Maple work session starts by using commands from the linear algebra package before it is loaded. The command vector( ) resides in the linear algebra package, but it also appears to be one of Maple's main commands that loads automatically. This is not true, however, of the addition command matadd( ). Notice Maple's response when this command is used. Output such as this is clear evidence that Maple does not understand.

```
**********************************
> vr:=<8|-3|-5>; wr:=<13|-6|7>;
```

$$vr := [8,\ -3,\ -5]$$
$$wr := [13,\ -6,\ 7]$$

```
> u1r:=vr+wr;
```

$$u1r := [21,\ -9,\ 2]$$

```
> u2r:=3*vr;
```

$$u2r := [24,\ -9,\ -15]$$

```
> u3r:=3*vr-5*wr;
```

$$u3r := [-41,\ 21,\ -50]$$

```
> vm:=vector([8,-3,-5]); wm:=vector([13,-6,7]);
```

$$vm := [8,\ -3,\ -5]$$
$$wm := [13,\ -6,\ 7]$$

```
> u1m:=matadd(vm,wm);
```

$$u1m := \mathrm{matadd}(vm,\ wm)$$

#Look at the output. It is a clear indication that Maple does not understand the command matadd( ).#

```
> with(linalg):
Warning, the protected names norm and trace have been redefined and
unprotected
> u1m:=matadd(vm,wm);
```

$$u1m := [21,\ -9,\ 2]$$

```
> u2m:=scalarmul(vm,-3);
```

$$u2m := [-24,\ 9,\ 15]$$

```
> u3m:=matadd(vm,wm,3,-5);
```

$$u3m := [-41,\ 21,\ -50]$$

```
> u4m:=3*vm-5*wm;
```

$$u4m := 3\ vm - 5\ wm$$

Notice how the above expression is left unevaluated. This will happen whenever the (+), (-), or (*) operator is used over matrices.#

```
> evalm(u4m);
```

$$[-41,\ 21,\ -50]$$

#In the following input statements, notice the way in which the vector expression is left unevaluated. Because of this, it can be used to create the symbolic form of a vector

$$\vec{v} = a\vec{i} + \vec{j} + c\vec{k}$$

that we are accustomed to seeing in print, where $\vec{i}, \vec{j}, \vec{k}$ are the standard unit vectors in the x, y, z direction, respectively. If these standard vectors, $\vec{i}, \vec{j}, \vec{k}$, are used, they must be defined. The linear combination $3\vec{v} - 5\vec{w}$ is computed a second time below by using this environment, and as you can see, Maple performs admirably. #

```
> i:=vector([1,0,0]); j:=vector([0,1,0]); k:=vector([0,0,1]);
```

$$i := [1, 0, 0]$$
$$j := [0, 1, 0]$$
$$k := [0, 0, 1]$$

```
> v:=8*i-3*j-5*k; w:= 13*i-6*j+7*k;
```

$$v2 := 8\,i - 3\,j - 5\,k$$
$$w2 := 13\,i - 6\,j + 7\,k$$

```
> 3*v-5*w;
```

$$-41\,i + 21\,j - 50\,k$$

```
> evalm(%);
```

$$[-41, 21, -50]$$

****

Before we present Maple's command for the length of a vector, it will help if we first discuss some background. Mathematicians have many ways to measure the "size" of a vector besides the geometric notion of length. The word **"norm,"** in mathematics, is used to **represent any one of these "size" measurements.** In vector spaces of dimension greater than three, and in more abstract vector spaces, there is no sense of vision anyway—except one that is mathematically created. Maple is meant to deal with many vector spaces besides the familiar two- and three-dimensional spaces in this manual, and so it has a command for all of the standard norms (or sizes) of vectors in all of these spaces. **If v represents the vector $\vec{v} = a\vec{i} + b\vec{j} + c\vec{k}$** in a Maple work session, and if **n is a positive integer,** then

$$\mathbf{norm}(v, n) = \sqrt[n]{|a|^n + |b|^n + |c|^n}.$$

A similar formula can be expressed for norm(v, n) of a two-dimensional vector $\vec{v}$. The geometric notion of the length of $\vec{v}$ is therefore just norm$(v,2)$. This "2-norm" might well be the only norm we ever use in this manual, but, when we compute the length of a vector, it will usually be done through this command, and so we must **remember to insert the number 2 as its second argument.**

Example 9.1 *Find a unit vector in direction $\vec{v} = 13\vec{i} + 4\vec{j} - 9\vec{k}$. Find a vector of length 18 in direction $\vec{v}$.*

Often, the first step in solving a problem involving vectors is to turn a vector into a unit vector. This frequently makes the vector much easier to deal with. Scalar multiplication does not change the direction of a vector, and so **scalar multiplying a vector by the reciprocal of its length will turn it into a unit vector** with the same direction. This is such a basic vector operation that the concept should be completely understood before a more direct command is used. Once the principle is understood, then **the command normalize( ) is a slightly more convenient way to turn a vector into a unit vector.**

****

```
> with(linalg):
Warning, the protected names norm and trace have been redefined and
unprotected
> v:=<113|4|-9>;
```

$$v := [113, 4, -9]$$

```
> u:=(1/norm(v,2))*v;
```

$$u := \left[\frac{113}{12866}\sqrt{12866}, \frac{2}{6433}\sqrt{12866}, -\frac{9}{12866}\sqrt{12866}\right]$$

```
> check:=norm(u,2);
```

$$check := 1$$

```
> w:=(18/norm(v,2))*v;
```

$$w := \left[\frac{1017}{6433}\sqrt{12866}, \frac{36}{6433}\sqrt{12866}, -\frac{81}{6433}\sqrt{12866}\right]$$

```
> check:=norm(w,2);
```

$$check := 18$$

```
> u1:=normalize(v);
```

$$u1 := \left[\frac{113}{12866}\sqrt{12866}, \frac{2}{6433}\sqrt{12866}, -\frac{9}{12866}\sqrt{12866}\right]$$

```
> u1;
```

$$u1$$

#Noticed that the vector $u1$ does not fully evaluate—it has turned into a matrix. We mentioned in the introduction that this might happen without our knowledge.#

****

An unassigned name in Maple is always regarded as a real (or complex) number. To create the name of an unassigned vector we use the vector( ) command. In order to treat $\vec{C}$ as an arbitrary (unassigned) vector of, for example, dimension 3, we enter the input statement:

```
> C:=vector(3);
```

The three components of $\vec{C} = [C_1, C_2, C_3]$ can be extracted in the usual way with square brackets. The second component, for example, would be extracted with the input notation C[2], and the output would be appear as C_2 in a Maple worksheet. As unassigned names, C_1, C_2, and C_3 represent arbitrary complex-valued numbers (real numbers are also complex numbers), and this complex-value assumption can cause complications throughout Maple.

Getting Maple to perform calculations on each component of a vector can be a tedious experience if it is not done thoughtfully. This is especially true when vectors in higher-dimensional spaces are involved, but even vectors in three-dimensional space can be tedious to work with, unless an attempt is made to use Maple's tools effectively. **Any repetitive step should only have to be entered once** on a computer. This is probably true of any computer application. If you find yourself repeating the same calculations more than once in a problem, you can probably find a way of doing it all with just one set of instructions. Your work will be less tedious, and it will certainly look more elegant. Initially, it may require some effort to incorporate clever shortcuts in your work, but some early effort in this direction will make vectors easier to work with later.

The next example is included for two important reasons. It shows how a simple "for ..., do ..., od;" loop can be used to do repetitive work on the components of a vector. Later in this section, we will present **map( )**, a wonderful new Maple command for doing **multiple applications**. The map( ) command can frequently (though not always) be used, instead of a program loop, to do repetitive work on the components of a vector. Program loops will still be useful.

There is another reason for this example. The solution is needlessly awkward. It would not be presented at all, except that it is very natural for the problem to be set up in this way. A better approach involves a short "pencil and paper" strategy session prior to a Maple solution. This is shown at the end of the example.

Example 9.2 *Let $\vec{f}$ be the vector $\vec{f} = x^2\vec{\imath} + (x-4)^2\vec{\jmath} + (9-2x)\vec{k}$, and suppose that $\vec{p} = \vec{f} + \vec{C}$ for some unknown constant vector $\vec{C}$. Determine a $\vec{C}$ such that $\vec{p} = 18\vec{\imath} - 33\vec{\jmath} + 4\vec{k}$ when $x = 7$.*

****

```
> with(linalg):
```

```
Warning, the protected names norm and trace have been redefined and
unprotected
```

```
> f:=<x^2|(x-4)^2|9-2*x>;
```

$$f := [x^2,\ (x-4)^2,\ 9-2\,x]$$

```
> f7:=subs(x=7,f);
```

$$f7 := [49,\ 9,\ -5]$$

```
> C:=vector(3);
```

$$C := \mathrm{array}(1..3,\ [])$$

```
> p:=f+C;
```

$$p := C + [x^2,\ (x-4)^2,\ 9-2\,x]$$

```
> pp:=<18|-33|4>;
```

$$pp := [18,\ -33,\ 4]$$

```
> f7+C;
```

$$C + [49,\ 9,\ -5]$$

```
> p7:=matadd(f7,C);
```

$$p7 := [49 + C_1,\ 9 + C_2,\ -5 + C_3]$$

```
> for j from 1 to 3 do C[j]:=solve(p7[j]=pp[j],C[j]); od; j:='j':
```

$$C_1 := -31$$
$$C_2 := -42$$
$$C_3 := 9$$

```
> evalm(p);
```

$$[-31 + x^2,\ -42 + (x-4)^2,\ 18 - 2\,x]$$

#There is a better way to solve this problem. In the above, we defined, as a Maple assignment, $\vec{p}$ in terms of an unknown vector $\vec{C}$, and then solved for $\vec{C}$. Maple, unfortunately, forced us to determine $\vec{C}$ component by component. A brief "pencil-and-paper" strategy session might suggest an easier approach. Because $\vec{p} = \vec{f} + \vec{C}$ for every x, and because $\vec{p}(7)$, and $\vec{f}(7)$ are known, it follows that $\vec{C} = \vec{p}(7) - \vec{f}(7)$. Thus after the setting up of the vectors $\vec{f7}$, and $\vec{pp}$, as we did above, the following brief work session would finish the problem. Certainly, the following approach is easier.#

```
> pp=f7+C;
```

$$[18, -33, 4] = C + [49, 9, -5]$$

```
> C:=pp-f7;
```

$$C := [-31, -42, 9]$$

```
> p;
```

$$[-31 + x^2, -42 + (x - 4)^2, 18 - 2x]$$

****

Each Maple work session in this chapter began by loading the linear algebra package. Hopefully, this is now recognized as an **automatic first step to perform vector operations** in Maple.

9.3 The Dot and Cross Products

The two most interesting vector operations are the dot product and the cross product. **The Maple commands dotprod( ) and crossprod( )** are very straightforward to use.

The important geometric rule

$$\vec{v} \cdot \vec{w} = |\vec{v}||\vec{w}| \cos(\theta)$$

can be used to determine the angle, θ, between $\vec{v}$ and $\vec{w}$. Undoubtedly, this is what Maple uses to define its own command, **angle( ), for finding the angle, θ, between $\vec{v}$ and $\vec{w}$.** There is a similar (and equally important) rule

$$|\vec{v} \times \vec{w}| = |\vec{v}||\vec{w}| \sin(\theta)$$

for cross products, but it is generally not a good idea to use this rule to determine the angle, θ, between $\vec{v}$ and $\vec{w}$. Finding the angle θ in this manner means writing this equation in the form $\sin(\theta) = p$ for θ, where $p \geq 0$ is the obvious product and quotient of lengths of vectors. **For every such p, however, there are two angles θ in the interval $0 \leq \theta \leq \pi$ that satisfy this equation.** Only one of these angles is an appropriate answer, and this is the root of the problem. A few sample calculations using these commands are shown below.

In the first part of our Maple work session, we discuss a **surprising complication involving the dot product of arbitrary vectors**. Maple delivers two versions of the dot product. For vectors having real-valued components, the two versions are equivalent, but for vectors with complex-valued (and non-real) components, they are different. When we use unassigned names, Maple assumes they represent complex numbers, and this assumption causes an undesirable evaluation of the dot product when we work with arbitrary vectors.

Maple's default version of the dot product is the standard one mathematicians use for vectors over the complex number system. Our work session begins with this definition.

****

```
> with(linalg):
```

```
Warning, the protected names norm and trace have been redefined and
unprotected
```

#We create two arbitrary vectors and then compute Maple's default version of the dot product.#

```
> v:=vector(3): w:=vector(3):
> dotprod(v,w);
```

$$v_1\,\overline{(w_1)} + v_2\,\overline{(w_2)} + v_3\,\overline{(w_3)}$$

#If the components of $\vec{v}$ and $\vec{w}$ are real numbers, the above expression would be the same as the familiar dot product, but the conjugation operations would be exceedingly awkward to deal with in our work. It is essential that we avoid this complication. By inserting the option "orthogonal", we can avoid this complication.#

```
> dotprod(v,w,orthogonal);
```

$$v_1\,w_1 + v_2\,w_2 + v_3\,w_3$$

This optional argument only has to be used when we work with arbitrary vectors having real-valued components. When we use it, the output will be in a useful form, but it is still awkward to be forced to insert this optional term in our imput statements. **There is an easy way out of this dilemma. We could create our own dot product.**

```
> dot:=(u,v)->dotprod(u,v,orthogonal);
```

$$dot := (u,\, v) \rightarrow \mathrm{dotprod}(u,\, v,\, orthogonal)$$

With this new personal command, we can easily compute any dot product. If $\vec{p}$ and $\vec{q}$ represent any two (arbitrary or specific) vectors over the real number system, then $\vec{p}\cdot\vec{q}$ is computed as follows.#

```
> p:=vector(3):q:=vector(3):
> dot(p,q);
```

$$p_1\,q_1 + p_2\,q_2 + p_3\,q_3$$

Make it a practice to begin every Maple work sheet involving vectors with real-valued components by loading the linear algebra package and defining our personal dot-product command. If we always use our personal dot-product command (even if we don't have to use it), then we will not be surprised by the sudden appearance of awkward conjugation symbols.

Let us now continue with a sample of calculations. To demonstrate, at least once, that our new personal dot-product command is not needed if we work with specific vectors, we use the dot-product command without the optional argument for our initial work.#

```
> v:=<2|-7|3>: w:=<8|1|-6>:
> dotprod(v,w);
```

$$-9$$

```
> crossprod(v,w);
```

$$[39, 36, 58]$$

```
> theta:=evalf(arccos(dotprod(v,w)/(norm(v,2)*norm(w,2))));
```
#This angle is in radian measure.#

$$\theta := 1.684775822$$

```
> alpha:=evalf(angle(v,w));
```
#Here is Maple's command for computing the same angle.#

$$\alpha := 1.684775822$$

#With Maple, it is easy to show that the cross product of two vectors in three-dimensional space is just the determinant of a certain 3×3 matrix. This is the familiar rule that we use to compute cross products by hand. We start with Maple's answer for computing, directly, the cross product of $\vec{v}$ and $\vec{w}$.#

```
> v:=vector(3):w:=vector(3):
> crossprod(v,w);
```

$$[v_2\, w_3 - v_3\, w_2,\ v_3\, w_1 - v_1\, w_3,\ v_1\, w_2 - v_2\, w_1]$$

#The command for computing the determinant of a matrix is det( ). Neither it nor the matrix command **matrix( )** will be used very much in this manual. Notice below, in the verification of a cross product as a determinant, that the vectors $\vec{i}, \vec{j}, \vec{k}$ are represented by unassigned names i, j, k. Maple has no idea that they represent the three unit coordinate vectors. Remember that unassigned names are regarded as real or complex numbers by Maple. Let's not tell Maple what we mean by i, j, k. It is enough to just hold on to this meaning in our minds.

The rows of a matrix must be entered as "lists," and so we have to **convert the vectors to lists** before we create our matrix.#

```
> r2:=convert(v,list);
```

$$r2 := [v_1, v_2, v_3]$$

```
> r3:=convert(w,list);
```

$$r3 := [w_1, w_2, w_3]$$

```
> A:=matrix([[i,j,k],r2,r3]);
```

$$A := \begin{bmatrix} i & j & k \\ v_1 & v_2 & v_3 \\ w_1 & w_2 & w_3 \end{bmatrix}$$

```
> det(A);
```

$$i\,v_2\,w_3 - i\,v_3\,w_2 - v_1\,j\,w_3 + v_1\,k\,w_2 + w_1\,j\,v_3 - w_1\,k\,v_2$$

```
> collect(%,{i,j,k});
```

$$(v_3\,w_1 - v_1\,w_3)\,j + (v_1\,w_2 - v_2\,w_1)\,k + i\,(v_2\,w_3 - v_3\,w_2)$$

****

Some of the algebraic rules governing the behavior of dot products and cross products are easy to justify. The dot product, for example, is obviously a commutative operation. That is to say, $\vec{u}\cdot\vec{v} = \vec{v}\cdot\vec{v}$, for any vectors $\vec{u}$ and $\vec{v}$. The less obvious rules can be proved quite easily by using Maple. One such rule is the *Distributive Law of Cross Product over Addition.* Its proof, and those of several other algebraic rules, are interesting problems that appear in the exercise set at the end of the chapter.

In the next Maple work session, we explore the proof of the algebraic rule $a(\vec{u}\times\vec{v}) = (a\vec{u})\times\vec{v}$.

Along the way, we will use **a significant new Maple command. The command map( ) is Maple's <u>m</u>ultiple <u>app</u>lication command. If *comd*() is a Maple command and *obj* is a Maple object (an expression, function, list, etc.), then**

```
> map(comd,obj);
```

applies the command *comd*() to each <u>operand</u> of *obj* and expresses the result as an object of the same data type as *obj*. If the command *comd*() takes **more than one argument** then these arguments appear as **additional arguments in the map( ) command**, starting with the third argument. Notice the arrangement of parentheses. **There is no left parenthesis after the name** *comd.* Finally, we underlined the word "operand" in the above definition. This word was discussed back on page 125. For example,

$$\text{if } a = \textstyle\sum_{j=1}^{10} a_j\text{,then map(factor,}a\text{)} = \sum_{j=1}^{10}\text{factor}(a_j),$$

$$\text{if } y = seq(y_j, j = 1..20)\text{, then map(diff,}y, x\text{)=seq(diff}(y_j, x), j = 1..20).$$

Many of Maple's commands operate only on real- (or complex-) valued expressions and consequently do not operate directly on vectors. This is where the map( ) command will prove to be useful. For example, if $v = [v_1, v_2, v_3]$ is a vector, then map(expand,v) = [expand(v_1),expand(v_2),expand(v_3)].

Many other commands will have to be used in a similar way in order to have them operate over vectors. The map( ) command resides in Maple's main environment. It is not just a vector command, but it will certainly prove to be useful in a wide variety of vector computations. It can frequently (though not always) be used instead of a more complicated "for..., do..., od;" loop, to do repetitive applications of a command. The map( ) command could have been used in Example 9.2 instead of the program loop to evaluate the vector $\vec{C}$.

Now we're ready to prove the rule $a(\vec{u} \times \vec{v}) = (a\vec{u}) \times \vec{v}$.

****

```
> with(linalg):

Warning, the protected names norm and trace have been redefined and
unprotected
> u:=vector(3):  v:=vector(3):
> L:=scalarmul(crossprod(u,v),a);
```

$$L := [a\,(u_2\,v_3 - u_3\,v_2),\ a\,(u_3\,v_1 - u_1\,v_3),\ a\,(u_1\,v_2 - u_2\,v_1)]$$

```
> R:=crossprod(scalarmul(u,a),v);
```

$$R := [a\,u_2\,v_3 - a\,u_3\,v_2,\ a\,u_3\,v_1 - a\,u_1\,v_3,\ a\,u_1\,v_2 - a\,u_2\,v_1]$$

Looking at the above expressions for $\vec{L}$ and $\vec{R}$, we can see that they are the same. Such conclusions are not always so obvious, however, so let us demonstrate some methods we could use to show that $\vec{L} = \vec{R}$.

```
> z:=matadd(L,-R); #This should be the zero vector.#
```

$$z := \big[a\,(u_2\,v_3 - u_3\,v_2) - a\,u_2\,v_3 + a\,u_3\,v_2,\ a\,(u_3\,v_1 - u_1\,v_3) - a\,u_3\,v_1 + a\,u_1\,v_3,\\ a\,(u_1\,v_2 - u_2\,v_1) - a\,u_1\,v_2 + a\,u_2\,v_1\big]$$

#The simplify( ) command is one of the commands that was redesigned in later versions of Maple to operate on vectors. #

```
> simplify(z);
```

$$z$$

#It appears that the vector $\vec{z}$ is too "closed." We open it up by using the evalm( ) command.#

```
> simplify(evalm(z));
```

$$[0, 0, 0]$$

This means that $\vec{L} = \vec{R}$, which proves the algebraic rule. Another way to **show equality of two vectors is to use the equal( ) command.#**

```
> equal(L,R);
```

$$false$$

#Why do we get this response? We already know that the answer should be *true*. This is more fuel to fan the fires of healthy skepticism! We could try to expand the vector $\vec{L}$ ($\vec{R}$ is already in expanded form).#

```
> L1:=expand(evalm(L));
```

The output is omitted—the command has no effect on $\vec{L}$. **A response like this on a vector should always prompt us to try the map( ) command.**

```
> L1:=map(expand,L);
```

$$L1 := [a\, u_2\, v_3 - a\, u_3\, v_2,\ a\, u_3\, v_1 - a\, u_1\, v_3,\ a\, u_1\, v_2 - a\, u_2\, v_1]$$

```
> equal(L1,R);
```

$$true$$

****

We used a simple strategy in the above work session, which deserves more attention. It can be visually taxing to see that two complicated vectors are equal to each other from just looking at them. We used an approach that was easier on our eyes. **To show that $\vec{L} = \vec{R}$, just show, instead, that $\vec{R} - \vec{L} = \vec{0}$. This approach can be used on almost any two Maple objects (vectors or otherwise) to show their equivalence. This is worth remembering!** We also saw in the last work session, that **Maple has a special command equal( ) for showing equality of vectors.** The command **equal( ) can only be used to compare vectors.**

Some of the familiar algebraic rules that hold for addition and multiplication of real numbers do not hold in vector algebra. **Cross product, for example, is not an associative operation.** In the next Maple work session, we investigate this curious behavior in the cross product. Almost any combination of three vectors $\vec{u}$, $\vec{v}$, and $\vec{w}$, will show that $\vec{u}\times(\vec{v}\times\vec{w})$ is different from $(\vec{u}\times\vec{v})\times\vec{w}$. (Some combinations satisfy associativity, but the chances are remote that a random choice would produce such a combination.)

****

```
> with(linalg):

Warning, the protected names norm and trace have been redefined and
unprotected
> u:=<2|-5|1>:  v:=<-1|8|2>:  w:=<6|1|3>:
> crossprod(u,crossprod(v,w));
  crossprod(crossprod(u,v),w);
```

$$[230, 120, 140]$$

$$[-26, 120, 12]$$

****

This proves that cross product is not associative. What, then, can we say about these two triple cross products? Among other things, the vector $(\vec{u}\times\vec{v})\times\vec{w}$ is perpendicular to $\vec{u}\times\vec{v}$, and so it is in the plane formed by $\vec{u}$ and $\vec{v}$ (if all of the vectors are based at the same point). This means that it is a linear combination of $\vec{u}$ and $\vec{v}$. Given arbitrary vectors, $\vec{u}$, $\vec{v}$, and $\vec{w}$, Maple can be used to find scalars a and b such that $(\vec{u}\times\vec{v})\times\vec{w} = a\vec{u}+b\vec{v}$. This and other interesting properties of this triple cross product will be explored in the exercises.

The operations we have been discussing in the last two sections can be used effectively in a large variety of mathematical problems. A sample of these standard mathematical problems are considered in the next section. The operations can also be used to solve very interesting and impressive applied problems. Some of these appear in the exercises at the end of this chapter. In particular, Captain Ralph reappears after a long vacation. Now that we are working in three-dimensional space, we will have the opportunity to accompany Captain Ralph more frequently on his adventures.

We end this section with one standard mathematical application. Frequently, vectors need to be decomposed into a sum of two vectors, one which is parallel to a given direction, the other perpendicular. For example, if $\vec{f}$ is a force vector being applied to an object traveling along a curve C, then it would make sense to express $\vec{f}$ in the form

$$\vec{f} = \vec{f_t} + \vec{f_n},$$

where $\vec{f_t}$ is tangent to the curve C, and $\vec{f_n}$ is normal to the curve C.

Example 9.3 *Express the vector* $\vec{v} = 13\vec{i} - 32\vec{j} + 7\vec{k}$ *as the sum of a vector* $\vec{w}_1$ *parallel to* $\vec{w} = -8\vec{i} + 9\vec{j} + 2\vec{k}$ *and a vector* $\vec{w}_2$ *perpendicular to* $\vec{w}$.

We project $\vec{v}$ onto $\vec{w}$ to get $\vec{w}_1$. Once we have $\vec{w}_1$, the vector $\vec{w}_2$ is easy to get from $\vec{v} = \vec{w}_1 + \vec{w}_2$ by subtraction.

Calculus texts always have a formula for the projection vector, but there is no need to memorize this formula. We first turn $\vec{w}$ into a unit vector $\vec{u}$—this is the direction of $\vec{w}_1$, or its opposite direction. Then $\vec{w}_1 = L\,\vec{u}$, where L is either the length of the projection vector or its negative (depending on whether $\vec{w}_1$ and $\vec{u}$ have the same or opposite directions). Elementary trigonometry easily implies that $L = |\vec{v}|\cos(\theta)$, where θ is the angle between $\vec{v}$ and $\vec{u}$. This number L automatically has the right sign as well as the right size.

****

```
> with(linalg):
```

Warning, the protected names norm and trace have been redefined and unprotected

```
> v:=<13|-32|7>:  w:=<-8|9|2>:
> u:=normalize(w);
```

$$u := \left[-\frac{8}{149}\sqrt{149}, \frac{9}{149}\sqrt{149}, \frac{2}{149}\sqrt{149}\right]$$

```
> u;
```

$$u$$

#The normalize( ) command above must have turned our row vector into a matrix, because the vector u did not fully evaluate.

In the next calculation, we use the name "COS" (it's just a name) to suggest $\cos(\theta)$, where θ is the angle between the unit vector $\vec{u}$ and $\vec{v}$. #

```
> COS:=dotprod(u,v)/norm(v,2);
```

$$COS := -\frac{21}{3427}\sqrt{149}\sqrt{138}$$

```
> w1:=evalm(norm(v,2)*COS*u);
```

$$w1 := \left[\frac{3024}{149}, \frac{-3402}{149}, \frac{-756}{149}\right]$$

```
> w2:=v-w1;
```

$$w2 := -w1 + [13, -32, 7]$$

```
> w2:=evalm(w2);
```

$$w2 := \left[\frac{-1087}{149}, \frac{-1366}{149}, \frac{1799}{149}\right]$$

Just to be sure, we check to see that $\vec{v}$ is the sum of $\vec{w1}$ and $\vec{w2}$. Surely $\vec{w1}$ will be parallel to $\vec{w}$, but we check to see that $\vec{w2}$ is perpendicular to $\vec{w}$. Both $w1$ and $w2$ are now matrices, so we use the matadd( ) command.#

```
> evalm(w1+w2);dotprod(w2,w);
```

$$[13, -32, 7]$$

$$0$$

****

9.4 Lines and Planes

In the previous section, we discussed a difficulty that can be encountered when the dotprod( ) command is applied to arbitrary vectors. One way to avoid the problem is to use our own dot-product command. Starting with this section, we will make a practice of defining this special dot product every time we open Maple. We might not always need it, but, when we do, we will not be surprised by unanticipated complications.

Example 9.4 *Find the equation of the plane through the points* $P(2,-5,-4)$, $Q(-1,6,7)$, $R(9,5,-6)$.

Mathematically, a point and a vector in three-dimensional space are equivalent. They are both just ordered triplets of numbers. The only difference is a point of view. This is frequently an advantage, but it can also cause confusion. In this one example, we will regard a point as a Maple "list" and a vector as a Maple "vector." We have discussed the difference between these two data types; hopefully, it can be used to help end this confusion. You might want to use this practice in your own work, if it helps.

After this example, we will simply treat both points and vectors, right from the start, as vectors.

****

```
> with(linalg):dot:=(u,v)->dotprod(u,v,orthogonal):
```

```
Warning, the protected names norm and trace have been redefined and
unprotected
```

```
> P:=[2,-5,-4]; Q:=[-1,6,7];R:=[9,5,-6];
```

$$P := [2,\ -5,\ -4]$$
$$Q := [-1,\ 6,\ 7]$$
$$R := [9,\ 5,\ -6]$$

```
> p:=<2|-5|-4>; q:=<-1|6|7>;r:=<9|5|-6>;
```

$$p := [2,\ -5,\ -4]$$
$$q := [-1,\ 6,\ 7]$$
$$r := [9,\ 5,\ -6]$$

```
> n:=crossprod(q-p,r-p);
```

$$n := [-132,\ 71,\ -107]$$

The point $X(x,y,z)$ is in the plane if and only if the vector from P to X is orthogonal to the normal vector $\vec{n}$.

```
> v:=<x|y|z>;
```

$$v := [x,\ y,\ z]$$

```
> eq:=dot(v-p,n)=0;
```

$$eq := -132\,x + 191 + 71\,y - 107\,z = 0$$

****

Example 9.5 *Find the distance between the line through* $A(4,-3,8)$, $B(-1,9,5)$ *and the line through* $C(5,7,-2)$, *and* $D(13,2,-6)$.

Imagine a plane which contains one of the lines and is parallel to the second line. Then the distance in question is just the perpendicular distance from a point (any point on the second line) to the plane. Most calculus books provide a formula for the distance from a point Q to a plane, but it is fairly easy to see geometrically. Just take any vector from a point in the plane to Q, project it onto the plane's normal vector, and compute the length of the projection.

This problem can also be done by minimizing a certain function of two variables, namely the function giving the distance between a variable point on one line and a variable point on the other line. We will return to this example in our study of functions of several variables.

****

```
> with(linalg):dot:=(u,v)->dotprod(u,v,orthogonal):
Warning, the protected names norm and trace have been redefined and
unprotected
> a:=<4|-3|8>:  b:=<-1|9|5>:  c:=<5|7|-2>:  d:=<13|2|-6>:
> u:=a-b; v:=c-d;
```

$$u := [5, -12, 3]$$

$$v := [-8, 5, 4]$$

```
> n:=crossprod(u,v);
```

$$n := [-63, -44, -71]$$

#A plane parallel to one line and containing the other has normal vector $\vec{n}$. To find the distance, we project the vector from A to C onto $\vec{n}$ and compute its length.#

```
> d:=abs(dot(c-a,n)/norm(n,2));
```

$$d := \frac{207}{10946}\sqrt{10946}$$

```
> d:=evalf(d);
```

$$d := 1.978529926$$

****

Example 9.6 *Find an equation for the line of intersection between the plane through* $A(5,1,-8)$, $B(-6,2,7)$, $C(3,9,1)$, *and the plane through* $P(-8,1,1)$, $Q(4,-7,-3)$, *and* $R(-5,-7,-1)$.

We will discuss Maple's three-dimensional plotting devices in the next section. They could be used to see the line of intersection, but they contribute little to this problem.

There are two basic ways of proceeding. In one approach, the equation of the line is found by solving a system of two equations in three unknowns. This problem is a part of a more general problem involving m equations in n unknowns, and it is one of the main topics of discussion in a linear algebra course, usually taken soon after a calculus sequence.

The approach we take is more geometrical. Let $\vec{n_1}$ and $\vec{n_2}$ be normal vectors for each of the two planes. The line of intersection lies in both planes, and so it is orthogonal to both $\vec{n_1}$ and $\vec{n_2}$. It follows that the vector $\vec{v} = \vec{n_1} \times \vec{n_2}$ is a direction vector for the line of intersection. All we need, then, is a point of intersection of the two lines.

****

```
> with(linalg):dot:=(u,v)->dotprod(u,v,orthogonal):

Warning, the protected names norm and trace have been redefined and
unprotected
> a:=<5|1|-8>:  b:=<-6|2|7>:  c:=<3|9|1>:
> p:=<-8|1|1>:  q:=<4|-7|-3>:  r:=<-5|-7|-1>:
> ab:=b-a; ac:=c-a; pq:=q-p; pr:=r-p;
```

$$ab := [-11, 1, 15]$$
$$ac := [-2, 8, 9]$$
$$pq := [12, -8, -4]$$
$$pr := [3, -8, -2]$$

```
> n1:=crossprod(ab,ac); n2:=crossprod(pq,pr);
```

$$n1 := [-111, 69, -86]$$
$$n2 := [-16, 12, -72]$$

```
> v:=crossprod(n1,n2);
```

$$v := [-3936, -6616, -228]$$

The vector $\vec{v}$ is the direction vector for the line. We discussed this in the introduction to this Maple work session.

```
> X:=<x|y|z>:
> ABC:=dot(X-a,n1)=0; PQR:=dot(X-p,n2)=0;
```

$$ABC := -111\,x - 202 + 69\,y - 86\,z = 0$$
$$PQR := -16\,x - 68 + 12\,y - 72\,z = 0$$

#All we need is to find one point on the line of intersection. We have two equations and three unknowns, so we can assign any number to one of the variables and solve for the remaining two variables. #

```
> X:=<x|y|z>:
> ABC:=dot(X-a,n1)=0; PQR:=dot(X-p,n2)=0;
```

$$ABC := -111\,x - 202 + 69\,y - 86\,z = 0$$
$$PQR := -16\,x - 68 + 12\,y - 72\,z = 0$$

```
> x:=-2:
> solve({ABC,PQR},{y,z});
```

$$\{z = \frac{-227}{328}, y = \frac{-189}{164}\}$$

```
> assign(%);
> X;
```

$$\left[-2, \frac{-189}{164}, \frac{-227}{328}\right]$$

```
> line:=X-t*v;
```

$$line := -t\,v + \left[-2, \frac{-189}{164}, \frac{-227}{328}\right]$$

```
> line:=evalm(line);
```

$$line := \left[3936\,t - 2,\ 6616\,t - \frac{189}{164},\ 228\,t - \frac{227}{328}\right]$$

Finally, the parametric equation of the line can be defined as follows. The punctuation marks around the letters x, y, and z are needed in this input statement. Why?

```
> 'x'=line[1],'y'=line[2],'z'=line[3];
```

$$x = 3936\,t - 2,\ y = 6616\,t - \frac{189}{164},\ z = 228\,t - \frac{227}{328}$$

****

Projection vectors were used frequently throughout this section, and we always went back to basic vector work to construct the projections. This is a good learning device, but eventually it is more efficient to **simply use the proc( ) command to create a special projection command.** This is a problem in the exercise set. Once such a command is created, it can be used to easily compute projection vectors.

If you create your own personal projection command (or any other personal command), and wish to use it on another worksheet, it would, of course, have to be entered as an input line on the new worksheet. Personal commands like this could be **saved on a separate file that you could regard as your "library." Be careful how you name such a file,** however. Stay away from names like "lib" and "maple.lib," which are used by Maple.

Such a file could also be **saved and stored using Maple's internal code**, or it could be **saved as a package** using Maple's internal code. This code is not "human readable," but it is a fast and effective way to get information onto a worksheet. A personal "package" would be used like any other package in Maple. To create a file in Maple's internal code, end the file name with the suffix ".m." A file named *"mylib"* would be saved in Maple's internal code if it were saved under the named *mylib.m.*

The use of Maple's internal code and the creation of special packages is presented only as a suggestion to someone who has an extraordinary interest in Maple, and the time to experiment with this idea. It is not at all a necessary part of this manual, nor is it important for a better understanding of mathematics. For those who wish to pursue this idea, however, look up the key words: "save, packages, Maples internal code" in a Maple Reference Manual or in Maple's on-line Help file. This should get you started on the right path.

9.5 Surfaces in Space

The graph of an equation in three variables is a surface in three-dimensional space. The graph of a function f of two variables x and y, is the graph of the equation $z = f(x, y)$. Traditionally, the letters x and y are used as independent variables and z is used as the dependent variable, but Maple has no particular preference for one letter choice over another. Maple has a variety of plotting commands for functions, expressions, and equations. This would be a good time to load the *plots* package with a semicolon (;) instead of a colon (:) at the end of the input statement, so that the names of all the plotting commands can be viewed. At this point, we will introduce two of the main plotting commands. A few others will be introduced towards the end of this section.

Maple's main three-dimensional plotting tool, **plot3d( )**, does not reside in the *plots* package. It is automatically loaded, like all of the other "central" commands, whenever Maple is activated. Typical input statements using this command would look like:

```
> plot3d(f,a..b,c,,d);
```

```
> plot3d(p(x,y),x=a..b,y=c,,d);
```

where ***f* is a function and *p(x, y)* is an expression in *x* and *y*.** The letters a, b, c, d in the ranges $a..b, c..d$, and $x = a..b, y = c..d$ represent given real numbers, but surprisingly enough, there is no requirement that $a < b$ or that $c < d$. This might seem odd, but it adds a measure of convenience to the plotting environment. **Be careful with the ranges. In the first input statement above, the variable names cannot appear in the range. In the second statement, the variables—in this case, *x* and *y*—must appear in the range.**

In the second plot statement, the plotting domain does not have to be a rectangle. In the second range, $c = c(x)$, and $d = d(x)$ can also be expressions that depend on the first variable x. This is one reason why it is a convenience to allow ranges to go from larger to smaller numbers.

The plot3d( ) command cannot be used to plot an equation. Consequently, even if f is a function, the first argument in the plot3d( ) command cannot be $z = f(x, y)$. The equality symbol (=) makes this an equation.

The Maple command for **plotting an equation in three variables is called implicitplot3d( )**, and it resides in the **plots package**. Remember that an **equation is a statement involving the equality (=) symbol.** A typical input statement would look like

```
> implicitplot3d(eqn,x=a..b,y=c..d,z=p..q);
```

where *eqn* is an equation in x, y, and z.

All three variables appear in the ranges. The variables in the first two ranges become horizontal variables and the **variable in the third range becomes the vertical variable.** Any choice of letters other than the traditional x, y, z can be used—Maple has no preference. In the above plot statement, the x-axis and y-axis will be horizontal and the z-axis will be vertical.

There is a **large number of options** that can be used along with any of the three-dimensional plot commands. After a plot command has been entered, and a plot appears on a worksheet, **put the mouse pointer on the plot and click. A rectangular border will appear around the plot, and a large battery of plot option buttons will appear at the top of the screen.** Experiment with each button, so that its function is understood. Other options are available by selecting items in the menu-bar. A large number of additional options (beyond button and menu items) are available as *names* that can be entered as arguments inside a plot command. **Take a look at all of these options.** Pull down the **Help Menu**, select **Full Text Search...**, and type **"plot options"** in the word(s) field.

Finally, before we attempt any Maple activity, realize that **plotting is a memory-intensive activity**. It is easy to run out of memory before a plot request is finished. **Maple might let you save before it forces you to quit, or it could force you to quit without saving!** It is a good practice, especially when using plotting commands, to **save your file before you enter a request for a memory-intensive activity**. The amount of memory that has been used is shown at the right-hand end of the Status Line, which is located at the bottom of the computer screen.

Maple's plotting commands are effective tools for understanding mathematics and solving problems. Now, let us experiment with these commands. The plots given below should be experienced first-hand, and so the plot figures have not been included. **Try to predict the basic shape of the surface before you enter the plot request—this is always a good experience.**

****

```
> f:=(x,y)->(9-x^2-y^2)*cos(abs(x)+abs(y)):
> plot3d(f(x,y),x=-10..10,y=-10..10);
```

To get a different view point, use the mouse to move the pointer to the plot window, and click. Then **hold the mouse button down and drag** the pointer. To get a cut-away view, enter a plot request with a different range.#

```
> plot3d(f(x,y),x=-10..10,y=0..10);
```

#To plot two or more expressions in the same three-dimensional coordinate system, we use a standard Maple practice. When two or more objects must be placed in the same argument of a command, group them together by using curly brackets.#

```
> g:=(x,y)->x^2+y^2-400:
> plot3d({f(x,y),g(x,y)},x=-10..10,y=-10..10);
```

#By now, if you are following along on a computer, you could be low on memory. If so, you will need to quit and restart to recover spent memory. Keep an eye on the memory window in the lower left-hand corner of the worksheet.

The implicitplot3d command is a particularly memory-hungry command. If it is possible, it is best to avoid using it, especially on a slow computer or one that has insufficient RAM. There are frequently alternative ways of getting such a plot.#

```
> with(plots):

Warning, the name changecoords has been redefined
> implicitplot3d(x^2+z^2+y^2/25=1,x=-2..2,y=-5..5,z=-2..2);
```

#The variable in the third range determines the axis that will be vertical.#

****

Example 9.7 *Draw the region bounded by the graphs of*

$$f(x,y) = 9 - \frac{y^2}{16} - \frac{x^2}{84}, \text{ and } g(x,y) = y^2 + \frac{x^2}{64} + 5.$$

There are some interesting ideas that come from this simple example. To plot both functions together, we group them together with curly brackets inside the plot command. What ranges should we use for x and y? If a standard range of $-10..10$ is used for each variable, the region bounded by the two surfaces will not quite fit in one direction, but it will still be **dwarfed by the large surfaces involved**. By a "hit or miss" procedure, these ranges could be adjusted to eventually produce a fairly large x range and a fairly small y range. However, each attempt to plot the region might require a long wait for the plot if a slow computer is being used, so this could be called a "hit, wait, and miss" procedure.

A better approach would be to look at the curve $f(x, y) = g(x, y)$, and choose a range on x and y so that the corresponding rectangle in the x, y-plane just covers this curve and its interior. After all, the region under consideration surely consists of all points $P(x, y, z)$ with $g(x, y) \leq z \leq f(x, y)$.

****

```
> f:=(x,y)->9-y^2/16-x^2/81:  g:=(x,y)->y^2+x^2/64+5:
> eq:=f(x,y)-g(x,y)=0;
```

$$eq := 4 - \frac{17}{16}y^2 - \frac{145}{5184}x^2 = 0$$

```
> a:=fsolve(subs(y=0,eq),x); b:=fsolve(subs(x=0,eq),y);
```

$$a := -11.95854910, 11.95854910$$

$$b := -1.940285000, 1.940285000$$

```
> a:=a[2]:  b:=b[2]:
> plot3d({f(x,y),g(x,y)},x=-a..a,y=-b..b);
```

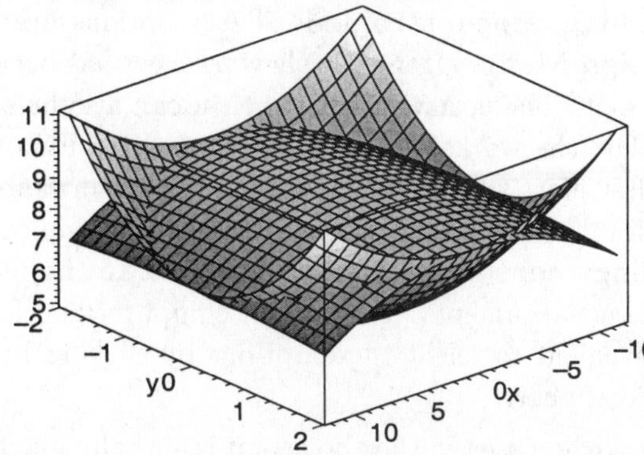

#The above plot adequately shows the region, especially if suitable options are used to enhance the plot, but **we can do much better**! Both surfaces are plotted over a rectangle in the x, y-plane, and so there are overlapping portions of both surfaces appearing in the plot that are not a part of the region under consideration.

What is really desired is not a plot of both surfaces over this rectangle, but rather a plot of both surfaces over the ellipse $f(x,y) - g(x,y) = 0$ and its interior.#

```
> s:=solve(eq,y);
```

$$s := \frac{1}{306}\sqrt{352512 - 2465\,x^2},\ -\frac{1}{306}\sqrt{352512 - 2465\,x^2}$$

```
> plot3d({f(x,y),g(x,y)},x=-a..a,y=s[1]..s[2]);
```

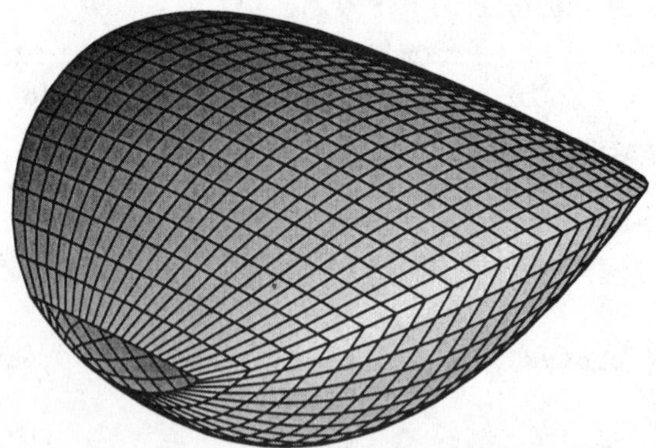

#We consider one last enhancement of the above plot. When Maple makes a two- or three-dimensional plot, it scales the axes of the dependent variables so that the graph fits nicely in the window. It takes even further liberties on three-dimensional plots to produce a nice fit. Maple feels **unconstrained** if it has the artistic liberty to size a plot to its own liking. If, on the other hand, Maple is forced to treat scales rigorously, so that the plot is like a photographic copy of the actual object, then Maple feels **constrained.** These are **"scaling" options**, and Maple's default setting is **scaling=unconstrained.** These options are available as menu items in the **Projection Menu**, after a Maple worksheet has been put into its "plot mode" by clicking a plot. The **constrained option** can also be selected by clicking the **[1:1] button** while the worksheet is in its plot mode. Finally, this option can be selected on the first plot, which avoids waiting for an unwanted plot to appear, making button choices, and waiting again for the desired plot to appear. Just add the option **scaling=constrained** as an argument to the plot3d( ) command, placing it after the range arguments. With a fast computer and lots of memory, it is probably just as convenient to select these options by clicking buttons and waiting for a second graph to appear.

The round and fat region seen in the above plot is actually much longer and leaner than it appears. **Click on the picture to bring up the plotting menu-bar. Click the [1:1] button in the menu-bar** to produce a "true-scale, photographic" picture of the actual region. In the picture shown below, we also clicked the menu button that looks like the the border of a square. #

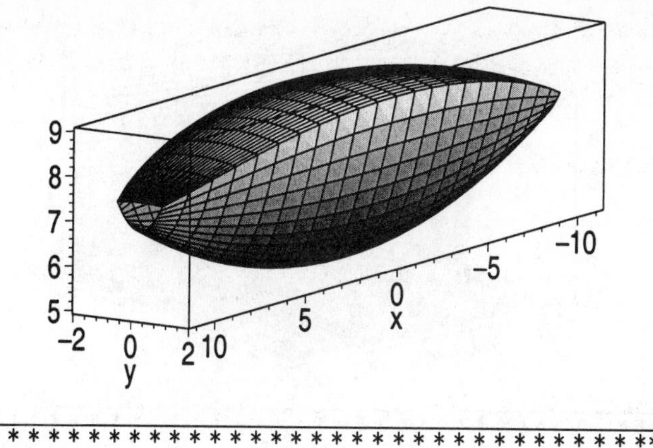

****

Example 9.8 *Plot the region bounded by the surfaces $y = 9 - x^2$, $99z - 10y = 900$, and the xy-plane.*

The xy-plane is just the graph of the equation $z = 0$. We could plot these three equations inside one implicitplot3d( ) command, using curly brackets. This command, however, uses so much memory that it is best to avoid using it if possible. The second equation could be expressed in the form $z = f(x, y)$, and we could then plot the second and third equations by plotting, instead, the expressions $f(x, y)$, and 0, using the faster plot3d( ) command. The first equation, however, cannot be expressed in the required form $z = g(x, y)$.

We can still use the plot3d( ) command, but we must use a very different approach. A point $P(x, y, z)$ in three-dimensional space is just an ordered triplet of numbers. Maple would view this point as the list $[x, y, z]$. The first entry in a plot3d( ) command can be a list of this sort. The surface $y = 9 - x^2$ can be viewed as the list $[x, y, z]$, where x and z are arbitrary and y depends on these variables according to the equation $y = 9 - x^2$. The list $[x, 9 - x^2, z]$ now depends only on x and z, and it can be plotted in this way by using the plot3d( ) command. In order to plot the two other equations in the same plot command, they must be viewed as lists that depend on **the same two variables x and z.** The second equation would be expressed in the form $[x, -90 + 99z/10, z]$. The third equation (the xy-plane) could be dropped—one can easily imagine where it would be in a plot. If the third equation is entered in the plot command, it could be expressed in the form $[x, -90 + 99z/10, 0]$. The third entry must be 0. The second entry cannot contain y, but it does not have to be expressed as it is. On the other hand, can you see why this might be a good way to express this second entry?

The plot appearing in the next Maple work session was moved around a bit with the mouse to get a better viewpoint. In addition, we **clicked a plot option button to produce the box shown in the picture.** It is easy to guess which button would produce this effect.

****

```
> plot3d({[x,9-x^2,z],[x,-90+99*z/10,z],
  [x,-90+99*z/10,0]},x=-10..10,z=0..10);
```

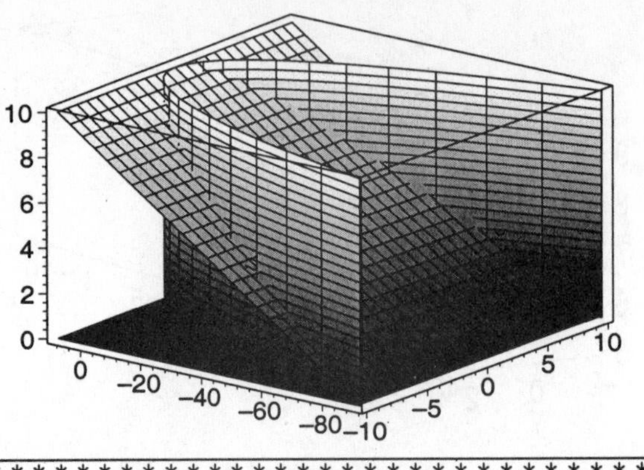

****

Frequently, a more complicated mixture of various plotting instructions is required, and the instructions cannot be expressed within one plotting command. In this situation, **several individual plots can be made, and they can be brought together with the command display( )**. The Maple command **display3d( )** is just another name for the same command.

To use this command, the **individual plots should be assigned names.** This is an interesting assignment. The value of such a name is not a new plot window, but rather a complex set of computer instructions used to draw the plot. An input statement such as

```
> p:=plot3d(f(x,y),x=a..b,y=c..d);
```

ending with a semicolon (;) rather than a colon (:), would load up a computer screen with pages of unwanted, incomprehensible computer code. **Definitely use a colon (:) to avoid the unwanted output.**

Example 9.9 *Plot, in the same coordinate system, the surface whose equation in cylindrical coordinates is*

$$r = 2\cos(2\theta), 0 \leq z \leq 10$$

and the surface whose equation in rectangular coordinates is

$$z = 10 + (4 - x^2)(4 - y^2), -2 \leq x, y \leq 2.$$

One of the reasons for selecting this particular example is to show that very different plotting environments can be brought together with the display( ) command.

****

```
> with(plots):

Warning, the name changecoords has been redefined
> r:=2*cos(2*theta):  f:=10+(4-x^2)*(4-y^2):
> p1:=cylinderplot(r,theta=0..2*Pi,z=0..10):
> p2:=plot3d(f,x=-2..2,y=-2..2):
> display({p1,p2});
```

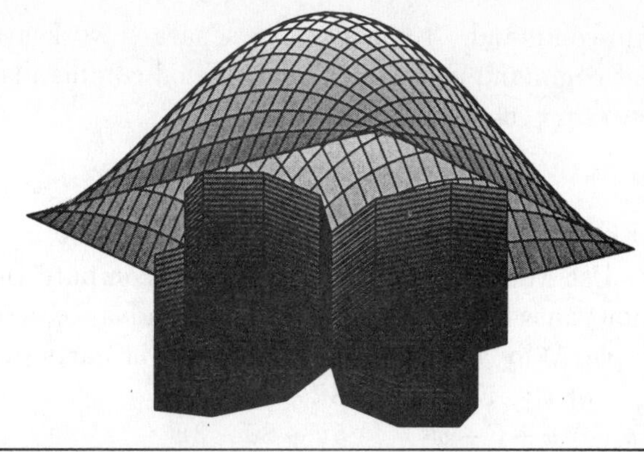

****

We used a new Maple command in the above work session. The three-dimensional plotting command **cylinderplot( ), which plots a surface in the cylindrical coordinate system** is very similar to the two-dimensional polarplot( ) command. The curve to be plotted can be entered into the command as a function, as an expression, or as a list. See Section 8.3 for further insight. This command, and the command **sphereplot( ) for plotting a surface in spherical coordinates**, will not be discussed further at this time, but they are mentioned to encourage their use when they are needed.

9.6 Exercise Set

New Maple Commands in this Chapter (and a few old commands as a reminder)

angle( )	crossprod( )	cylinderplot( )	det( )	display( )
dotprod( )	equal( )	evalf( )	evalm( )	implicitplot( )
implicitplot3d( )	map( )	matadd( )	matrix( )	norm(,2)
normalize( )	plot3d( )	proc( )	scalarmul( )	seq( )
sphereplot( )	vector( )	whattype( )	with( )	

Remember to load the *linalg* package for any work involving vectors and to load the *plots* package when it is needed. Make it a practice to define and use our special dot-product command to avoid unanticipated complications when working with arbitrary vectors.

The norm( ) command for the length of a vector can be awkward to manipulate when it is applied to unassigned vectors. Maple has difficulty getting past the absolute-value bars that appear in such expressions. It turns out that the elementary identity $|x|^2 = x^2$ does not hold for complex numbers.

If you experience such difficulties in working with the norm( ) command, take advantage of the vector identity

$$|\vec{v}|^2 = \vec{v} \cdot \vec{v}$$

and use $\sqrt{\vec{v} \cdot \vec{v}}$ (instead) for the length of $\vec{v}$. This avoids problems with the absolute-value bars. For added convenience, use the following input statement to create your own personal norm-command. Keep it on a separate worksheet along with our special dot product command and any other personal commands you wish to keep.

```
> mynorm:=v->sqrt(dot(v,v)):
```

1. Let $\vec{u}$, $\vec{v}$, $\vec{w}$ be the vectors defined by $\vec{u} = 7\vec{i} - 11\vec{j} - 9\vec{k}$, $\vec{v} = -13\vec{i} + 5\vec{j} - \vec{k}$, and $\vec{w} = 2\vec{i} + 3\vec{j} + 6\vec{k}$. Use Maple's vector commands to compute the following. Some of them have obvious answers. Try to catch them in advance, and state the answer before you use Maple. Why are parentheses not used in parts (m) and (n)?

a) $\vec{u} + \vec{v}$	b) $\vec{w} - \vec{v}$	c) $3\vec{u} + 7\vec{w}$
d) $7\vec{v} - 12\vec{w}$	e) $\vec{u} + \vec{v} + \vec{w}$	f) $5\vec{u} + 8\vec{v} - 3\vec{w}$
g) $\vec{u} \cdot \vec{u}$	h) $\vec{u} \times \vec{v}$	i) $\vec{u} \times (\vec{v} \times \vec{w})$
j) $(\vec{u} \times \vec{v}) \times \vec{w}$	k) $(\vec{u} \times \vec{v}) \times \vec{v}$	l) $\vec{w} \times (13\vec{w})$
m) $\vec{u} \cdot \vec{v} \times \vec{w}$	n) $\vec{v} \cdot \vec{v} \times \vec{w}$	o) $(3\vec{u} + 7\vec{w}) \cdot (7\vec{v} - 12\vec{w})$

2. Let $\vec{u}$, and $\vec{v}$ be the vectors $\vec{u} = 8\vec{i} - 3\vec{j} - 6\vec{k}$, $\vec{v} = 2\vec{i} + 5\vec{j} + 7\vec{k}$.
   a) Compute the lengths of $\vec{u}$ and $\vec{v}$.
   b) Compute the length of $5\vec{u} - 8\vec{v}$.
   c) Compute the length of $\vec{u} \times \vec{v}$.
   d) Find a unit vector in the direction of $2\vec{u} - 3\vec{v}$.
   e) Find a vector of length 7 in the direction of $\vec{u} \times \vec{v}$.
   f) Find the angle between $\vec{u}$, and $\vec{v}$ in degree measure.
   g) Find the projection of $\vec{v}$ onto $\vec{u}$.
   h) Express $\vec{v}$ in the form $\vec{v} = \vec{v_1} + \vec{v_2}$, where $\vec{v_1}$ is parallel to $\vec{u}$ and $\vec{v_2}$ is orthogonal to $\vec{u}$.

3. Create a projection command—name it "proj"—using the proc( ) command, so that, if u and v are Maple names for the vectors $\vec{u}$ and $\vec{v}$, then proj(u, v) is the projection of $\vec{u}$ onto $\vec{v}$. (This can also be defined without the proc( ) command, by using the same technique we used to define our special dot( ) command earlier in this chapter. You might want to save it in your own library, so that you can use it in the future.) Use it to compute the projection of $\vec{a} = 6\vec{i} - \vec{j} - 613\vec{k}$ onto $\vec{b} = 2\vec{i} + 11\vec{j} - 5\vec{k}$ and the projection of $\vec{b}$ onto $\vec{a}$.

4. Use Maple to prove the identity

$$(\vec{u} - \vec{v}) \cdot (\vec{u} + \vec{v}) = |\vec{u}|^2 - |\vec{v}|^2.$$

5. Use Maple to prove the *Distributive Law of Dot Product Over Addition.* Here is a point that might not be noticed unless Maple is used in this proof. The addition operation is different on each side of the equation: It is vector addition on one side and ordinary addition of real numbers on the other side.

6. Use Maple to prove that $\vec{u}$ and $\vec{v}$ are orthogonal to $\vec{u} \times \vec{v}$.

7. Use Maple to prove the identity

$$|\vec{u} \times \vec{v}|^2 = |\vec{u}|^2|\vec{v}|^2 - (\vec{u} \cdot \vec{v})^2.$$

Look at the terms involved, and appreciate how hard it would be to prove it by hand. This algebraic result is important, because it leads immediately to a proof of the fundamental geometric rule concerning the length of a cross product, namely,

$$|\vec{u} \times \vec{v}| = |\vec{u}||\vec{v}|\sin(\theta)$$

where θ $(0 \leq \theta \leq 2\pi)$ is the angle between $\vec{u}$, and $\vec{v}$.

8. For reasons of orthogonality that were discussed on page 219, it follows that $(\vec{u} \times \vec{v}) \times \vec{w}$ is a linear combination of $\vec{u}$ and $\vec{v}$. Find scalars a and b such that

$$(\vec{u} \times \vec{v}) \times \vec{w} = a\vec{u} + b\vec{v}.$$

Identify the scalars a and b that are found as **certain dot products**, and state a very interesting **identity** involving this **triple cross product**.

9 Use Maple to prove the rule

$$(\vec{u} \times \vec{v}) \times \vec{w} = \vec{u} \times (\vec{v} \times \vec{w}) + (\vec{u} \times \vec{w}) \times \vec{v}.$$

Use this rule to create a nontrivial example of vectors $\vec{u}$, $\vec{v}$, and $\vec{w}$ for which $(\vec{u} \times \vec{v}) \times \vec{w} = \vec{u} \times (\vec{v} \times \vec{w})$. (By nontrivial, we mean that the final products should be nonzero.)

10. Use Maple to prove the *Distributive Law of Cross Product Over Addition.*

11. Consider the parallelepiped formed by the three vectors $\vec{u} = 2\vec{i} - 7\vec{j} + 5\vec{k}$, $\vec{v} = 4\vec{i}+2\vec{j}+5\vec{k}$, $\vec{w} = 6\vec{i}+9\vec{j}-3\vec{k}$ based at a common vertex P. Determine the angle (in degree measure) between the main diagonal from P (to the opposite corner of the parallelepiped) and the diagonals from P along each of the three faces adjacent to P.

12. Let $\vec{u}$, $\vec{v}$, and $\vec{w}$, be any three noncollinear vectors in three-dimensional space. Let R be the parallelepiped formed by $\vec{u}$, $\vec{v}$, $\vec{w}$ based at a common vertex P. Let $\vec{f1}$, $\vec{f2}$, and $\vec{f3}$ be the diagonal vectors from P along the three faces of R adjacent to P, and, finally, let Rf be the parallelepiped formed by $\vec{f1}$, $\vec{f2}$, and $\vec{f3}$. Use Maple to show that the volume of Rf is exactly twice the volume of R. (This can also be shown by hand in just a few lines by using the algebraic properties of the vector operations.)

13. Let $\vec{u}$, $\vec{v}$ $\vec{w}$ be the vectors defined by

$$\vec{u} = \frac{3}{\sqrt{14}}\vec{i} - \frac{1}{\sqrt{14}}\vec{j} + \frac{2}{\sqrt{14}}\vec{k},$$

$$\vec{v} = \frac{1}{\sqrt{3}}\vec{i} + \frac{1}{\sqrt{3}}\vec{j} - \frac{1}{\sqrt{3}}\vec{k},$$

$$\vec{w} = \frac{-1}{\sqrt{42}}\vec{i} + \frac{5}{\sqrt{42}}\vec{j} + \frac{4}{\sqrt{42}}\vec{k}.$$

Show that $\vec{u}$, $\vec{v}$ $\vec{w}$ are mutually orthogonal unit vectors. (Such a collection is called an orthonormal set). Then use Maple to show that any vector $\vec{p}$ in three-dimensional space can be expressed in the form

$$\vec{p} = a\vec{u} + b\vec{v} + c\vec{w}$$

where a, b, and c are real numbers. In the process, find formulas for a, b, and c in terms of the components of $\vec{p}$.

The numbers a, b, and c can also be found by hand with very little computation, by an algebraic manipulation using the rules associated with the various vector operations. As a fitting conclusion to this problem, explain how a, b, and c can be found in this way. (The answer is obvious, once it is seen, but it can be puzzling at first glance. If you already have the formula generated by Maple, it might provide some insight.)

14. Find the equation of the plane through the points $A(5,-9,1)$, $B(16,-2,31)$, $C(-8,53,-17)$.

15. Find the equation of the plane that contains the point $A(7,15,-8)$ and the line defined parametrically by $x = 4+2t, y = -6+5t, z = 7-3t$.

16. Find the equation of the plane that contains the point $A(-4,9,6)$ and that is parallel to, and 2 units away from, the line L through the points $B(7,5,-3)$, and $C(-6,4,13)$.

17. Find the parametric equation of the line of intersection between the plane passing through $A(2,1,-5)$, $B(-4,11,8)$, $C(12,3,-2)$ and the plane passing through $P(14,23,18)$, $Q(-9,-1,-32)$, $R(-1,19,-6)$.

18. How would you define the angle between a plane and a nonparallel line? Define this term. Find the angle between the plane whose equation is $5x - 8y + 2z = 17$ and the line with equation $x = 7-t, y = 2+9t, z = -6+5t$. Maple has a command, angle( ), for computing the angle between two vectors, but it is easy enough to get along without it.

19. Create a Maple procedure that computes the distance between two lines. Be careful with names. Maple owns the name "distance." "LineDist" would work. Create the command so that, if a,u, b, and v are Maple names (expressed as vectors) for the line through A with direction $\vec{u}$ and the line through B with direction $\vec{v}$, then LineDist(a,u,b,v) is the distance between the two lines. Use it to compute the distance between the line through $A(8,1,-4)$, $B(-2,5,3)$ and the line through $P(14,5,-6)$, $Q(-3.-17,2)$.

20. Captain Ralph has spotted a small but deadly object at position

$$P(232.627, 632.879, 3275.68)$$

traveling at great speed in a straight line with direction vector

$$\vec{v} = 1.2764\vec{i} - 0.26271\vec{j} + 1.7014\vec{k}$$

towards a spherical space station of radius 250 meters centered at

$$Q([258.257, 627.413, 3309.599]).$$

Will it hit the space station? If not, how close will it come? The coordinate system is in kilometers.

21. A damaged space ship from United Earth Federation (UEF) is hiding from the dreaded Lizzard Warriors and attempting to make radio contact with its headquarters without being observed by the enemy. A UEF observer at position $P(73.1, -19.8, -42.6)$ has determined that a brief radio signal was broadcast from somewhere along the direction of the vector $\vec{u} = 3.2\vec{\imath} + 1.2\vec{\jmath} - 5.1\vec{k}$ based at P. At the same time, another UEF observer at position $Q(-1.4, 81.9, -13.7)$ received a radio signal broadcast from somewhere along the direction of the vector $\vec{v} = 9.4\vec{\imath} - 3.0\vec{\jmath} - 10.5\vec{k}$ based at Q. The coordinate system is expressed in kilometers.

Naturally there are slight errors in these measurements. Captain Ralph will launch a fast rescue mission only if the location data are accurate enough to justify the risk. The two direction lines used to locate the damaged ship should meet, but, as long as they are less than 1 kilometer apart, the data will be considered reliable enough to begin the operation. Captain Ralph, always ready to spring into action, awaits your call. Should he rescue, or should he not? Approximate the coordinates of the damaged ship. The answer is not unique. Several different strategies are available. (Use your imagination.)

22. Plot the following surfaces. **Try to predict the basic shape roughly before you ask Maple for a plot**. After all, how else would you know whether Maple is behaving?

Choose a suitable range of values. If infinite discontinuities are involved, Maple will choose a large window and lose the surface in the process, unless you tell Maple what window to use. The idea is to show the discontinuities without losing the surface.

How do you plot an equation like $x = f(y, z)$, for example? Actually, it is easy, if you do not mind that the x-axis is vertical—Maple does not mind. If you want the z-axis to be vertical (we usually do), then the surface has to be graphed more carefully. In the following, plot the appropriate surfaces both ways.

Recall Maple's behavior concerning fractional powers of negative numbers. Some of the following plots will require the use of the surd( ) command.

a) $z = \sqrt[3]{x^2 + y^2}$ b) $z = \sqrt[3]{x^2 - y^2}$ c) $z = \sqrt[3]{(x - y)^2}$
d) $z = \sqrt[3]{x - y}$ e) $z = \cos(xy)$ f) $z = \ln(x^2 + y^2)$
g) $z = \frac{x^2+y^2}{x^2-y^2}$ h) $y = e^{x-z}$ i) $x = 4y^2 + 16z^2$

23. Plot the region bounded by the surfaces $z = 5 + \frac{100\cos(|x|+|y|)}{10+x^2+y^2}$ and $z = |x| + |y| - 10$.

24. Plot the region bounded by the surfaces $f(x, y) = 200 - 9x^2 - 4y^2 - 18x - 8y$ and $g(x, y) = 4x^2 + 9y^2 + 8x - 18y - 200$.

a) Use a simple rectangular range of suitable size.

b) Use an exact range that shows the boundary exactly, without showing any overlapping parts of either surface. Try to enhance this plot to show it in the best possible way.

Project: Just for Fun

Find a three-dimensional shape (and a corresponding mathematical formula) that you find to be visually interesting and appealing. Make a Maple plot and enhance it "to the max." Pull down the **Help Menu**, click on **Full Text Search...**, and type "plot options". If your computer has enough RAM, use, for example, more points to generate a plot. Arrange for a color print of your "art." Frequently, color printers are on the campus network, in which case your file can be sent to a printer from your computer over the network. Send your print to the Metropolitan Museum of ...—no, on second thought, never mind. Hang it on your door, and enjoy.

Project: Captain Ralph Shoots at a Mirror

A planar circular mirror with a radius of 10 meters is built, and great effort is taken to make the mirror absolutely flat. The mirror is then mounted to the ends of three parallel, variable-length rods that project from the outer wall of a space station. A heavy, fixed, mounting device from the wall of the space station is attached to the center of the mirror's underside, which secures the center of the mirror in a fixed position, and the mirror is then allowed to pivot on its center in a way that is determined by the lengths of the three adjustment rods. The center of the mirror is fixed, so the lengths of any two of the adjustable rods determine a unique setting of the mirror, and the length of the third rod is then adjusted to provide a secure position for the mirror.

The *mirror's target vector* is defined to be the unit vector normal to the planar surface of the mirror that, if based at the mirror's center, would point away from the space station.

The mirror must be mounted so that its target vector is exactly the same as the unit vector with direction

$$\vec{n} = \sqrt{3}\vec{\imath} + \sqrt{7}\vec{\jmath} + 42\sqrt{2}\vec{k}$$

(relative to a certain three-dimensional coordinate system discussed below), and the lengths of the rods are to be adjusted to product this direction. The adjustment must be made with great accuracy.

Captain Ralph happens to be in the right region, on his way back to the space station for lunch, and he has been asked to help out. He is told to maneuver his space ship to a position roughly in line with the mirror's target vector and about 30 kilometers away. Here he is to shoot a beam of light from a laser gun in his ship at the mirror, adjust his position and shoot again, and continue doing this until the beam is reflected by the mirror and returned to exactly its original position at the gun. A record will then be made of the gun's coordinate position at this time, and this information will be used to adjust the mirror's position.

Being the best pilot in the fleet, Captain Ralph does the job in short order. While maneuvering his ship, he takes a test shot here and there, and soon sees the tell tale reflection of the laser beam appearing back in the gun's sight. As instructed, he marks his position (actually, the gun's position) as

$$P(873.57838, 1334.4355, 29957.581).$$

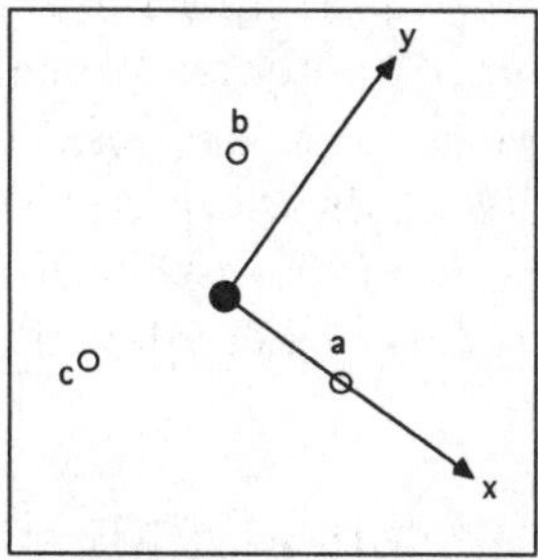

Here is some information concerning the rods and the coordinate system. All units are measured in *meters.* The coordinate system is arranged so that the fixed center of the mirror is at (0,0,0). The z-axis is parallel to the adjustment rods with a positive direction pointing away from the space station. The positive x-axis and y-axes are shown in the figure, along with cross sections of the three rods, which are labeled a, b, and c. The adjustment rods are 9.5 *meters* away from the z-axis and are equally distributed around the z-axis, as suggested in the picture.

1. Exactly where the laser light beam strikes the mirror is unknown, and so the center of the mirror is used as the point where the light beam strikes. What should the adjustment be at each adjustment rod in order to produce a mirror position whose target vector has direction $\vec{n}$? How many centimeters should each rod be lengthened or shortened?

Remember that the coordinate system is in meters, and the adjustment specifications are to be made in centimeters. Good luck. This has to be accurate!

2. The laser light beam could, of course, strike any point on the mirror, not necessarily its center, and this could cause an error in positioning the mirror. Determine the size of the error in the following way. If the final setting of the mirror results in a target vector $\vec{n_0}$, then determine, in degree measure, the largest possible angle between $\vec{n_0}$ and the vector $\vec{n}$.

Chapter 10

Vector-Valued Functions

Vector-valued functions of a real variable t play an important role in mathematics and its applications. Now that we have broken out of a one-dimensional world, we can look forward to a variety of really interesting problems. If done by hand, calculations can quickly overwhelm us, but fortunately, Maple stands ready to provide help.

10.1 Basic Calculations

Our study begins with a mixture of different ways to create vector-valued expressions and vector-valued functions. This is a confusing issue to discuss, because there are many different ways to create these data types. There are subtle but important differences between them. Furthermore, calculations are likely to cause changes from one data type to another.

It was said in the previous chapter that the safest way to work with vectors is to treat them always as **matrices**, by liberal use of the vector( ) command. That comment could be repeated in this chapter. With this approach, there would be fewer surprises in store for us, and Maple would seem to be more predictable. With this approach, however, we would have to use the less intuitive matadd( ) and scalarmul( ) commands for addition and scalar multiplication. There is much to be said in favor of a consistent matrix approach.

On the other hand, we have also noticed how much more intuitive it is to work with **row vectors** and the operators (+), (-), and (*). In this chapter, we will continue to use the more intuitive approach offered by row vectors. We will accept the risk of dealing with occasional surprises. Actually, we can often deal with surprising behavior by using the evalm( ) (evaluate as a matrix) command. Of course, this command will turn our output into a matrix.

Many of Maple's commands work only on real-valued expressions, and so they must be applied to the components of a vector rather than to the whole vector. For this reason, the opportunity to use **the map( ) command (multiple application command) will be ever present** throughout this chapter. This command was introduced in the last chapter, but it will be used much more frequently in the current chapter.

Our approach to the confusing issues raised in our first work session will be to lean toward the use of **vector-valued expressions** rather than **vector-valued functions**. Nevertheless, let us bring these issues regarding vector-valued functions to the forefront.

****

```
> with(linalg):
```

```
Warning, the protected names norm and trace have been redefined and
unprotected
```

```
> rm:=vector([3*t^2,cos(t),exp(-5*t)]);rr:=<3*t^2|cos(t)|exp(-5*t)>;
```

$$rm := \left[3\,t^2,\ \cos(t),\ e^{(-5\,t)}\right]$$

$$rr := \left[3\,t^2,\ \cos(t),\ e^{(-5\,t)}\right]$$

\# These are the two main ways of entering a vector-valued expression $\vec{r} = \vec{r}(t)$. We used the suffixes "m" and "r" as a reminder that rm represents $\vec{r}$ as a matrix and rr represents $\vec{r}$ as a row-vector.

It is a straightforward matter to substitute $t = 2$ into either expression, but notice that, because rm is a matrix, it does not fully evaluate until we use the evalm( ) command. #

```
> subs(t=2,rm),subs(t=2,rr);
```

$$rm,\ \left[12,\ \cos(2),\ e^{(-10)}\right]$$

```
> subs(t=2,evalm(rm));
```

$$\left[12,\ \cos(2),\ e^{(-10)}\right]$$

\# Let us now try to enter these vector-valued expressions as functions. Again, the suffixes "m" and "r" are used to remind us of the underlying (matrix or row vector) data type. We use the unapply( ) command to turn an expression into a function. To find out why we use such a strange word for this command, look it up in the index. It is easier to remember names of commands if they seem reasonable. The unapply( ) command operates only on real-valued (and complex-valued) expressions, so we must **use the map( ) command.** #

```
> fm:=map(unapply,rm,t);fr:=map(unapply,rr,t);
```

$$fm := \left[t \to 3\,t^2,\ \cos,\ t \to e^{(-5\,t)}\right]$$

$$fr := \left[t \to 3\,t^2,\ \cos,\ t \to e^{(-5\,t)}\right]$$

\# Look at the preceding output statements. It appears that fm and fr are both exactly what we want. Each seems to be three separate real-valued functions of t, put together to form one vector-valued function. **Watch, however, what happens when we call for an evaluation of fm and fr at $t = 2$.** #

```
> fm(2);fr(2);
```

$$\left[12,\ \cos(2),\ e^{(-10)}\right]$$

$$\left[t \to 3\,t^2,\ \cos,\ t \to e^{(-5\,t)}\right](2)$$

\# Unfortunately, ***fr* is not a vector-valued function**, so we will not be able to use this approach to create vector-valued functions.

We could have used the row vector rr to create a function if we had used evalm(rr) instead of rr as shown in the next line. Because evalm(rr) evaluates as a matrix, however, the function f produced in the next line is no different from fm. #

```
> f:=map(unapply,evalm(rr),t);
```

$$f := \left[t \to 3\,t^2,\ \cos,\ t \to e^{(-5\,t)}\right]$$

```
> f(2);
```

$$\left[12,\ \cos(2),\ e^{(-10)}\right]$$

The vector-valued function fm defined above is, in some sense, an "ideal" vector-valued function. Not only does it allow for an easy evaluation of $fm(t)$ for any t, but

$$fm = [fm[1], fm[2], fm[3]]$$

actually consists of three separate real-valued functions: $fm[1]$, $fm[2]$, $fm[3]$.

Here is another way to create a vector-valued function. It has some advantages and some disadvantages. The suffixes "m" and "r" are used in the same way. #

```
> gr:=t-><3*t^2|cos(t)|exp(-5*t)>;
  gm:=t->vector([3*t^2,cos(t),exp(-5*t)]);
```

$$gr := t \to \langle 3\,t^2 \mid \cos(t) \mid e^{(-5\,t)}\rangle$$

$$gm := t \to \left[3\,t^2,\ \cos(t),\ e^{(-5\,t)}\right]$$

```
> gr(2);gm(2);
```

$$\left[12,\ \cos(2),\ e^{(-10)}\right]$$

$$\left[12,\ \cos(2),\ e^{(-10)}\right]$$

This time, gr seems to work as well as gm as a vector-valued function. Additionally, the vector-valued function gr has the convenience of evaluating at t as the row-vector $gr(t)$.

The major disadvantage of both gm and gr is that **neither one represents three separate real-valued functions**. Maple sees only **one** "$\to$" in each of these functions. Consequently

$$gm \neq [gm[1], gm[2], gm[3]], \quad gr \neq [gr[1]gr[2], gr[3]].$$

In some sense, this is no cause for concern as long as we operate on the functions gr and gm as expressions $gr(t)$ and $gm(t)$. Problems arise only when we operate on the functions gr and gm themselves.

In the following, we differentiate the three "respectable" functions we have created so far, using the diff( ) command, which operates on expressions and returns expressions. We use p (p for prime) in our names. Notice that we **must use the map( ) command.**

```
> gmp:=map(diff,gm(t),t);grp:=map(diff,gr(t),t);
  fmp:=map(diff,fm(t),t);
```

$$gmp := \left[6\,t,\ -\sin(t),\ -5\,e^{(-5\,t)}\right]$$

$$grp := \left[6\,t,\ -\sin(t),\ -5\,e^{(-5\,t)}\right]$$

$$fmp := \left[6\,t,\ -\sin(t),\ -5\,e^{(-5\,t)}\right]$$

The differentiation command D( ) operates on a real-valued (or complex-valued) function and returns a function as output. It must be used in conjunction with the map( ) command. The D( ) command is a convenient tool in our arsenal, but from our preceding comments, we can predict how it will perform. **We can expect it to perform well on *fm*, which is three separate functions, but not on *gm* or *gr*, each of which is only one function.**

```
> fmp:=map(D,fm);gmp:=map(D,gm);grp:=map(D,gr);
```

$$fmp := \left[t \to 6\,t,\ -\sin,\ t \to -5\,e^{(-5\,t)}\right]$$

$$gmp := \mathrm{D}(gm)$$

$$grp := \mathrm{D}(gr)$$

```
> fmp(2);
```

$$\left[12,\ -\sin(2),\ -5\,e^{(-10)}\right]$$

```
> gmp(2),grp(2);
```

$$\mathrm{D}(gm)(2),\ \mathrm{D}(gr)(2)$$

We can see that D(gm) and D(gr) have both failed.

****

What should our conclusions be from this discussion about vector-valued functions? It appears that fm, gr, and gm are useful ways to define vector-valued functions, but they are all different from each other, and each has its advantages and disadvantages.

As long as there is not a call for too many evaluations, it might be more convenient to simply use vector-valued expressions rather than functions. That will be our approach as we continue, and, because we have grown accustomed to the notational advantage of using row vectors, we will continue to enter our vector-valued expressions as row vectors.

Example 10.1 *Plot the curve* Γ *defined by the vector-valued position function*

$$r(\vec{t}) = \sin(2t)\vec{i} + (\sin(t) - t + 2\cos(3t))\,\vec{j} + (2t - 3\cos(t))\,\vec{k},\ -2\pi \le t \le 2\pi.$$

Find the equation of the plane that passes through the point $r(\vec{3})$ *on* Γ *and that is parallel to both* $\vec{r}'(3)$ *and* $\vec{r}''(3)$.

If $\vec{r}(t)$ describes the position of an object at time t, then $\vec{r}'(t)$, and $\vec{r}''(t)$ turn out to be the velocity and acceleration vectors of the object at time t. We will study velocity and acceleration vectors soon enough, but, for now, notice, just for the sake of interest, that the plane under consideration seems to describe, in a sense, the plane where all of the "action" is taking place near time $t = 3$ and near the point $r(\vec{3})$.

To plot the curve Γ, we use a **new command, spacecurve( ),** which resides in the plots package. Only the curve and a range for the parameter must be entered into the command. In the plot shown next, we have added several options. Some of these options could have been accessed from the menu-bar. After a plot appears, we can gain access to the plotting menu-bar by clicking the picture.

****

```
> with(linalg):dot:=(u,v)->dotprod(u,v,orthogonal):
```

```
Warning, the protected names norm and trace have been redefined and
unprotected
```

```
> with(plots):
```

```
Warning, the name changecoords has been redefined
```

```
> r:=<sin(2*t)|sin(t)-t+2*cos(3*t)|2*t-3*cos(t)>;
```

$$r := [\sin(2\,t),\, \sin(t) - t + 2\cos(3\,t),\, 2\,t - 3\cos(t)]$$

```
> spacecurve(r,t=-2*Pi..2*Pi,axes=BOXED,thickness=2,numpoints=200);
```

```
Error, (in spacecurve) first argument must be array, list or set
```

There is no cause for alarm here. Simply use evalm(r) instead of r in this command to turn r into a matrix (array).

```
> spacecurve(evalm(r),t=-2*Pi..2*Pi,
  axes=BOXED,thickness=2,numpoints=200):
```

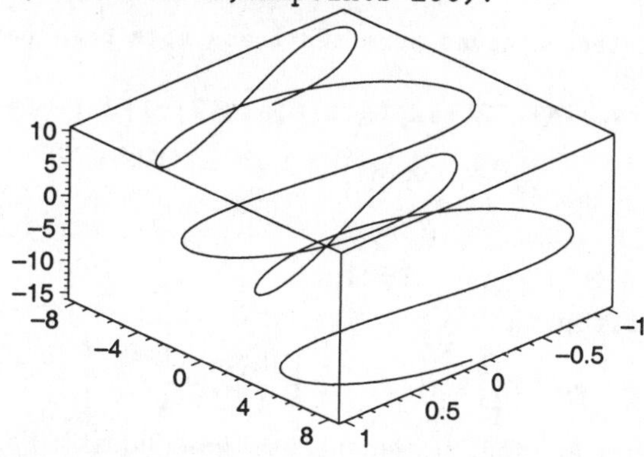


```
> rp:=map(diff,r,t);
```

$$rp := [2\cos(2\,t),\, \cos(t) - 1 - 6\sin(3\,t),\, 2 + 3\sin(t)]$$

```
> rpp:=map(diff,rp,t);
```

$$rpp := [-4\sin(2\,t),\, -\sin(t) - 18\cos(3\,t),\, 3\cos(t)]$$

```
> v:=evalf(subs(t=3,rp));a:=evalf(subs(t=3,rpp));
```

$$v := [1.920340573,\, -4.462703408,\, 2.423360024]$$

$$a := [1.117661993,\, 16.25922470,\, -2.969977490]$$

```
> n:=evalf(crossprod(v,a));
```

$$n := [-26.14782650,\, 8.411865670,\, 36.21104286]$$

```
> p:=evalf(subs(t=3,r));
```

$$p := [-.2794154982,\, -4.681140516,\, 8.969977490]$$

```
> X:=<x|y|z>:
> plane:=dot(n,X-p)=0;
```

$$plane := -26.14782650\,x - 292.7412221 + 8.411865670\,y + 36.21104286\,z = 0$$

****

Example 10.2 *Find a vector-valued function $\vec{r}(t)$ such that*

$$\vec{r}''(t) = \cos(3t)\vec{i} + (4+t^2)\vec{j} - e^{4t}\vec{k},$$

$$\vec{r}'(0) = 2\vec{i} - \vec{j} + 2\vec{k},\ \vec{r}(0) = \vec{i} + 6\vec{j} - \vec{k}.$$

Each time we antidifferentiate, we pick up another constant of integration. No big surprise there, but remember that, in the current vector setting, the **constant of integration is vector-valued.** Recall that Maple leaves it up to us to attach the constant of integration. These constants, denoted by C and K below, are just unassigned names, but **they must be set up as vectors, or Maple would regard them as real numbers.** Evaluating these constants can be a tedious experience if Maple is not used effectively. (This issue was discussed separately in Example 9.2.)

In the next work session, we follow our customary practice of using names like *rp* (p for prime) for the derivative and *rpp* for the second derivative of $\vec{r}$. Other unassociated names are used for intermediate objects of no final importance.

****

```
> with(linalg):
Warning, the protected names norm and trace have been redefined and
unprotected
> rpp:=<cos(3*t)|4+t^2|-exp(4*t)>;a:=<2|-1|2>; b:=<1|6|-1>;
```

$$rpp := \left[\cos(3\,t),\ 4+t^2,\ -e^{(4\,t)}\right]$$

$$a := [2,\ -1,\ 2]$$

$$b := [1,\ 6,\ -1]$$

```
> gp:=map(int,rpp,t);
```

$$gp := \left[\frac{1}{3}\sin(3\,t),\ 4\,t+\frac{1}{3}\,t^3,\ -\frac{1}{4}\,e^{(4\,t)}\right]$$

#Without entering an input statement, use these "ideas": $\vec{rp} = \vec{gp} + \vec{C}$ for some constant vector $\vec{C}$, and $\vec{rp} = \vec{a}$ when $t = 0$, to get $\vec{C}$ directly, as follows.#

```
> C:=a-subs(t=0,gp);
```

$$C := \left[2,\ -1,\ \frac{9}{4}\right]$$

```
> rp:=gp+C;
```

$$rp := \left[\frac{1}{3}\sin(3\,t)+2,\ 4\,t+\frac{1}{3}\,t^3-1,\ -\frac{1}{4}\,e^{(4\,t)}+\frac{9}{4}\right]$$

```
> g:=map(int,rp,t);
```

$$g := \left[-\frac{1}{9}\cos(3\,t)+2\,t,\ 2\,t^2+\frac{1}{12}\,t^4-t,\ -\frac{1}{16}\,e^{(4\,t)}+\frac{9}{4}\,t\right]$$

#Again, use the ideas that $\vec{r} = \vec{g} + \vec{K}$ for some constant vector $\vec{K}$ and that $\vec{r} = \vec{b}$ when $t = 0$ to get the following.#

```
> K:=b-subs(t=0,g);
```

$$K := \left[\frac{10}{9}, 6, \frac{-15}{16}\right]$$

```
> r:=g+K;
```

$$r := \left[-\frac{1}{9}\cos(3\,t) + 2\,t + \frac{10}{9}, 2\,t^2 + \frac{1}{12}\,t^4 - t + 6, -\frac{1}{16}\,e^{(4\,t)} + \frac{9}{4}\,t - \frac{15}{16}\right]$$

Just to check our results, we compute $\vec{r}(0)$ and $\vec{r}'(0)$:#

```
> subs(t=0,r);subs(t=0,map(diff,r,t));
```

$$[1, 6, -1]$$

$$[2, -1, 2]$$

****

10.2 Motion in Space

If $\vec{r}(t)$ is the position of an object at time t, then $\vec{v}(t) = \vec{r}'(t)$ is the object's velocity, and $\vec{a}(t) = \vec{v}'(t)$ is its acceleration at time t. The speed of the object at time t is $s(t) = |\vec{v}(t)|$. Maple can be an excellent tool to use in the solution of motion problems. We begin with a fairly straightforward example and follow that with a more interesting one.

Example 10.3 *A golf ball is hit with an initial speed of 140 feet per second and with an angle of inclination of 34°. The ball travels down a field that is level for the first 374 feet and then rises at an angle of inclination of 22°. How far up the hill does the ball hit, and what is its speed at impact?*

This can be realized as a two-dimensional problem with a vertical y-axis. If $\vec{g}$ is the gravitational vector and $\vec{v_0}$ is the initial velocity vector, then the velocity of the ball at time t is $\vec{v}(t) = t\vec{g} + \vec{v_0}$, and the position of the ball at time t is $\vec{r}(t) = t^2\vec{g}/2 + t\vec{v_0}$. At the moment of impact,

$$\vec{r}(t) = 374\vec{\imath} + s(\cos(22°)\vec{\imath} + \sin(22°)\vec{\jmath},$$

where s is the distance up the hill, and t is the time at impact.

****

```
> with(linalg):

Warning, the protected names norm and trace have been redefined and
unprotected

> g:=<0|-32>;
```

$$g := [0, -32]$$

```
> v0:=<140*cos(34*Pi/180)|140*sin(34*Pi/180)>;
```

$$v0 := \left[140\cos(\frac{17}{90}\,\pi), 140\sin(\frac{17}{90}\,\pi)\right]$$

```
> v:=t*g+v0;
```

$$v := t\,[0, -32] + \left[140\cos(\frac{17}{90}\,\pi), 140\sin(\frac{17}{90}\,\pi)\right]$$

Look at the preceding output. It appears that we have a problem with this scalar multiplication. Maple does not think of this as scalar multiplication, because it does not know that t is a real number. Using the input command

>assume(t,real);

does not seem to help. There are some fancy ways around the problem, but perhaps the best strategy is to **use row vectors only as long as they are convenient.** An easy application of the familiar evalm( ) command eliminates the problem, but of course it turns the output into a matrix vector. #

```
> v:=evalm(t*g+v0);
```

$$v := \left[140\cos(\frac{17}{90}\pi),\ -32\,t + 140\sin(\frac{17}{90}\pi)\right]$$

```
> r:=evalm(t^2/2*g+t*v0);
```

$$r := \left[140\,t\cos(\frac{17}{90}\pi),\ -16\,t^2 + 140\,t\sin(\frac{17}{90}\pi)\right]$$

The hill is defined by the following vector-valued expression.

```
> p:=<374|0>+<s*cos(20*Pi/180)|s*sin(20*Pi/180)>;
```

$$p := \left[374 + s\cos(\frac{1}{9}\pi),\ s\sin(\frac{1}{9}\pi)\right]$$

```
> fsolve({r[1]=p[1],r[2]=p[2]},{s,t},t=0..10);
```

$$\{s = 127.3106976,\ t = 4.253063511\}$$

```
> assign(%);
> HillDistance:=s; ImpactTime:=t; ImpactSpeed:=evalf(norm(v,2));
```

$$HillDistance := 127.3106976$$
$$ImpactTime := 4.253063511$$
$$ImpactSpeed := \sqrt{13471.14461 + |32.\,t - 78.28700650|^2}$$

Maple's failure to evaluate the last term is somewhat of a surprise. Ordinarily, if an expression A involving a variable t is defined in a worksheet, and a value is assigned to t later in the work session, then A is automatically evaluated at the assigned value of t. What we see in the above line is that this full evaluation doesn't take place when matrix vectors are involved. This could well be another consequence of a (matrix) vector being "too closed."#

```
> ImpactSpeed:=subs('t'=t,ImpactSpeed);
```

$$ImpactSpeed := 129.6659529$$

Recall that single quotes postpone evaluation. They were needed here, because t had a value.

****

Example 10.4 *An enemy space ship is spotted traveling along a wide sweeping arc. Captain Ralph hits the attack button on his console and his computer determines the enemy ship's current path, relative to a certain coordinate system expressed in kilometers. The curve appears on his computer screen, defined parametrically as*

$$x = 3.1t + 85\cos(.01t) + 400,\ \ y = -12.8t,\ \ z = 13\cos(.03t) - 4.1t,$$

where t is time in seconds since he hit the attack button. The computer also determines the current path of his own fighter ship and displays the curve on his computer screen parametrically as

$$x = 340 - 2.1t, \ y = 197\sqrt{t+1} - .015t^2 - 2300, \ z = 0.032t^2 + 3.6t - 608.$$

"What a great computer," Ralph says to himself, as he prepares to launch a missile and attack the enemy ship.

First he sets the missile launch time to $t = 40$ seconds, and then he tries to determine the missile launch direction at $t = 40$. The missile will launch with a speed of 20.8 kilometers per second (relative to Ralph's ship). The missile has no power of its own, and so it will travel in a straight line with a constant velocity vector until it bumps (hopefully) into the enemy ship and explodes. What direction vector should Captain Ralph give the missile so that it will knock out the enemy ship? At what time will the missile reach the enemy ship, and how far will it be from Captain Ralph at that moment? Ralph, unfortunately, is a "shoot-from-the-hip" kind of guy, and not a very good student of calculus. He really needs our help.

We let $\vec{e}(t)$, $\vec{r}(t)$, and $\vec{m}(t)$, denote the vector-valued position functions for the enemy ship, Ralph's ship, and the missile respectively. These letters will be prefixed with v to denote respective velocities. If Ralph points the missile in the direction of some unit vector $\vec{u}$, then the velocity vector of the missile will be

$$v\vec{m}(t) = 20.8\vec{u} + v\vec{r}(40).$$

The reasoning here is that the missile will acquire not only the launch velocity relative to Ralph, but the velocity of Ralph's ship at the firing time as well. We need to determine the vector $\vec{u}$ (three unknowns), and the time t (a fourth unknown) when $\vec{m}(t) = \vec{e}(t)$, when their destinies collide.

****

```
> with(linalg):

Warning, the protected names norm and trace have been redefined and
unprotected
> e:=<3.1*t+85*cos(.01*t)+400|-12.8*t|13*cos(.03*t)-4.1*t>;
```

$$e := [3.1\,t + 85\cos(.01\,t) + 400,\ -12.8\,t,\ 13\cos(.03\,t) - 4.1\,t]$$

```
> r:=<340-2.1*t|197*sqrt(t+1)-.015*t^2-2300|0.032*t^2+3.6*t-608>;
```

$$r := \left[340 - 2.1\,t,\ 197\sqrt{t+1} - .015\,t^2 - 2300,\ .032\,t^2 + 3.6\,t - 608\right]$$

```
> u:=<a|b|c>;
```

$$u := [a,\ b,\ c]$$

We could have used u=vector(3) to create the unknown direction vector (a standard approach), but the output would have been in the form of a matrix vector.

```
> vr:=map(diff,r,t);
```

$$vr := \left[-2.1,\ \frac{197}{2}\frac{1}{\sqrt{t+1}} - .030\,t,\ .064\,t + 3.6\right]$$

```
> r40:=subs(t=40,r);vr40:=subs(t=40,vr);
```

$$r40 := \left[256.0,\ 197\sqrt{41} - 2324.000,\ -412.800\right]$$

$$vr40 := \left[-2.1,\ \frac{197}{82}\sqrt{41} - 1.200,\ 6.160\right]$$

```
> vm:=20.8*u+vr40;
```

$$vm := \left[20.8\,a - 2.1,\ 20.8\,b + \frac{197}{82}\sqrt{41} - 1.200,\ 20.8\,c + 6.160\right]$$

```
> m:=evalm((t-40)*vm+r40);
```

$$m := \Big[(t-40)\,(20.8\,a - 2.1) + 256.0,$$

$$(t-40)\,(20.8\,b + \frac{197}{82}\sqrt{41} - 1.200) + 197\sqrt{41} - 2324.000,$$

$$(t-40)\,(20.8\,c + 6.160) - 412.800\Big]$$

Notice our use of the evalm( ) command in the above input line. Because t is an unassigned name, Maple will not recognize $(t-40) * vm$ as scalar multiplication. We were forced to turn this expression into a matrix vector. Recall that this problem was encountered in the previous example as well.

We set the missile's position equal to the enemy ship's position, to get three equations in four unknowns (t, and the three unknowns in the direction vector u).#

```
> for j from 1 to 3 do eq[j]:=m[j]=e[j];od;
```

$$eq_1 := (t-40)\,(20.8\,a - 2.1) + 256.0 = 3.1\,t + 85\cos(.01\,t) + 400$$

$$eq_2 := (t-40)\,(20.8\,b + \frac{197}{82}\sqrt{41} - 1.200) + 197\sqrt{41} - 2324.000 = -12.8\,t$$

$$eq_3 := (t-40)\,(20.8\,c + 6.160) - 412.800 = 13\cos(.03\,t) - 4.1\,t$$

#A fourth equation comes from setting the length of u (a unit vector) equal to one.#

```
> eq[4]:=u[1]^2+u[2]^2+u[3]^2=1;
```

$$eq_4 := a^2 + b^2 + c^2 = 1$$

```
> fsolve({eq[1],eq[2],eq[3],eq[4]},{a,b,c,t});
```

$$\{a = .9955856911,\ b = -.08023467754,\ c = .04869833936,\ t = 61.74999724\}$$

```
> assign(%);
> direction:=u;
```

$$direction := [.9955856911,\ -.08023467754,\ .04869833936]$$

```
> ExplosionTime:=t;
```

$$ExplosionTime := 61.74999724$$

```
> ExplosionDistance:=norm(r-e,2);
```

$$ExplosionDistance := 450.4991844$$

****

The last evaluation is interesting, because it was fully evaluated. Compare this evaluation to our first attempt to evaluate ImpactSpeed in the solution to Example 10.3. In that example we were evaluating a matrix vector. Matrices are more "closed." They do not evaluate fully until we use the evalm( ) command. In the evaluation here, the expression $\vec{r} - \vec{e}$ is a row vector, which evaluates fully—that is one of the advantages of using rowvectors.

10.3 Geometry of Curves

The ideas of slope and concavity that we used to describe the geometry of the graph of a real-valued function of a real variable are ideas that can be used to describe the geometry of a parametrically defined curve in two- or three-dimensional space in much the same way.

The geometry is described by the so-called "curvature" of a curve and by three vectors: the unit tangent vector $\vec{T}$, the principal unit normal vector $\vec{N}$, and the binormal vector $\vec{N}$. For the sake of brevity, they are introduced without explanation or interpretation. Visit our web site for a more thorough explanation of these ideas, or consult your main calculus text.

Let Γ be a smooth curve with arc length L, defined by the vector-valued function

$$\vec{r} = \vec{r}(t), \quad a \le t \le b. \tag{10.1}$$

The first vector we meet hardly needs an introduction. The **unit tangent vector** is defined to be

$$\vec{T} = \vec{T}(t) = \frac{1}{|\vec{r}\,'(t)|}\vec{r}\,'(t) = \frac{\vec{r}\,'(t)}{|\vec{r}\,'(t)|}. \tag{10.2}$$

By definition of a smooth curve, $\vec{r}\,'(t)$ exists and is never the zero vector, so the unit tangent vector is well defined.

The **principal unit normal vector** (or just unit normal vector) to the curve Γ defined by (10.1) is

$$\vec{N} = \vec{N}(t) = \frac{1}{|\vec{T}\,'(t)|}\vec{T}\,'(t) = \frac{\vec{T}\,'(t)}{|\vec{T}\,'(t)|}.$$

The last of the three vectors, the **binormal vector**, is defined by

$$\vec{B} = \vec{T} \times \vec{N}.$$

For each t, the vectors $\vec{T}$, $\vec{B}$, and $\vec{N}$ are mutually orthogonal unit vectors (a so-called orthonormal set). Why? Actually, the answer is immediate—no computations are required to verify this result. **The collection $\{\vec{T}, \vec{B}, \vec{N}\}$, of vectors based at the point $\vec{r}(t)$ on the curve Γ is called the frame (or TNB frame) based at $\vec{r}(t)$.** This is a frame of mutually orthogonal unit vectors that moves along the curve with the point.

In order to define the curvature of the curve Γ, it is useful to think of Γ as being parametrized in terms of the arc length variable s $(0 \le s \le L)$. We can think of the position vector $\vec{r}$ as a function of t or as a function of s. A formula for $\vec{r}$ as a function of s exists as a concept, but it is usually impossible to find. It is, however, useful to think of this as a concept, even though we are forced to use the given formula $\vec{r} = \vec{r}(t)$ in order to perform calculations. With this in mind, we define the **curvature** of the curve Γ as

$$\kappa = \left|\frac{dT}{ds}\right| = \left|\frac{dT}{dt}\frac{dt}{ds}\right| = \frac{1}{|\vec{r}\,'(t)|}\left|\frac{dT}{dt}\right|.$$

The curvature is a measure of how quickly the unit tangent vector is turning. We do not want this to depend on the speed of a point moving along a curve, and so curvature is defined in terms of the rate of change of $\vec{T}$ with respect to s rather than t. Notice that $\frac{ds}{dt} = |\vec{r}\,'(t)|$ is just the speed of the point $\vec{r}(t)$ moving along the curve. The formula

$$\vec{N} = \frac{1}{\kappa}\frac{dT}{ds} = \frac{1}{\kappa}\frac{dT}{dt}\frac{dt}{ds} = \frac{1}{\kappa}\frac{dT}{dt}\frac{1}{|\vec{r}\,'(t)|}$$

is sometimes useful.

The radius of curvature is $\rho = 1/\kappa$. If Γ is the curve defined by (10.1), and if $\vec{T}$, $\vec{N}$, $\vec{B}$, κ, and ρ are all evaluated at the point $\vec{r}_0 = \vec{r}(t)$ on Γ, then the circle of curvature is the circle having radius ρ and center (the point) $\vec{r}_0 + \rho\vec{N}$, which lies in the plane through $\vec{r}_0$ with normal vector $\vec{B}$ (the plane formed by $\vec{T}$, and $\vec{N}$).

Incidentally, if formula (10.1) defines the motion of an object along a curve Γ, then the most useful way to present the acceleration vector, $\vec{a}$, acting on the object at time t is to express it in the form

$$\vec{a} = a_T\vec{T} + a_N\vec{N}.$$

The tangential component of acceleration, a_T, changes the speed of the object, and the normal component of acceleration, a_N, changes the direction of the object. Finally, the object's acceleration is related to the force acting on it by *Newton's Second Law*: $\vec{F} = m\vec{a}$. These issues will be raised in the exercise set.

Example 10.5 *Let Γ be the curve defined by the vector-valued position function*

$$\vec{r} = \sin(t)\vec{i} + (\cos(t) - \sin(t))\vec{j} + \cos(3t)\vec{k},\ \ 0 \le t \le 2\pi.$$

Compute the vectors $\vec{T}$, $\vec{N}$, $\vec{B}$ at the point on Γ corresponding to $t = \pi/4$. Plot Γ along with the frame $\{\vec{T}, \vec{N}, \vec{B}\}$ based at $\vec{r}(\pi/4)$. Finally, determine the curvature of Γ at $t = \pi/4$.

Some attention has to be focused on the evaluation procedure. It is clear from these formulas that we must compute $\vec{T}$ as an expression (or a function) in t, so that we can differentiate $\vec{T}$ to get $\vec{N}$. This is where computations frequently get out of hand. Once we differentiate $\vec{T}$, there is no longer a need to maintain t as a variable, and so we immediately evaluate everything at $t = \pi/4$ to avoid complications.

On another matter, it is always possible to experience **problems manipulating expressions involving Maple's norm( ) command.** The norm( ) command involves the use of absolute-value bars. You might already have noticed that **Maple behaves in a clumsy way when it manipulates expressions involving absolute-value bars.** Maple does not seem to know, for example, that $|a|^2 = a^2$. Actually, Maple knows quite well that this is true when a is a real number, but it doesn't make such rash assumptions unless it is told otherwise. (When a is allowed to be a complex number, the equation $|a|^2 = a^2$ is false, so Maple's behavior is understandable.) We could eliminate some of the problems if we used the assume( ) command. We could also simplify by using the option **assume = real**. In Maple 7, the convenient and less permanent "assuming" facility can be used.

It is easy enough to avoid the norm( ) command all together by taking advantage of the vector identity

$$|\vec{v}|^2 = \vec{v} \cdot \vec{v}$$

and so by using $\sqrt{\vec{v} \cdot \vec{v}}$, instead, for the length of $\vec{v}$.

Finally, as a convenience, we can **create our own length command**. The word *length* is already used by Maple for something else, so a name like **mynorm** might be a more appropriate length command. We could have use for it in the future, as well. If you are building a personal library of special commands, this might be a good one to include.

Actually, in this problem, Maple's norm( ) command works just as well as our newly created mynorm( ) command. Nevertheless, it could only be a matter of time before difficulties with the norm( ) command are experienced. To make an example out of this, we use our newly created "mynorm" command in our Maple worksheet.

The details of Maple's solution to this problem are **available at our web site**, but they are omitted from this manual to save space.

10.4 Exercise Set

New Maple Commands in this Chapter (and a few old commands as a reminder)

applyop( )	assume( )	assuming	convert( )	D( )
diff( )	evalm( )	map( )	norm(,2)	spacecurve( )
unapply( )	vector( )			

1. An object moves along a curve with position vector

$$\vec{r} = (3 - \sin(4t))\,\vec{i} + 2t\cos(t)\vec{j} + 5\sin(3t)\vec{k}, \;\; 0 \le t \le 10.$$

Plot the curve. It might not look very good unless a larger number of points is used to form the plot (the default value is only 50 points). To plot more points, say 200, for example, use **numpoints=200** as a plot option. Use the mouse to move the plot around so that it can be viewed from several directions. Find, in decimal form, at time $t = 7$, the velocity, acceleration, and speed of the object, the angle between the velocity and acceleration vectors, and the projection of the acceleration vector onto the velocity vector. Using just this decimal information at time $t = 7$, decide whether the object is speeding up or slowing down at this time.

2. Let γ be a circle in the xy-plane, having radius r and center at the point $A(0, R)$, where $0 < r < R$. A torus is formed by rotating γ about the x-axis. Two circular axes are naturally associated with the torus: One is γ itself, and the other is the circle β formed by rotating the point $A(0, R)$ about the x-axis.

If m and n are relatively prime integers (m, n have no common prime factor other than 1), then the following curves wrap m times around the torus in one "circular" way and n times around the torus in the other "circular" way. Set $r = 1$ and $R = 4$, and experiment with this family of curves, by using a variety of n and m values. Which of the integers, n or m, controls the number of times the curves wrap in the direction of γ, and which controls the number of times the curves wrap in the direction of β? Use the plot option **numpoints=200**, or some other suitably large number, to produce curves that look nice.

$$\begin{aligned} x &= r\sin(2\pi mt) \\ y &= (R + r\cos(2\pi mt))\cos(2\pi nt) \\ z &= (R + r\cos(2\pi mt))\sin(2\pi nt) \end{aligned}$$

3. In a rectangular coordinate system with units in feet, a projectile is shot from a gun at position $P(2, -3, 6)$ into a constant force field with acceleration vector $\vec{a} = 4.2\vec{\imath} - 1.8\vec{\jmath} + 29.7\vec{k}$ in ft/sec^2. No other forces are acting on the projectile. The muzzle velocity of the gun (the initial speed of the projectile) is 2,000 ft/sec^2, and the gun is pointed in the direction of the vector $\vec{g} = 7.2\vec{\imath} + 12.4\vec{\jmath} + 8.82\vec{k}$. Find the velocity and position of the projectile at time t in seconds. How close does the projectile come to a target located at $Q(120, 213, 89)$? The muzzle velocity of the gun is a constant 2,000 ft/sec^2, but its direction can be altered. Find a direction (expressed as a unit vector) so that the projectile will hit the target.

4. A force (in pounds) of

$$\vec{F} = 8\sqrt{t}\,\vec{\imath} - 3t\vec{\jmath} - \frac{7}{t^2+4}\vec{k}, \quad 0 \le t \le 60$$

is exerted on an object of mass $m = 12$ *slugs*, where t is time in seconds. The position and velocity of the object at time $t = 0$ are

$$\vec{v}_0 = 2\vec{\imath} + 7\vec{\jmath} + 13\vec{k}, \quad \vec{r}_0 = 8\vec{\imath} - 33\vec{\jmath} - 9\vec{k}.$$

Find the velocity and position of the object for time t in the interval $[0, 60]$. (Force is related to acceleration by *Newton's Law*, $\vec{F} = m\vec{a}$. When the force $\vec{F}$ is expressed in pounds, and mass m is expressed in slugs, then acceleration $\vec{a}$ is expressed in $^{ft}/_{sec^2}$.)

5. Let $\vec{f}(t)$ and $\vec{g}(t)$ be vector-valued functions. State a *Product Rule* for the derivative of $\vec{f}(t) \cdot \vec{g}(t)$, and use Maple to verify the rule.

6. Let $\vec{f}(t)$ and $\vec{g}(t)$ be vector-valued functions. State a *Product Rule* for the derivative of $\vec{f}(t) \times \vec{g}(t)$, and use Maple to verify the rule.

7. Let $\vec{f}(t)$, be vector-valued function, and let $c(t)$ be a scalar (real-valued) function. State a *Product Rule* for the derivative of $c(t)\vec{f}(t)$, and use Maple to verify the rule.

8. A projectile is shot from the top of a 100-ft-tall building, at an angle of elevation of 27° (above horizontal) and at an initial speed of $243^{ft}/_{sec}$. How high does the projectile go? What is its speed and angle at impact? How much time does it take to land, and how far down-range is it at impact? Assume a level field surrounding the building.

9. A baseball is struck 3 ft above ground, down the third base line, at an angle of elevation of 26°, and at an initial speed of $140^{ft}/_{sec}$. Will it clear a 30-ft-high fence that is 340 ft from home plate down the third base line?

10. Let Γ be the spiral

$$\vec{r} = \vec{r}(t) = a\cos(2\pi nt)\vec{\imath} + a\sin(2\pi nt)\vec{\jmath} + 2\pi nbt\vec{k},\ 0 \le t \le 1,$$

where a and b are positive constants and n is a positive integer. Show that Γ is a curve of constant curvature κ and determine the value of κ. Find the center of the circle of curvature corresponding to the point $\vec{r}(t)$ on Γ. Before asking Maple to perform these computations, try to guess their expected values.

11. Determine the frame vectors $\vec{T}$, $\vec{N}$, and $\vec{B}$ at the point $t = 2/3$ on the curve Γ defined by

$$x = \sin(4\pi t),\ y = (4 + \cos(4\pi t))\cos(10\pi t),\ z = (4 + \cos(4\pi t))\sin(10\pi t)$$

The curve Γ is one of the curves discussed in Problem 2. Determine the curvature, the radius of curvature, and the center of the circle of curvature at the point $t = 2/3$. If t is time, the parametric representation for Γ describes the motion of an object along the curve. Find the tangential, a_T, and normal, a_N, components of acceleration at $t = 2/3$.

Project: Slamming Sam's Home Run Attempt

A baseball field is shown in the following figure. The angle at home plate (labeled H in the picture) is 90°, and the three outfield fences are all 30 ft. tall. All of the other information concerning the field is shown in the picture. Slamming Sam strikes the ball 2.8 ft above ground, at an angle of elevation of 25°, and at an initial speed of 152 $ft/_{sec}$. From a "bird's eye" view, far above the playing field, the ball travels in a direction exactly midway between second and third base, as shown in the picture. Does Slamming Sam have another home run? If not, describe, in very exact language, how close it is to being a home run.

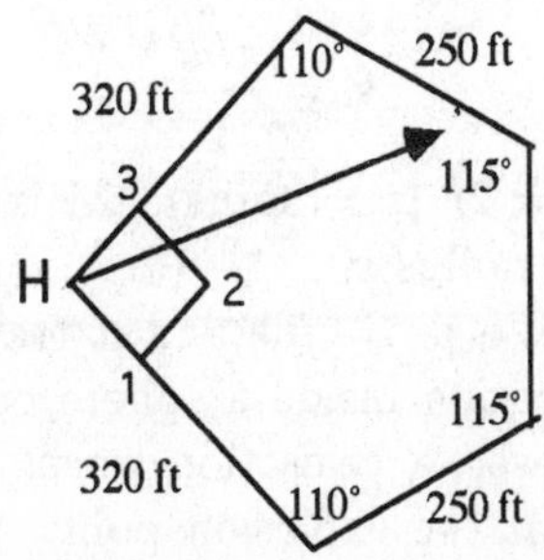

Project: Captain Ralph's Computer Failure (Version 1)

Captain Ralph is on a routine patrol a few thousand miles away from his home space station, where the origin of a three-dimensional coordinate system is located. (Coordinates are expressed in miles.) Looking for action, he grins as he spots an enemy ship. His display monitor shows that his coordinate position is $P(405, -3201, 95)$. He hits the attack button on his computer, and his ship begins to travel, starting at $t = 0$, along the attack curve

$$\vec{r}(t) = (405 - t^2)\vec{i} + (5t - 3201)\vec{j} + (95 + t^3)\vec{k}$$

(t is in seconds), firing off a projectile at the enemy ship $t = 2$ seconds later. The projectile travels along a straight line having the same direction as Captain Ralph's velocity vector at time $t = 2$, with a constant acceleration vector of 1 $mile/sec^2$ (naturally, in the same direction as Captain Ralph's velocity vector at time $t = 2$). The projectile has an initial speed that is 14 miles per second faster than Captain Ralph's speed at the firing time.

Unfortunately, Ralph's computer makes a mistake, and it programs the projectile to explode prematurely (4 seconds later). When the projectile explodes, it will annihilate everything within a 100 mile radius, and, in the meantime, Captain Ralph has no alternative to continuing along the same attack curve for another 5 seconds, before he regains control of his ship. Will he make it? Will he be 100 miles away from the projectile when it explodes? Come on, Captain Ralph! Weeeeeeeeeee ... neeeeeeeeeeed ... youuuuuuuuuu!

Project: Captain Ralph's Computer Failure (Version 2)

This version is the same as the first, except for the way that the bomb explodes. As every fighter pilot for the United Earth Federation knows, bombs do not always explode in spherical patterns. Acceleration warps and flattens the explosive region.

By a circular ellipsoid, we mean the surface generated by revolving an ellipse about its major or minor axis. The axis of the circular ellipsoid is the line segment on the axis of rotation terminating at points on the surface. The center of the ellipse is the midpoint of the axis. The diameter of the ellipsoid is the diameter of the circle formed by intersecting the ellipsoid with a plane passing through the center that is perpendicular to the axis. It is a straightforward exercise to show that, if d is the diameter of and l is the length of the axis of a circular ellipsoid, then the volume of the enclosed region is

$$V = \frac{\pi l d^2}{6}.$$

As you can see, when $l = d = 2r$ (r =radius), this turns into the familiar formula for the volume of a sphere of radius r.

When Captain Ralph fires a projectile and it travels at a constant velocity, its region of destruction is the region inside a sphere of radius 100 miles centered at the point of explosion. However, acceleration warps the region of destruction into a circular ellipsoid centered at the explosion point, having an axis parallel to the acceleration vector. The volume of the region of destruction remains the same, but it flattens so that l decreases, and d increases, according to the rule

$$l = \frac{d}{1 + 2|\vec{a}|},$$

where $\vec{a}$ denotes the acceleration vector expressed in $miles/sec^2$.

Outside of this change in the explosive pattern, this version of the problem is the same as the first. Determine whether Captain Ralph survives this computer failure.

You have all the tools needed to solve this problem, but this is definitely a nontrivial project. A planar ellipse is defined in terms of the sum of the distances between a point on the ellipse and its two focal points. This definition can be used very effectively to decide whether a point P is inside or outside an ellipse. Now try to turn this problem into a planar problem.

Chapter 11

Multivariate Differential Calculus

11.1 Functions of Several Variables

In a pencil-and-paper mathematical environment, expressions and functions are often treated as if they were equivalent, even though there is, philosophically, a difference between the two ideas. Maple, however, not only appreciates the difference between a function and an expression, but will not allow us to get careless and treat these ideas as being the same.

The same is true of functions and expressions in two or more variables. We will work with both of these Maple data types, but their differences must be appreciated. If f is a Maple function of x and y, then $f(x, y)$—a Maple expression—is the value of f at (x, y). If g is a Maple expression in x and y, then g itself is the value of (what we're thinking of as) the function at (x, y), and $g(x, y)$ makes no sense to Maple.

The methods used to introduce functions of several variables are quite similar to the methods used for introducing functions of one variable. A few examples are shown in the following Maple work session. The **unapply( ) command** that we are accustomed to using for turning an expression in one variable into a function of one variable can also be used to turn an **expression in several variables into a function of several variables**. More complicated functions can be introduced by using the **proc( ) command**, but there are usually more convenient ways to define even complicated functions. Remember that a **Maple procedure** must be terminated with the word **end**. Actually, regardless of how a function is introduced, it ultimately becomes a Maple procedure. The proc( ) command is mentioned here only because it is versatile enough that it can be used when more appropriate command names are forgotten.

****

```
> f:=(x,y)->x^2*sin(y);
```

$$f := (x,\ y) \to x^2 \sin(y)$$

```
> f(3,2*Pi/3); ln(f(y,x*y));
```

$$\frac{9}{2}\sqrt{3}$$

$$\ln(y^2 \sin(x\,y))$$

```
> g:=(u,v,w,x)->u+v^2+w^3+x^4;
```

$$g := (u,\, v,\, w,\, x) \to u + v^2 + w^3 + x^4$$

```
> g(4,3,2,1); g(u,v,w,f(u,v+w));
```

$$22$$

$$u + v^2 + w^3 + u^8 \sin(v + w)^4$$

```
> f:=x^2+y^3;
```

$$f := x^2 + y^3$$

```
> g:=unapply(f,x,y);
```

$$g := (x,\, y) \to x^2 + y^3$$

```
> h:=unapply(f,y,x);
```

$$h := (y,\, x) \to x^2 + y^3$$

```
> g(x,y);h(x,y);
```

$$x^2 + y^3$$

$$y^2 + x^3$$

Notice the order of appearance of the variables in the two unapply commands. The functions g and h are quite different.

```
> p:=(x,y)->piecewise(y<=9-x^2,25-x^2-y^2,y>0,10*sqrt(y),0);
```

$$p := (x,\, y) \to \text{piecewise}(y \leq 9 - x^2,\, 25 - x^2 - y^2,\, 0 < y,\, 10\sqrt{y},\, 0)$$

```
> p(x,y);
```

$$\begin{cases} 25 - x^2 - y^2 & y \le 9 - x^2 \\ 10\sqrt{y} & 0 < y \\ 0 & \textit{otherwise} \end{cases}$$

```
> p(2,5), p(2,6), p(4,-8), p(4,-6);
```

$$-4,\ 10\sqrt{6},\ -55,\ 0$$

```
> plot3d(p(x,y),x=-5..5,y=-5..5);
```

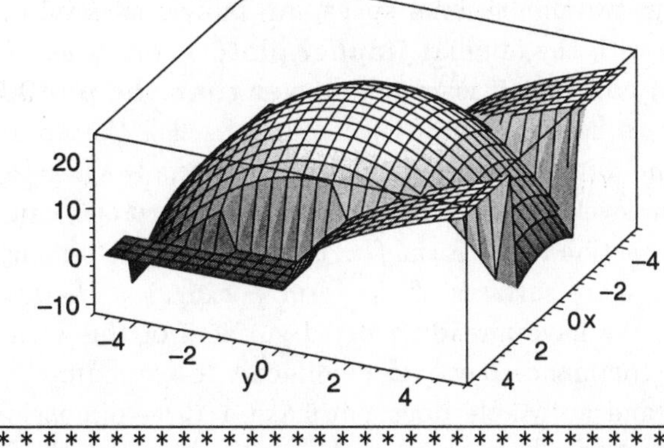

****

The plot demonstrates how the conditions are used in the piecewise( ) command. In particular, $p(x, y)$ takes the second value $10\sqrt{y}$ only if the **second condition $y > 0$ is true and the first condition $y \le 9 - x^2$ is false.**

We have already plotted expressions in two variables; there is no need to amplify on the above plot3d( ) command. Notice how the range was expressed in the last plot command. **Expressions depend on letter choices, functions do not.** Using the letters x, y in the range of the last plot command would have made no more sense than using any two other letters.

We cannot graph a function of three or more variables, at least not visually. We can still define the graph of a function of n variables as a certain subset of $(n+1)$-dimensional space, and such a graph can be studied mathematically, but, obviously, we can see it with our eyes only when $n = 1$ or 2. Nevertheless, there are a few "tricks", most notably level sets and sections, that can be used to bring our sense of vision to bear on some of these graphs. These techniques are even useful to use on the three-dimensional graph of a function of two variables. Plan on loading the **plots package** for much of this work.

Graph a function f of two variables x and y. Move the mouse pointer to the plot and click. As we have seen before, this activates the plot tool bar. With a three-dimensional plot, however, the buttons are different, and there are more of them. To find out more about these buttons, use a method we have used before. Click one of the buttons and hold the mouse key down. An explanation of the button's function will appear as a statement in the Status Line at the bottom of the computer screen. Slide the mouse pointer off the button before releasing the mouse key, and the button will not be activated. With the plot tool bar activated, click the **Contour button**, or choose **Contour** from the **Style Menu**. Then click the **[R] button to redraw the graph**. The curves that appear are the level curves for the function. Now use the mouse to turn the plot around so that it is being viewed exactly from above (the frame box will turn into a rectangle when you are in the right position). What will now be seen, instead of a three-dimensional graph, is a two-dimensional graph of a family of level curves for f, or, in other words, several curves of the form $f(x,y) = C$ corresponding to different values of C.

The **level curves of a function of two variables can be plotted** directly by using the command **contourplot( )**, which resides in the **plots package.** This allows us to view a graph living in three-dimensional space, by looking at certain parts of the graph in two-dimensional space. An individual level curve $f(x,y) = C$ can also be plotted with the familiar **implicitplot( )** command. It turns out that the **contourplot( ) command is much slower than the plot3d( ) command.** It is usually better to use the plot3d( ) command, click the appropriate buttons, turn the plot around with the mouse pointer, and finally click [R] to redraw the graph. A plot of the level curves of a function of two variables can be very useful.

The graph of a function of the form $f(x,y,z)$ lives in four-dimensional space, but the level surfaces of f are surfaces of the form $f(x,y,z) = C$, which live in three-dimensional space. We have already plotted surfaces of this form, using Maple's **implicitplot3d( )** command. It would be difficult to see more than one such level surface at a time, and so Maple does not have a three-dimensional analogue of its contourplot command. It is also worth mentioning that the implicitplot3d( ) command is time and memory intensive.

Another effective way of reducing dimensions and looking at parts of a graph is to plot **sections.** The graph of $z = f(x,y)$ lives in three-dimensional space, but the intersection of this graph with a vertical plane is a curve that can be plotted as a two-dimensional graph. Such plots are frequently called **sections of a graph.** The most convenient sections to plot are those of the form $z = f(x,C)$, or $z = f(C,y)$, where C is an arbitrary constant. These sections are the intersections of $z = f(x,y)$ which vertical planes parallel to the x or y axes.

Sections of $g(x,y,z)$ are particularly useful, because the graph of a function of three variables lives in four-dimensional space. The three-dimensional sections of the form $w = g(x,y,C)$, $w = g(x,C,z)$, $w = g(C,y,z)$ can be be viewed with the plot3d( ) command, and the two-dimensional sections of the form $w = g(x,C_1,C_2)$, $w = g(C_1,y,C_2)$, $w = g(C_1,C_2,z)$ can be viewed with the plot( ) command.

A motion picture of the sections of $z = f(x,y)$ can be created by Maple's animate( ) command. For example,

```
> animate(f(x,C),x=a..b,C=p..q);
```

will draw the intersection of $z = f(x,y)$ with the plane $y = C$ for 16 values (the default number) of C over the interval $[p,q]$. The three-dimensional sections of $w = g(x,y,z)$ can be viewed by entering an input statement similar to

```
> animate3d(g(x,y,C),x=a..b,y=c..d,C=p..q);
```

When the plot tool bar is activated on an animation plot (by clicking the plot), the buttons in the tool bar become animation buttons. A little experimentation will explain their functions. **These are memory-intensive activities.** The first entry in the animate and animate3d commands must be an expression (or a set of expressions enclosed in curly brackets), and so these commands cannot be used to plot level curves or level surfaces. However, animations of this sort could be initiated by using the implicitplot or implicitplot3d commands along with the animated versions of Maple's display( ) and display3d( ) commands.

Look at the output created by the following work session. The animation needs to be experienced first-hand, of course, so its output is not included.

****

```
> with(plots):

Warning, the name changecoords has been redefined

> f:=(x,y)->(25-x^2-y^2)*sin(x^2+y^2):
> contourplot(f(x,y),x=-2..2,y=-2..2);
```

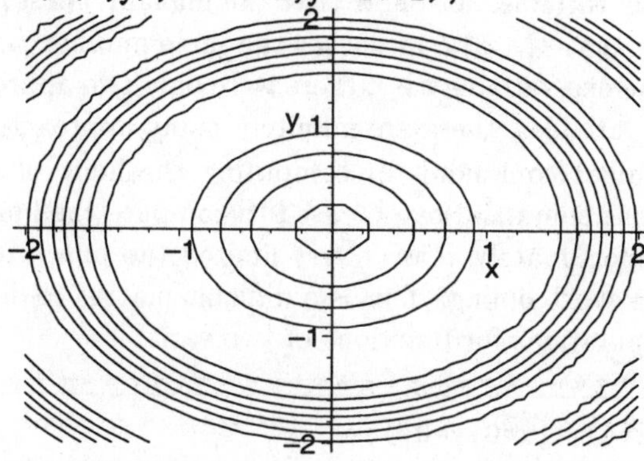


```
> animate(f(x,c),x=-5..5,c=0..5);
> implicitplot(f(x,y)=1,x=-5..5,y=-5..5);
```

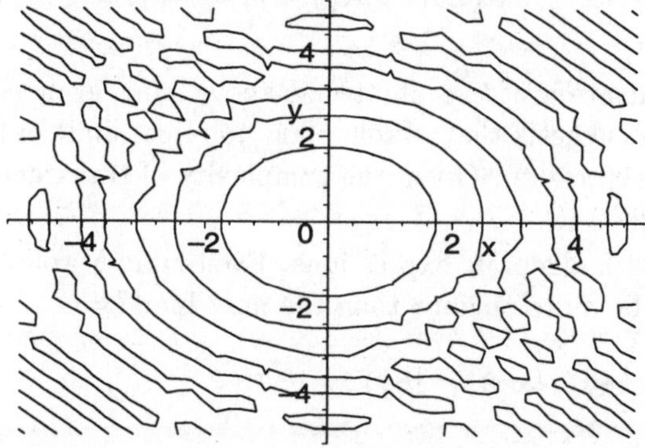

#This plot is clearly wrong! The function f becomes increasingly chaotic as the point (x, y) moves further away from the origin. Maple seems to be able to handle this function reasonably well for x and y between -2 and 2, but not for x and y in the larger range. This **should have been expected**, if not for x and y between -5 and 5, then at least in some slightly larger range. Why should this behavior by Maple be expected? **Think about this question. It is important to understand the limitations of computer software.#**

```
> g:=(x,y,z)->(10-z)*x^2+z*y^2:
> animate3d(g(x,y,c),x=-20..20,y=-20..20,c=-10..25);
```

****

11.2 Limits with Two or More Variables

Limits of functions are frequently easy to compute by observation. We are all familiar with the circumstances that allow for the immediate evaluation of a limit of a function of one variable, and these same circumstances allow for the immediate evaluation of a limit of a function of several variables. It is only when a function has one of the so-called indeterminate forms at a point that its limit becomes more difficult to compute. In the case of a function of one variable, *L'Hôpital's Rule* can often be used to compute a limit. **This rule, unfortunately, does not extend to functions of more than one variable.** Establishing the existence of such a limit is fundamentally more difficult for a function of two or more variables than it is for a function of just one variable. To show that the limit of $f(x, y)$ as $(x, y) \to (a, b)$ exists, we must show that $f(x, y)$ approaches the same number L, regardless of how $(x, y) \to (a, b)$. In the one-variable case, there were only two approach paths. In the case of two or more variables, there are infinitely many approach paths.

Maple has a familiar command for computing the limit of a function of one variable, and this same command can be used to compute the limit of a function of several variables. Unfortunately, it is of very limited (no pun intended) use. Don't blame Maple for this shortcoming. This is a difficult matter to deal with. Here is a sample of limit computations for functions of two variables.

****

```
> limit(sin(x*y), {x=0,y=0});
```

$$\text{limit}(\sin(x\,y),\ \{y = 0,\ x = 0\})$$

#Maple makes no attempt to evaluate the above limit, even though it is obvious. This trivial example suggests that Maple plans to reject limit evaluations involving any transcendental expression. Given the complexity of transcendental expressions, this is not surprising.

That leaves us with algebraic expressions. Even here, Maple balks at some easy limits. What could be more obvious than the next limit?#

```
> limit(sqrt(x+y),{x=9,y=16});
```

$$\text{limit}(\sqrt{x+y},\ \{y = 16,\ x = 9\})$$

#It appears that Maple's multivariate limit command is restricted to rational functions. Next, we set up a rational expression with a 0/0 indeterminate at $(x, y) = (4, -1)$. We can see at a glance how the common zero producing terms in the numerator and denominator cancel, giving us a limit with an obvious value. This makes it easy to check Maple's performance.#

```
> p:=(x-4)*(y+1)*(x*y-2):  q:=(x-4)*(y+1)*(x+3*y):
> r:=expand(p)/expand(q);
```

$$r := \frac{x^2\,y^2 - 6\,x\,y + x^2\,y - 2\,x - 4\,x\,y^2 + 8\,y + 8}{x^2\,y + 3\,x\,y^2 + x^2 - x\,y - 12\,y^2 - 4\,x - 12\,y}$$

```
> limit(r,{x=4,y=-1});
```

$$-6$$

#Maple did well with the above limit. The next 0/0 indeterminate is not so obvious.#

```
> f:=(y+3*x-13)/(y-x-5);
```

#Notice that f is undefined everywhere on a line passing through the point $(x, y) = (2, 7)$, where we plan to take the limit.#

$$f := \frac{y + 3\,x - 13}{y - x - 5}$$

```
> limit(f,{x=2,y=7});
```

undefined

#Maple has given us a clear answer. The limit does not exist. Rather than check Maple's work, let's go on to the next example, where some verification will be attempted.#

```
> f:=(x-2)*(y-7)/(y-x-5);
>
```

$$f := \frac{(x-2)\,(y-7)}{y - x - 5}$$

```
> limit(f,{x=2,y=7});
```

$$\mathrm{limit}(\frac{(x-2)\,(y-7)}{y - x - 5}, \{y = 7,\, x = 2\})$$

#Does the limit fail to exist, or has Maple simply given up? It is hard to tell at this point. To investigate this matter further, we let $(x, y) \to (2, 7)$ along the line of slope m, where m is an arbitrary real number.#

```
> f1:=simplify(subs(y=m*(x-2)+7,f));
```

$$f1 := \frac{m\,(x-2)}{m - 1}$$

```
> limit(f1,x=2);
```

$$0$$

#This shows that the limit is 0 along every linear path leading to (2,7), except along the line corresponding to $m = 1$, a line which is not in the domain of f (its denominator is 0 on this line). This would suggest that the limit exists, and equals 0 (at least if we restrict (x, y) to the domain of the expression). It is important to realize, however, that the above work **does not prove that the limit exists.** In fact, it turns out, as we show next, that **the limit does not exist.**

In our next calculation, we let $(x, y) \to (2, 7)$ along a path which pushes (x, y) closer to the boundary line $y = x + 5$, where f is undefined. We let $(x, y) \to (2, 7)$ along a quadratic passing through (2,7), which has slope 1 at $x = 2$ (the same slope as $y = x + 5$). It is easy to create such quadratics. Just add a term of the form $a(x - 2)^2$, where a is an arbitrary constant, to the line $y = x + 5$.#

```
> f2:=simplify(subs(y=x+5+a*(x-2)^2,f));
```

$$f2 := \frac{a\,x - 2\,a + 1}{a}$$

```
> limit(f2,x=2);
```

$$\frac{1}{a}$$

#Now we have a clear proof that the limit does not exist. By using different values for a, we see that the limit varies as we let $(x, y) \to (2, 7)$ along different quadratic paths.#

****

Example 11.1 *Show that the limit does not exist at the origin for the function f defined by*

$$f(x, y) = \frac{sin(x^2 y)}{sin(x^3 + y^3)}$$

When a limit does not exist, it is usually a straightforward process to confirm this. All we have to do is show that $f(x, y)$ tends to different values as $(x, y) \to (0, 0)$ along two different paths. In this example, we approach the origin along the line $y = mx$, where m is any fixed real number.

****

```
> f:=sin(x^2*y)/sin(x^3+y^3):
> limit(f,{x=0,y=0});
```

$$\text{limit}(\frac{\sin(x^2\,y)}{\sin(x^3 + y^3)},\ \{x = 0,\ y = 0\})$$

```
> g:=subs(y=m*x,f);
```

$$g := \frac{\sin(x^3\, m)}{\sin(x^3 + m^3\, x^3)}$$

```
> limit(g,x=0);
```

$$\frac{m}{1+m^3}$$

#This limit value is clearly different for different values of m, and so the full limit does not exist.#

****

On the other hand, suppose we show that the limit of a function $f(x, y)$ as $(x, y) \to (a, b)$ is the same number L along a variety of paths. If the full limit exists, its value is obviously L, but have we shown that the full limit exists? **Clearly, the answer is no**! We saw dramatic evidence, in a previous work session, that such a limit can fail to exist, regardless of how many paths were used in the analysis. Showing that the value of a limit is path independent can be very difficult for a function of more than one variable, if it has an indeterminate form at the point.

On that less than happy note, we move on to other matters. Actually, these limit complications are seldom encountered. It is enough to be aware of the circumstances that create this unresolved limit problem. In addition, limits frequently have directions attached to them, so that path independence is not even an issue.

11.3 Partial Derivatives

We already have a Maple command for computing a partial derivative of an expression. If f is an expression, the familiar Maple input statement

```
> diff(f,x);
```

that we have been using differentiates the expression f with respect to x and treats all unassigned letters in f as constants. In other words, it computes the partial derivative of f with respect to x. The output, recall, is a Maple expression (rather than a Maple function).

If f is a Maple function of one variable, then D(f) is the derivative of f in the form of a Maple function. We have all experienced how convenient it is to have the derivative of a function returned as a function rather than as an expression. Maple, fortunately, has similar commands for partial derivatives. If f is a function in the variables $x_1, x_2, \ldots, x_n$, then D[j](f) represents—in the form of a Maple function—the partial derivative of f with respect to x_j. In addition, D[j, k](f) represents—in the form of a function—the second partial derivative of f, first with respect to x_j and then x_k. Higher-order partials are handled in a similar way.

Example 11.2 *Compute the first-order and second-order partial derivatives of the function f defined by*

$$f(x,y) = \frac{x^2 y}{4+y^2}.$$

Verify the first-order partials by using the definition of a partial derivative as the limit of a difference quotient.

Notice how the inert form Diff( ) of the differentiation command diff( ) is used to create a visually pleasing display.

****

```
> f:=(x,y)->x^2*y/(4+y^2):
> Diff(f,x)=diff(f(x,y),x);
```

$$\frac{\partial}{\partial x} f = 2\frac{x\,y}{4+y^2}$$

The above output display looks nice, but it is **awkward to use in subsequent calculations**, and so it is not recommended except as a final display. Assignments like the following are much more convenient.

```
> fy:=diff(f(x,y),y);
```

$$fy := \frac{x^2}{4+y^2} - \frac{2\,x^2\,y^2}{(4+y^2)^2}$$

```
> fyx:=diff(f(x,y),y,x);
```

$$fyx := 2\frac{x}{4+y^2} - \frac{4\,x\,y^2}{(4+y^2)^2}$$

```
> fxy:=diff(f(x,y),x,y);
```

$$fxy := 2\frac{x}{4+y^2} - \frac{4\,x\,y^2}{(4+y^2)^2}$$

```
> fxx:=diff(f(x,y),x,x);
```

$$fxx := 2\frac{y}{4+y^2}$$

```
> fyy:=diff(f(x,y),y,y);
```

$$fyy := -6\frac{x^2\,y}{(4+y^2)^2} + \frac{8\,x^2\,y^3}{(4+y^2)^3}$$

```
> qx:=(f(x+h,y)-f(x,y))/h;
```

$$qx := \frac{\dfrac{(x+h)^2\,y}{4+y^2} - \dfrac{x^2\,y}{4+y^2}}{h}$$

```
> qx1:=simplify(qx);
```

$$qx1 := \frac{y\,(2\,x+h)}{4+y^2}$$

The limit of $qx1$ as $h \to 0$ is now obvious.#

```
> Diff(f,x)=limit(qx1,h=0);
```

$$\frac{\partial}{\partial x} f = 2 \frac{x\,y}{4+y^2}$$

```
> qy:=(f(x,y+k)-f(x,y))/k;
```

$$qy := \frac{\frac{x^2\,(y+k)}{4+(y+k)^2} - \frac{x^2\,y}{4+y^2}}{k}$$

```
> qy1:=simplify(qy);
```

$$qy1 := -\frac{x^2\,(-4+y^2+y\,k)}{(4+y^2+2\,y\,k+k^2)\,(4+y^2)}$$

The limit of $qy1$ as $k \to 0$ is now obvious.#

```
> Diff(f,y)=limit(qy1,k=0);
```

$$\frac{\partial}{\partial y} f = -\frac{(-4+y^2)\,x^2}{(4+y^2)^2}$$

* *

The next example is offered just to demonstrate the convenience of using the command D( ) to compute a partial derivative.

Example 11.3 *Let f be the function defined by $f(x,y) = \cos(2x-3y)$. Determine the polynomial function p defined by*

$$p(x,y) = f(0,0) + f_x(0,0)x + f_y(0,0)y + \frac{1}{2}f_{xy}(0,0)xy$$
$$+\frac{1}{2}f_{yx}(0,0)xy + \frac{1}{2}f_{xx}(0,0)x^2 + \frac{1}{2}f_{yy}(0,0)y^2.$$

Evaluate $f(x,y)$, $p(x,y)$, and $e(x,y) = f(x,y) - p(x,y)$ at $(x,y) = (0.2, 0.2)$.

By using Maple's D( ) command, we can compute the polynomial directly in one step. For example, $D[2,1](f)$ is a function, so we can compute its value, $D[2,1](f)(0,0)$, at the origin, directly, without any intermediate calculations.

The function f and the polynomial p agree at $(x,y) = (0,0)$. All of the first-order and second-order partials of f agree with those of p at $(x,y) = (0,0)$ as well. This is evident by observation, but it is easy to verify with Maple. Not surprisingly, with all this agreement, the polynomial $p(x,y)$ is a good approximation of $f(x,y)$ for values of (x,y) close to $(x,y) = (0,0)$. Look carefully, and you can see a similarity here to the notion of a Taylor polynomial of a function of one variable.

****

```
> f:=(x,y)->cos(2*x-3*y):
> p:=f(0,0)+D[1](f)(0,0)*x+D[2](f)(0,0)*y+D[1,2](f)(0,0)*x*y/2
  +D[2,1](f)(0,0)*x*y/2+D[1,1](f)(0,0)*x^2/2+D[2,2]( f)(0,0)*y^2/2;
```

$$p := 1 + 6\,x\,y - 2\,x^2 - \frac{9}{2}\,y^2$$

```
> p:=unapply(p,x,y);
```

$$p := (x,\,y) \rightarrow 1 + 6\,x\,y - 2\,x^2 - \frac{9}{2}\,y^2$$

```
> e:=(x,y)->f(x,y)-p(x,y);
```

$$e := (x,\,y) \rightarrow \mathrm{f}(x,\,y) - \mathrm{p}(x,\,y)$$

```
> evalf(f(.2,.2)); evalf(p(.2,.2)); evalf(e(.2,.2));
```

$$.9800665778$$

$$.9800000000$$

$$.0000665778$$

****

Let *eqn* (or *expr*= 0) denote an equation in two or more variables. It is natural to think of solving the equation for one of the letters (w, let's say) in terms of the other letters. If the solution exists, we say that the equation *eqn* defines w **implicitly** as a function of the other variables. If we actually find the solution, then we have w **explicitly** as a function of the other variables, but most of the time the equation *eqn* is **too difficult to actually find the solution.** Graphical considerations suggest, however, that solutions almost always exist, even though we might not be able to find them. We will look at some of this graphical evidence in a moment.

Implicit differentiation is a technique that can be used to compute a derivative (or partial derivative) of such a function without finding a formula for the function. This technique was used early in the manual, when we assumed that an equation *eqn* in x and y defined y implicitly as a function of x, and we computed the derivative, $\frac{dy}{dx}$, using this method. The situation is much the same when *eqn* is an equation in more than two variables. The only real difference is that the derivatives we compute become partial derivatives.

Maple has a command, implicitdiff( ), for computing a derivative implicitly. It is an impressive way to get an answer quickly, but it will not lead to any further understanding of the basic processes involved in the problem. It is much more instructive to differentiate implicitly without this command, by using Maple's diff( ) command instead. We use both methods in the next example. It may help to review Example 2.1.

Example 11.4 *Assume that the equation* $x^2y + e^{xz}y^2 + 4xz - 2yz^2 = 8$ *defines* z *implicitly as a differentiable function* $z = f(x, y)$ *of* x *and* y *in some neighborhood of the point* $(x_0, y_0) = (5, -2)$. *Use implicit differentiation and Maple's diff( ) command to compute the partials of* z *with respect to* x *and* y *at* $(5, -2)$. *The equation could define, implicitly, more than one such function* $z = f(x, y)$ *in a neighborhood of the point* $(5, -2)$. *In this case, compute the partials for each such function. Use Maple's implicitdiff( ) command to confirm the values of all of the partials.*

Intuitively, each function $z = f(x, y)$ represents one of the solutions obtained by solving the equation for z in terms of x and y. There could be several such functions defined by the equation. The assumption that they exist and are differentiable is really quite reasonable. The graph of the given equation is a surface, which could overlap the (x, y) plane several times. If we graph the equation over a small enough rectangle R in the (x, y) plane centered at the point (5,-2), we will see the parts of the whole surface which lie over the small rectangle R. It is reasonable to expect that each one of these parts (if there is more than one) represents the graph of a differentiable function of the form $z = f(x, y)$. These are the graphs of the functions that are defined implicitly in a neighborhood of $(x_0, y_0) = (5, -2)$ by the given equation.

We begin our Maple work session by forming this graph. To do this, pick (by guesswork) a small x-range and y-range centered about the point (5,-2). All that is left to do is pick a range of z-values for the plot. This requires a little thought, for we don't want to exclude parts of the surface that could represent the kinds of functions we are looking for. How many solutions are there, and where are they likely to occur?

It helps to look at the equation in z, which is obtained by letting $x = 5$ and $y = -2$. This is denoted by *eq0* in the Maple work session below. Clearly the term $4z^2$ dominates this equation for negative values of z, and the term e^{5z} quickly dominates the equation for positive values of z. The left-hand side of *eq0* is clearly larger (much larger) than 8 for z outside the interval $-10 < z < 1$, and so any solution must be in this interval. We add some wiggle room to this range of z values and plot the surface accordingly. Surely, with this range of z-values, we are not excluding any part of the surface that could generate another solution to this problem.

****

```
> eq:=x^2*y+exp(x*z)*y^2+4*x*z-2*y*z^2=8:
> eq0:=subs(x=5,y=-2,eq);
```

$$eq0 := -50 + 4\,e^{(5\,z)} + 20\,z + 4\,z^2 = 8$$

```
> with(plots):
```

```
Warning, the name changecoords has been redefined
> implicitplot3d(eq,x=3..7,y=-4..0,z=-20..10);
```

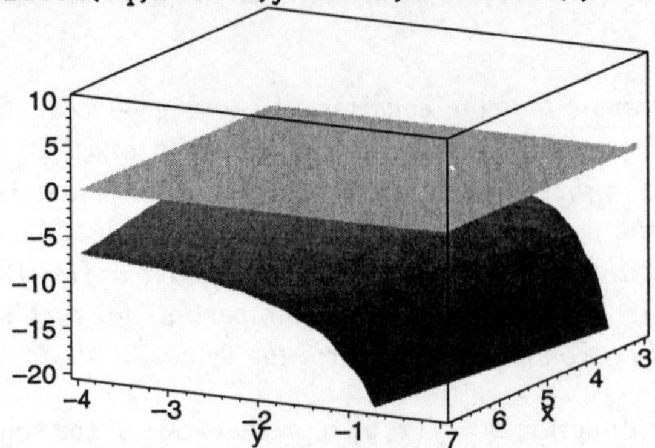

#Notice that **two distinct surfaces** have appeared. Each one is the graph of a function $z = f(x, y)$ defined implicitly by the given equation.

Finding the two values of z corresponding to $x = 5$ and $y = -2$ is next on our agenda. Maple will **never** give us more than one solution to the equation *eq0* (if that) without help. Can you look at the above graph and supply Maple with a range of z-values containing either one of the two solutions? Perhaps, but here is a more reliable method. Write the equation *eq0* in the form *lhs(eq0)-rhs(eq0)=0*. The left-hand side of this new equation is an expression in z that can be graphed. The solutions we are looking for are the z-intercepts of this graph.#

```
> plot(lhs(eq0)-rhs(eq0),z=-10..1);
```

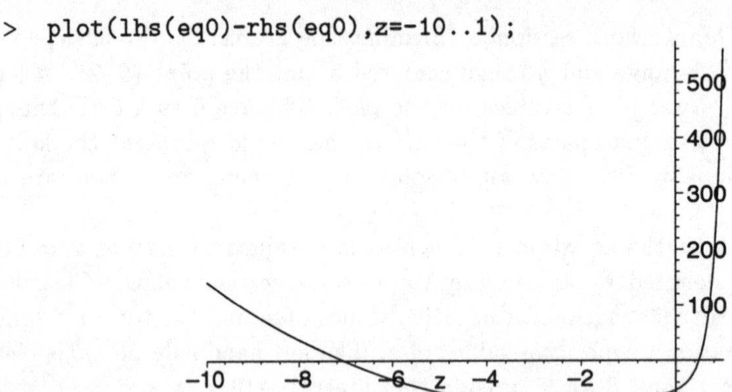

#From this plot, we see that there is a solution between 0 and 1 and a solution between -8 and -6.#

```
> z0:=fsolve(eq0,z);
```

$$z0 := .4934388706$$

```
> z1:=fsolve(eq0,z,z=-8..-6);
```

$$z1 := -7.055216790$$

Before we differentiate, we must tell Maple that z depends on x and y. Unassigned letters are always independent of each other. Maple will simply return 0 for diff(z, x), and diff(z, y) unless it is told that z depends on x and y.#

```
> imp_eq:=subs(z=z(x,y),eq);
```

$$imp\_eq := x^2\,y + e^{(x\,\mathrm{z}(x,\,y))}\,y^2 + 4\,x\,\mathrm{z}(x,\,y) - 2\,y\,\mathrm{z}(x,\,y)^2 = 8$$

Notice that we used a substitution rather than an assignment in the last input statement. The assignment $z := z(x, y)$ might seem like a reasonable strategy , but **it will not work.** Assigning z to an expression involving z is a "circular" assignment and definitely prohibited as long as z is initially unassigned.#

```
> deqx:=diff(imp_eq,x); deqy:=diff(imp_eq,y);
```

$$deqx := 2\,x\,y + (\mathrm{z}(x,\,y) + x\,(\frac{\partial}{\partial x}\mathrm{z}(x,\,y)))\,e^{(x\,\mathrm{z}(x,\,y))}\,y^2 + 4\,\mathrm{z}(x,\,y) + 4\,x\,(\frac{\partial}{\partial x}\mathrm{z}(x,\,y)) - 4\,y\,\mathrm{z}(x,\,y)\,(\frac{\partial}{\partial x}\mathrm{z}(x,\,y)) = 0$$

$$deqy := x^2 + x\,(\frac{\partial}{\partial y}\mathrm{z}(x,\,y))\,e^{(x\,\mathrm{z}(x,\,y))}\,y^2 + 2\,e^{(x\,\mathrm{z}(x,\,y))}\,y + 4\,x\,(\frac{\partial}{\partial y}\mathrm{z}(x,\,y)) - 2\,\mathrm{z}(x,\,y)^2 - 4\,y\,\mathrm{z}(x,\,y)\,(\frac{\partial}{\partial y}\mathrm{z}(x,\,y)) = 0$$

#Now that we have the partials, we simplify these equations with more manageable names.#

```
> deqx2:=subs(diff(z(x,y),x)=dzx,z(x,y)=z,x=5,y=-2,deqx);
  deqy2:=subs(diff(z(x,y),y)=dzy,z(x,y)=z,x=5,y=-2,deqy);
```

$$deqx2 := -20 + 4\,(z + 5\,dzx)\,e^{(5\,z)} + 4\,z + 20\,dzx + 8\,z\,dzx = 0$$

$$deqy2 := 25 + 20\,dzy\,e^{(5\,z)} - 4\,e^{(5\,z)} + 20\,dzy - 2\,z^2 + 8\,z\,dzy = 0$$

```
> dzx:=solve(deqx2,dzx); dzy:=solve(deqy2,dzy);
```

$$dzx := -\frac{-5 + e^{(5\,z)}\,z + z}{5\,e^{(5\,z)} + 5 + 2\,z}$$

$$dzy := \frac{1}{4}\,\frac{-25 + 4\,e^{(5\,z)} + 2\,z^2}{5\,e^{(5\,z)} + 5 + 2\,z}$$

```
> z=z0; Diff(z,x)=evalf(subs(z=z0,dzx));
  Diff(z,y)=evalf(subs(z=z0,dzy));
```

$$z = .4934388706$$

$$\frac{\partial}{\partial x} z = -.02018602924$$

$$\frac{\partial}{\partial y} z = .08718250218$$

```
> z=z1; Diff(z,x)=evalf(subs(z=z1,dzx));
  Diff(z,y)=evalf(subs(z=z1,dzy));
```

$$z = -7.055216790$$

$$\frac{\partial}{\partial x} z = -1.323231950$$

$$\frac{\partial}{\partial y} z = -2.045790885$$

#To finish this problem, we check our answers with Maple's **very impressive implicitdiff( ) command.#**

```
> zx:=implicitdiff(eq,z,x);
```

$$zx := -\frac{2\,x\,y + e^{(x\,z)}\,y^2\,z + 4\,z}{e^{(x\,z)}\,y^2\,x + 4\,x - 4\,y\,z}$$

```
> zy:=implicitdiff(eq,z,y);
```

$$zy := -\frac{x^2 + 2\,e^{(x\,z)}\,y - 2\,z^2}{e^{(x\,z)}\,y^2\,x + 4\,x - 4\,y\,z}$$

```
> Diff(z,x)=evalf(subs(x=5,y=-2,z=z0,zx));
  Diff(z,y)=evalf(subs(x=5,y=-2,z=z0,zy));
```

$$\frac{\partial}{\partial x} z = -.02018602921$$

$$\frac{\partial}{\partial y} z = .08718250216$$

```
> Diff(z,x)=evalf(subs(x=5,y=-2,z=z1,zx));
Diff(z,y)=evalf(subs(x=5,y=-2,z=z1,zy));
```

$$\frac{\partial}{\partial x} z = -1.323231950$$

$$\frac{\partial}{\partial y} z = -2.045790885$$

* *

The *Chain Rule* for functions of several variables bears a striking resemblance to its one-variable form, but it is somewhat more involved. It can be written in matrix form as one rule, but it is usually presented in calculus as a list of rules. Maple understands how to use the *Chain Rule* even in its most abstract setting.

Example 11.5 *Let $f(u,v,w)$ be a differentiable function of u, v, and w, and let $u = u(x,y)$,$v = v(x,y)$,and $w = w(x,y)$, be differentiable functions of x and y, so that*

$$h(x,y) = f(u(x,y), v(x,y), w(x,y))$$

is a differentiable function of x and y. Compute the partials of $h(x,y)$ with respect to x and y.

* *

```
> h:=f(u(x,y),v(x,y),w(x,y)):
> Diff('h',x)=diff(h,x);
```

$$\frac{\partial}{\partial x} h = D_1(f)(\mathrm{u}(x,y), \mathrm{v}(x,y), \mathrm{w}(x,y))\,(\frac{\partial}{\partial x}\mathrm{u}(x,y)) \\ + D_2(f)(\mathrm{u}(x,y), \mathrm{v}(x,y), \mathrm{w}(x,y))\,(\frac{\partial}{\partial x}\mathrm{v}(x,y)) \\ + D_3(f)(\mathrm{u}(x,y), \mathrm{v}(x,y), \mathrm{w}(x,y))\,(\frac{\partial}{\partial x}\mathrm{w}(x,y))$$

```
> Diff('h',y)=diff(h,y);
```

$$\frac{\partial}{\partial y} h = D_1(f)(\mathrm{u}(x, y), \mathrm{v}(x, y), \mathrm{w}(x, y)) \left(\frac{\partial}{\partial y} \mathrm{u}(x, y)\right) + D_2(f)(\mathrm{u}(x, y), \mathrm{v}(x, y), \mathrm{w}(x, y)) \left(\frac{\partial}{\partial y} \mathrm{v}(x, y)\right) + D_3(f)(\mathrm{u}(x, y), \mathrm{v}(x, y), \mathrm{w}(x, y)) \left(\frac{\partial}{\partial y} \mathrm{w}(x, y)\right)$$

****

11.4 Directional Derivatives and the Gradient

The Maple command for computing the gradient, ∇f, of a real-valued expression f of several variables is grad( ). Not surprisingly, we must tell Maple what variables should be used in the differentiation process, but in this command there is an added twist. When we compute $\nabla f(x, y, z)$ by hand, we have a common understanding that x is the first variable, y is the second, and z is the third. This ordering of the variables is a critical issue in the definition of the gradient vector. Maple has no order preference for the variables used in an expression, so we must **specify not only the variables, but also their ordering, by enclosing the variables in square brackets to create an ordered list.**

Like all of the other vector commands, the gradient command resides in the linear algebra package.

****

```
> with(linalg):
```

```
Warning, the protected names norm and trace have been redefined and
unprotected
```

```
> grad(f(x,y,z),[x,y,z]);
```

$$\left[\frac{\partial}{\partial x} \mathrm{f}(x, y, z), \frac{\partial}{\partial y} \mathrm{f}(x, y, z), \frac{\partial}{\partial z} \mathrm{f}(x, y, z)\right]$$

```
> grad(f(x,y),[x,y]);
```

$$\left[\frac{\partial}{\partial x} \mathrm{f}(x, y), \frac{\partial}{\partial y} \mathrm{f}(x, y)\right]$$

```
> grad(f(x,y,z),[y,z,x]);
```

$$\left[\frac{\partial}{\partial y} \mathrm{f}(x, y, z), \frac{\partial}{\partial z} \mathrm{f}(x, y, z), \frac{\partial}{\partial x} \mathrm{f}(x, y, z)\right]$$

****

Maple does not have a command for computing the directional derivative of a real-valued function. It hardly seems necessary to create one, because it is determined so easily from the gradient vector. In three dimensions, the directional derivative of a real-valued function f of x, y, and z at a point $P_0 = (x_0, y_0, z_0)$ in the direction of the unit vector $\vec{u}$ is

$$\mathcal{D}_{\vec{u}}(f)(P_0) = \nabla f(P_0) \cdot \vec{u}.$$

A similar formula holds for a function of two variables, or, for that matter, for a function of any number of variables.

Example 11.6 *An engineer is laying out the direction for a road that must be built through a mountainous area. A flat map of the region is being studied, and a two-dimensional grid has been placed over the map with units in miles, with the positive x-axis pointing east and the positive y-axis pointing north. The area being studied extends roughly 5 miles in each direction from the center of the map. The elevation h, in feet above sea level,at the point (x, y) on the map is*

$$h = 200000\frac{\cos(x + 0.2 * y + 2)\cos(0.3 * y - 0.1 * x - 1)}{50 + x^2 + y^2} + 3200.$$

The grade of the road must be less than 4% at each point. Starting at a scenic point located at $P_0(-2.1, -0.4)$, determine the range of directions which will produce an upward grade of between 2% and 4%. Express the answer as an angle from the positive x-axis (in degree measure).

The elevation function, h, is expressed in feet, and x and y are in miles. These are appropriate units for elevation, so we will not express h in terms of miles. At the appropriate point, however, an adjustment to miles will have to be made.

```
         ***********************************
> h:=(x,y)->200000*cos(x+0.2*y+2)*cos(0.3*y-0.1*x-1)/

  (50+x^2+y^2)+3200:
> plot3d(h(x,y),x=-5..5,y=-5..5);
```

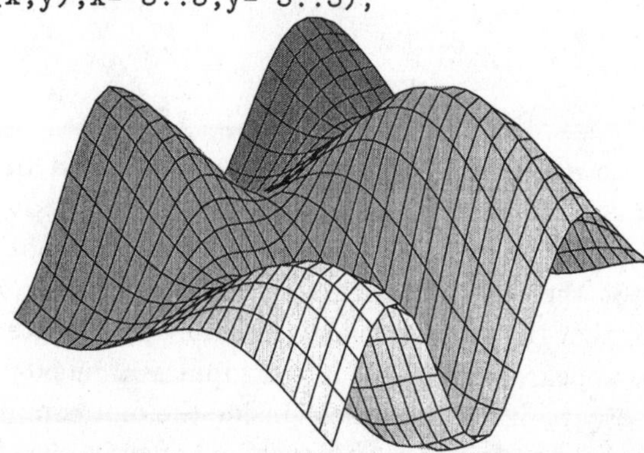

#This plot is not a necessary part of the problem, but it is interesting to look at the mountains that the road is running through. Remember that **we have not yet turned elevation from feet to miles**, and, even if we did, **it would have no**

effect on the plot, because Maple would adjust the picture for "best fit." The mountains are not nearly as dramatic as they appear to look on the plot, but, as you can see from the evaluations below, there are still some striking changes in elevation. In a moment, we will suggest a way to look more accurate at the hills, but, before we do, the plot, with its "magnifying glass" on elevation, is probably the best way to see the twists and turns. Turn the plot around to get a good look at the hills. Click some of the buttons in the plot tool bar, such as the "contour" button and the "box" button. Adjust your view point to get a "bird's eye view" looking straight down on the land. From this point of view, looking at the level curves, you can see approximately where the high and low points are on the land. The plot is shown below. Evaluate h at some of these points, and you will see some striking changes in elevation.#

```
> plot3d(h(x,y),x=-5..5,y=-5..5);
```

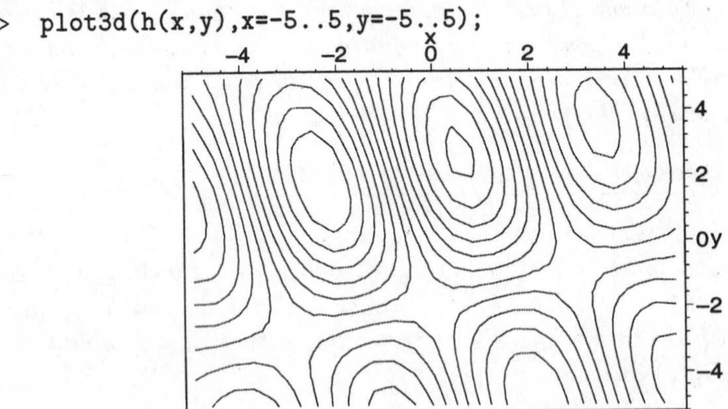


```
> evalf(h(-2.4,1.6)); evalf(h(.4,2.2)); evalf(h(-2.1,-.4));
```

6485.259604

−24.541822

5413.047040

#Now, let's get an **accurate picture of the land area involved**. We must **plot $h(x,y)/5280$**. In addition, we **must tell Maple to provide a true scale**, rather than a "best fit" picture. This can be done from the plot window, by clicking the [1:1] button in the plot tool bar, or by choosing "unconstrained" from the Projection Menu. Rather than make a plot, click the appropriate buttons, and wait for a second plot to appear, we elected, in the input statement below, to bring up the desired plot directly by inserting options inside the plot command. Surprisingly, the plot looks somewhat flat for land with such a change in elevation. It could help to remember that the map covers 100 square miles of land. #

```
> plot3d(h(x,y)/5280,x=-5..5,y=-5..5, scaling=constrained);
```

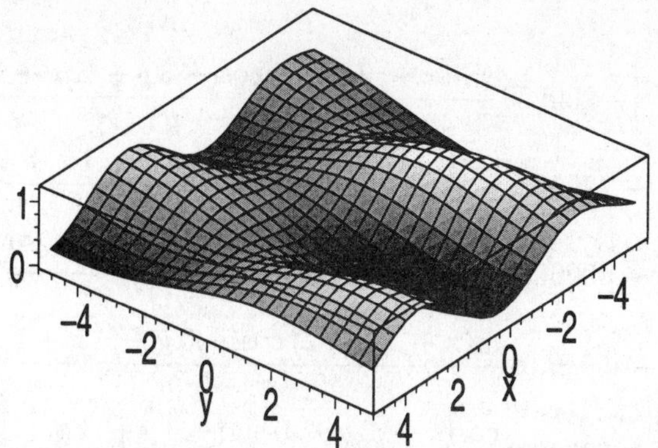

#By looking directly down on a contour map, it becomes clear that much of the elevation change occurs near a line through (-4,0) and (4,4), or, in other words, along the line $x = -4 - 2y$. An interesting final look at this land can be obtained by plotting the section $h(x, y)/5280$, with $x = -4 - 2y$. Maple must be told to maintain the same scale on the elevation axis. The plot is omitted, but its input statement should look like the following.

```
> plot(h(-4-2*y,y)/5280,y=-5..5,-5..5);
```

The grade of the road is its vertical rise over its horizontal run, or, in other words, just the slope of a certain tangent line to the surface—the tangent line determined by the direction of the road. To specify the direction of the road, simply project this tangent line, with its forward direction, down onto the x, y-plane, and let $\vec{L}$ denote a direction vector for this line with its forward direction. Then the unit vector

$$\vec{u} = \frac{1}{|\vec{L}|}\vec{L} = \cos(\theta)\vec{\imath} + \sin(\theta)\vec{\jmath},$$

where θ is the angle from $\vec{\imath}$ to $\vec{L}$ (or $\vec{u}$), will be called the (forward) direction of the road. The grade of the road is, essentially, the directional derivative of h at P_0 in the direction of $\vec{u} = \cos(\theta)\vec{\imath} + \sin(\theta)\vec{\jmath}$. Naturally, this number must be divided by 5280, which turns elevation, h, from feet into miles. It is convenient to express $\vec{u}$ in terms of θ.

A visual idea of what to expect for a range of allowable directions can be found by looking at our contour plot of the region. At the point $P_0(-2.1, -0.4)$, the tangent line to the contour through P_0 has the direction of a 0% grade. Consequently, a slight turn up the hill from the tangent line is the kind of direction we can expect from our calculations. #

```
> with(linalg):  dot:=(u,v)->dotprod(u,v,orthogonal):

Warning, the protected names norm and trace have been redefined and
unprotected

> g:=grad(h(x,y),[x,y]);
```

$$g := \left[-200000\,\frac{\sin(x+.2\,y+2)\cos(-.3\,y+.1\,x+1)}{50+x^2+y^2} - 20000.0\,\frac{\cos(x+.2\,y+2)\sin(-.3\,y+.1\,x+1)}{50+x^2+y^2} - 400000\,\frac{\cos(x+.2\,y+2)\cos(-.3\,y+.1\,x+1)\,x}{(50+x^2+y^2)^2},\right.$$
$$\left. -40000.0\,\frac{\sin(x+.2\,y+2)\cos(-.3\,y+.1\,x+1)}{50+x^2+y^2} + 60000.0\,\frac{\cos(x+.2\,y+2)\sin(-.3\,y+.1\,x+1)}{50+x^2+y^2} - 400000\,\frac{\cos(x+.2\,y+2)\cos(-.3\,y+.1\,x+1)\,y}{(50+x^2+y^2)^2}\right]$$

```
> g2:=subs(x=-2.1,y=-.4,evalm(g));
```

$$g2 := [-3665.017408\sin(-.18)\cos(.91) - 366.5017408\cos(-.18)\sin(.91) + 282.0794046\cos(-.18)\cos(.91), -733.0034816\sin(-.18)\cos(.91) + 1099.505222\cos(-.18)\sin(.91) + 53.72941040\cos(-.18)\cos(.91)]$$

```
> g3:=map(evalf,g2);
```

$$g3 := [288.3555068, 967.0236334]$$

```
> u:=vector([cos(theta),sin(theta)]);
```

$$u := [\cos(\theta), \sin(\theta)]$$

```
> grade:=dot(g3,u)/5280;
```

$$grade := .05461278538\cos(\theta) + .1831484154\sin(\theta)$$

```
> plot(grade,theta=-Pi..Pi);
```

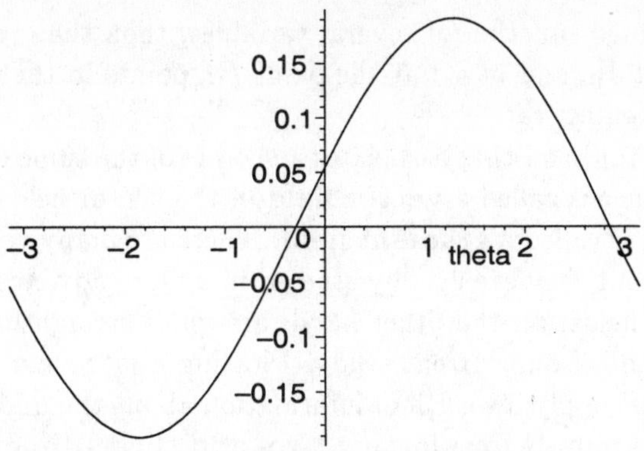

#By looking at this plot, we can see approximately what angles θ will return a grade between 2% and 4% percent.#

```
> a:=fsolve(grade=.02,theta,theta=-1..0);
```

$$a := -.1849545973$$

```
> b:=fsolve(grade=.04,theta,theta=-1..0);
```

$$b := -.07893993753$$

```
> c:=fsolve(grade=.04,theta,theta=2..3);
```

$$c := 2.640944168$$

```
> d:=fsolve(grade=.02,theta,theta=2..3);
```

$$d := 2.746958828$$

#To build a road with a grade between 2% and 4%, we must choose a direction angle (in degrees) between the two entries in the following list (or in the second list which follows).#

```
> Direction_Angle_1:=evalf([a*180/Pi,b*180/Pi]);
```

$$\mathit{Direction\_Angle\_1} := [-10.59711782, -4.522925256]$$

```
> Direction_Angle_2:=evalf([c*180/Pi,d*180/Pi]);
```

$$\mathit{Direction\_Angle\_2} := [151.3149547, 157.3891473]$$

****

If f is a real-valued function of several variables, then the gradient vector, ∇f, evaluated at a point P_0, and based at the point P_0, points in the direction in which f increases at its greatest rate.

A vector-valued function that associates a vector (of the same dimension) at each point in a vector space is called a **vector field**; if the vector field is of the form ∇f, then the vector field is called a **gradient field**. There are many vector fields that are not gradient fields, but they are usually studied in a more advanced course in vector calculus. Gradient fields, on the other hand, are quite straightforward and are an important component of our current study. Plotting a large sample of vectors in a vector field, ∇f, can supply a wealth of information about the underlying function f itself. **Maple's commands for plotting two- and three-dimensional gradient fields are gradplot( ) and gradplot3d( )**, respectively.

Example 11.7 *Plot the gradient field for* $f(x, y) = -4 + 6x + 4y - 2xy$.

****

```
> f:=-4+6*x+4*y-2*x*y:
> with(plots):
Warning, the name changecoords has been redefined
> gradplot(f,x=-10..10,y=-10..10);
```

10
y 5
-10 -5 5 x 10
-5
-10

****

Gradient fields are used frequently in applications. Heat energy, for example, flows in the direction of the greatest rate of decrease in temperature. If $T(x, y, z)$ is the temperature at the point (x, y, z) in a region, then the gradient field $-\nabla T$ models the flow of heat energy in the region.

The gradient ∇f always points in the direction of the greatest rate of increase in the values of $f(x, y)$, so it follows that ∇f is always orthogonal to the level sets of f. Thus, if S is a surface described by an equation of the form $f(x, y, z) = 0$, then the vector $\nabla f(x_0, y_0, z_0)$ based at (x_0, y_0, z_0) is orthogonal to the surface S. A similar statement can be made about a curve defined by an equation of the form $f(x, y) = 0$. The graph of a function $f(x, y)$ of two variables can be viewed as the graph of the equation $F(x, y, z) = z - f(x, y) = 0$. This level set of $F(x, y, z)$ has ∇F as a normal vector.

Example 11.8 *Find the equation of the tangent plane to the graph of*

$$f(x, y) = e^{3x} \cos(2y)$$

at the point on the graph corresponding to $(x, y) = (0.7, 2.8)$. *Graph* $f(x, y)$ *along with its tangent plane.*

****

```
> f:=(x,y)->exp(3*x)*cos(2*y):

> with(linalg): dot:=(u,v)->dotprod(u,v,orthogonal):

Warning, the protected names norm and trace have been redefined and
unprotected

> Q:=vector([0.7,2.8,f(0.7,2.8)]):

> F:=z-f(x,y):
```

#This turns the graph of f into the level set $F = 0$.#

```
> g:=grad(F,[x,y,z]);
```

$$g := \left[-3\, e^{(3\,x)} \cos(2\,y),\, 2\, e^{(3\,x)} \sin(2\,y),\, 1\right]$$

```
> n:=evalf(subs(x=Q[1],y=Q[2],evalm(g)));
```

$$n := [-19.00020823, -10.31006125, 1.]$$

```
> P:=vector([x,y,z]): QP:=matadd(P,-Q):
```

#If P is an arbitrary point in the plane, then the vector from Q (a fixed point in the plane) to P is orthogonal to the normal vector n.#

```
> plane:=dot(n,QP)=0;
```

$$plane := -19.00020823\, x + 35.83491452 - 10.31006125\, y + 1.\, z = 0$$

```
> h:=solve(plane,z);
```

$$h := 19.00020823\, x - 35.83491452 + 10.31006125\, y$$

```
> plot3d({f(x,y),h},x=0..1,y=0..4);
```

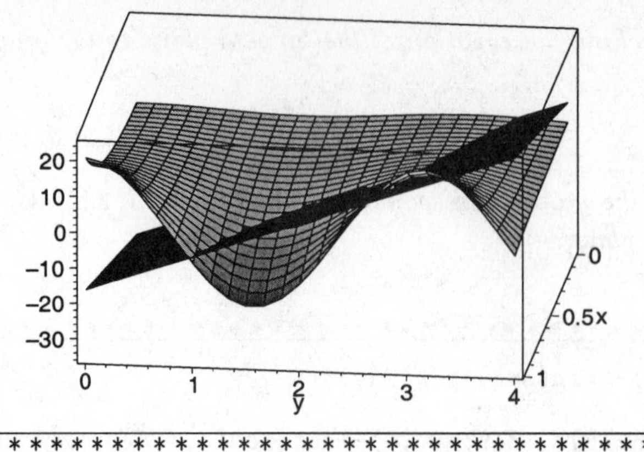

****

Example 11.9 *Find the equation of the plane tangent to the surface defined by the equation*

$$3x^2 - y^2 + 2z^2 = 8x + 5y - 7z + 15$$

at the point $P(-3, 10, 6)$ on the surface.

****

```
> f:=(x,y,z)->3*x^2-y^2+2*z^2-8*x-5*y+7*z-15:
> f(-3,10,6);
```

$$0$$

This verifies that the point is on the surface.

```
> with(linalg): dot:=(u,v)->dotprod(u,v,orthogonal):
```

```
Warning, the protected names norm and trace have been redefined and
unprotected
```

```
> g:=grad(f(x,y,z),[x,y,z]);
```

$$g := [6\,x - 8,\ -2\,y - 5,\ 4\,z + 7]$$

```
> n:=subs(x=-3,y=10,z=6,evalm(g));
```

$$n := [-26,\ -25,\ 31]$$

```
> v:=vector([x+3,y-10,z-6]);
```

#If (x, y, z) is an arbitrary point in the plane, then the vector $\vec{v}$ from $(-3, 10, 6)$ (a fixed point in the plane) to (x, y, z) is orthogonal to the normal vector $\vec{n}$.#

$$v := [x + 3,\ y - 10,\ z - 6]$$

```
> plane:=dot(n,v)=0;
```

$$plane := -26\,x - 14 - 25\,y + 31\,z = 0$$

****

11.5 Optimization

Let f be a function of two or more variables. If f has a local maximum or local minimum at a point P in its domain, then the partials of f at P either are zero or fail to exist. This provides us with an excellent means of finding the coordinates of the local maxima and minima.

Example 11.10 *Find the critical points of $f(x,y) = x^2 - 4xy^2 + 2x + 5y - 8y^3$, and, in each case, decide whether it is a local maximum point, a local minimum point, or a saddle point.*

We simply compute the partial derivatives and set them equal to zero. This gives us a system of two nonlinear equations in the unknowns x and y. Maple is happy to solve such a system, but it will give us only one solution with the fsolve( ) command. Are there others? Where are they? If we supply an approximate location for another solution, Maple's fsolve( ) command can be used to solve the system and produce that solution. By just looking at a system of equations, however, it is usually hard to tell how many solutions exist and hard to approximate their location. Plotting tools can frequently be used to supply these answers. In the present example, it is easier to simply solve one equation for x in terms of y, and substitute that into the other equation. This process of elimination, as long as it can be done, is probably the most reliable way to gather all of the solutions.

If (a,b) is a critical point of $f(x,y)$, then we can decide whether (a,b) is a local maximum point, local minimum point, or saddle point by looking at the sign of $f_{xx}(a,b)$, and the sign of $M(a,b)$, where

$$M(x,y) = \frac{\partial^2 f}{\partial x^2}(a,b)\frac{\partial^2 f}{\partial y^2}(a,b) - \left(\frac{\partial^2 f}{\partial x \partial y}(a,b)\right)^2.$$

A graph of f can also be used to make these decisions, but sometimes graphical evidence can be quite subtle. The so-called *Second Derivative Test* might be more reliable.

****

```
> f:=(x,y)->x^2-4*x*y^2+2*x+5*y-8*y^3:
> xeq:=diff(f(x,y),x)=0; yeq:=diff(f(x,y),y)=0;
```

$$xeq := 2\,x - 4\,y^2 + 2 = 0$$

$$yeq := -8\,x\,y + 5 - 24\,y^2 = 0$$

```
> a:=solve(xeq,x);
```

$$a := 2\,y^2 - 1$$

```
> eq:=subs(x=a,yeq);
```

$$eq := -8\,(2\,y^2 - 1)\,y + 5 - 24\,y^2 = 0$$

```
> s:=fsolve(eq,y);
```

$$s := -1.686597979,\ -.3471427509,\ .5337407301$$

```
> pt1:=subs(y=s[1],a),s[1]; pt2:=subs(y=s[2],a),s[2];
  pt3:=subs(y=s[3],a),s[3];
```

$$pt1 := 4.689225486,\ -1.686597979$$

$$pt2 := -.7589838210,\ -.3471427509$$

$$pt3 := -.4302416660,\ .5337407301$$

```
> fxx:=D[1,1](f);M:=D[1,1](f)*D[2,2](f)-D[1,2](f)^2;
```

$$fxx := 2$$

$$M := 2\,((x,\,y) \to -8\,x - 48\,y) - ((x,\,y) \to -8\,y)^2$$

```
> M(pt1); M(pt2); M(pt3);
```

$$-95.16941730$$

$$37.75692749$$

$$-62.58751011$$

```
> relmin:=[pt2,f(pt2)]; saddle1:=[pt1,f(pt1)];
  saddle2:=[pt3,f(pt3)];
```

$$relmin := [-.7589838210, -.3471427509, -1.977102118]$$

$$saddle1 := [4.689225486, -1.686597979, 7.95991927]$$

$$saddle2 := [-.4302416660, .5337407301, 1.267182843]$$

#A plot3d( ) image of $f(x,y)$ is not very revealing. The behavior of $f(x,y)$ near these special points is quite subtle. More striking graphical evidence verifying these results can be achieved by graphing the level curves of $f(x,y)$. We chose a range that just barely includes all three critical points.#

```
> with(plots):
Warning, the name changecoords has been redefined
> contourplot(f(x,y),x=-1..5,y=-2..1,axes=framed,contours=20);
```

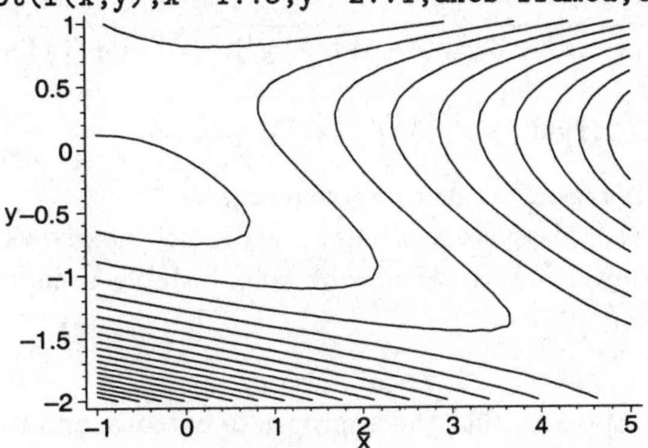

The above plot could have been produced from the plot3d( ) command. Simply use the plot option buttons in the plot menu, and use the mouse to move the view point to one that looks straight down on the graph.#

****

Notice the shape of the level curves near the local minimum and near the saddle points. **Notice how the level curves "split" near the saddle points.**

In closing, notice that the local minimum point is not an absolute minimum, because $f(0,y)$ clearly takes on large negative values when y is large and positive.

Example 11.11 *Find the critical points of*

$$f(x,y) = \frac{\ln(4 + x^2 + 3y^2)}{4 + 3x^2 + y^2},$$

and in each case determine whether it is local maximum point, a local minimum point, or a saddle point. Find the absolute maximum and absolute minimum if they exist.

The function f clearly is positive-valued and goes to 0 as x or y goes to ∞ (or as both do). For this reason, the absolute minimum does not exist. By the same token, the absolute maximum must exist. If D is a large closed disk, then f, being continuous on a closed and bounded set, must have an absolute maximum on D, and, because f is small on the bounding circle, the absolute maximum must be at an interior point of D, a local maximum. This maximum is clearly the absolute maximum over the entire (x, y) plane.

Following the standard practice, we set the partials equal to zero and try to solve the resulting system of equations.

****

```
> f:=(x,y)->ln(4+x^2+3*y^2)/(4+3*x^2+y^2):
> fx:=diff(f(x,y),x); fy:=diff(f(x,y),y);
```

$$fx := 2\,\frac{x}{(4+x^2+3\,y^2)\,(4+3\,x^2+y^2)} - 6\,\frac{\ln(4+x^2+3\,y^2)\,x}{(4+3\,x^2+y^2)^2}$$

$$fy := 6\,\frac{y}{(4+x^2+3\,y^2)\,(4+3\,x^2+y^2)} - 2\,\frac{\ln(4+x^2+3\,y^2)\,y}{(4+3\,x^2+y^2)^2}$$

```
> fsolve({fx=0,fy=0},{x,y});
```

Error, (in fsolve/gensys) did not converge

#As we shall see, this system of equations is not nearly as complicated as it looks, but Maple, nevertheless, needs our help. Maple even missed the obvious solution: $x = 0$, $y = 0$.#

```
> plot3d(f(x,y),x=-2..2,y=-2..2);
```

#After this plot appears, click the appropriate buttons, and turn the plot around to form a 2-dimensional graph of level curves.#

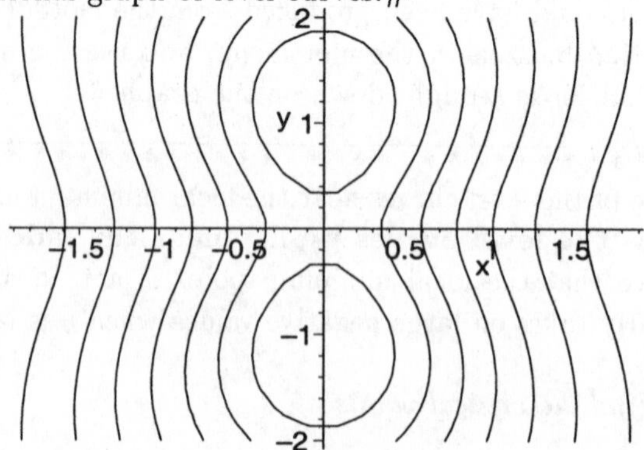


```
> fsolve({fx=0,fy=0},{x,y}, x=-0.5..0.5,y=0.5..1.5);
```

$$\{x = 0,\ y = 1.109275300\}$$

```
> a:=0:  b:=1.109275300:
```

#The three points (0,0), (a, b), and $(a, -b)$ are critical points. To decide whether they correspond to local maxima, local minima, or saddle points, we next form the function M used in the *Second Derivative Test.* Notice that it is introduced as a function. The output is so complicated that we suppressed its output by using a colon to close the input statement.#

```
> M:=D[1,1](f)*D[2,2](f)-D[1,2](f)^2:
```

```
> M(0,0);
```

$$\left(\frac{1}{8} - \frac{3}{8}\ln(4)\right)\left(\frac{3}{8} - \frac{1}{8}\ln(4)\right)$$

```
> evalf(%);
```

$$-.07964855383$$

#As the contour plot suggested, the point (0,0) is a saddle point.#

```
> M(a,b);
```

$$.05693603570$$

```
> D[1,1](f)(a,b);
```

$$-.3977112071$$

#It follows that the point (a, b), and hence the point $(a, -b)$, are local maximum points. It appears that these two points also correspond to absolute maximums as well, but have we done enough to draw such a conclusion?

The apparent complexity of the equations $fx = 0$, $fy = 0$ really does call for further investigation. Fortunately, the equations are not as complicated as they appear to be. If there is a solution with both x and y nonzero, then the common x and y terms in these equations can be canceled. The resulting two equations have many similar terms, and it is easy to show by hand that the system has no solution. It follows that the only solutions have an x or y coordinate which is zero. To search for all possible solutions with $x = 0$ or with $y = 0$, we plot the following two expressions. We could pursue the matter further, but this is solid evidence that we have found all of the solutions.#

```
> plot(subs(x=0,fy),y=-10..10);
```

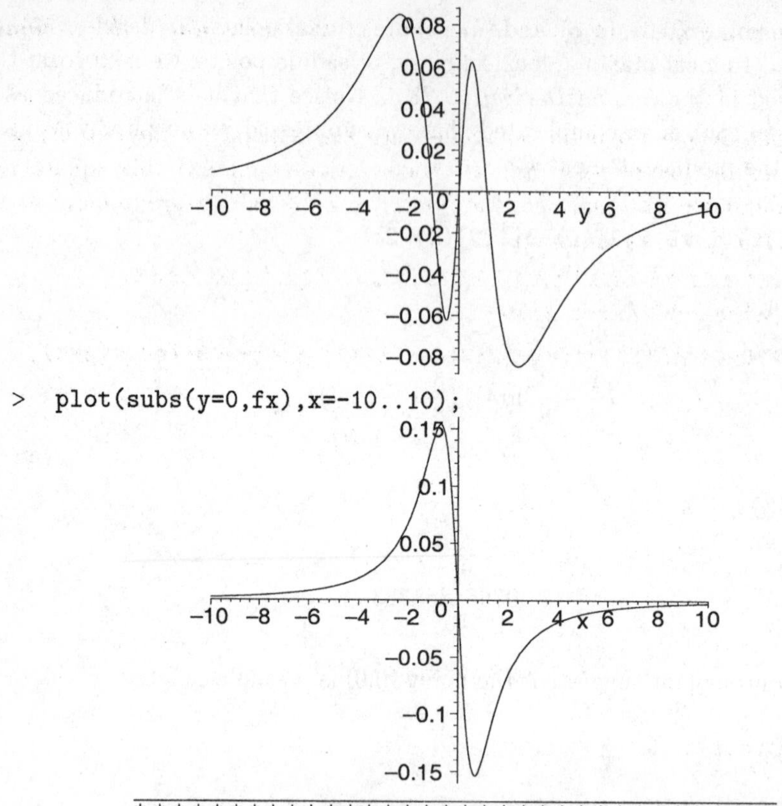


```
> plot(subs(y=0,fx),x=-10..10);
```

****

It is easy to imagine a system of equations of the form $fx = 0,\ fy = 0$ whose complete solution would not be available in such an elementary fashion. Each system must be considered on its own merits. Maple's fsolve( ) command frequently needs our help to get close to a desired solution. As an alternate strategy, we could have, for example, plotted the two equations, using the implicitplot( ) command, and then located the intersection points between the two curves by eyesight. Then, in much the same way as in the above solution, Maple's fsolve( ) command could be given the extra information it could need to produce a more accurate solution.

To find the local or absolute maximum or minimum points of a function of more than two variables is intrinsically more difficult. There is a version of the second derivative test similar to the so-called M-Test that we used above, but it is difficult to use when more than two variables are involved. Maple might have to struggle in order to solve a system of three or more equations in three or more unknowns. Showing that we have all of the solutions, or at least the desired solution, can be challenging. It is more difficult to use Maple's plotting tools to get useful information.

Example 11.12 *Find the minimum distance between the surface*

$$x = 4 - \frac{y^2}{10} - \frac{z^2}{30}$$

and the curve defined parametrically by

$$x = 16 + 4\sin(t),\ \ y = 9 + 2\cos(3t),\ \ z = -5 + 7\cos(2t).$$

Locate the point on the surface and the point on the curve where the minimum occurs.

The distance between a point on the surface and a point on the curve can be expressed as a function of y, z, and t. Rather than minimize this square root function, we minimize its square instead. Surely $\sqrt{expr}$ will be a minimum at the same point that *expr* is a minimum.

****

```
> f:=(y,z)->4-y^2/10-z^2/30:
> p:=t->16+4*sin(t); q:=t->9+2*cos(3*t); r:=t->-5+7*cos(2*t);
```

$$p := t \to 16 + 4\sin(t)$$

$$q := t \to 9 + 2\cos(3\,t)$$

$$r := t \to -5 + 7\cos(2\,t)$$

```
> with(plots):
Warning, the name changecoords has been redefined
> p1:=plot3d([f(y,z),y,z],y=-20..20,z=-20..40):
> p2:=spacecurve([p(t),q(t),r(t)],t=0..2*Pi,thickness=2):
> display({p1,p2});
```

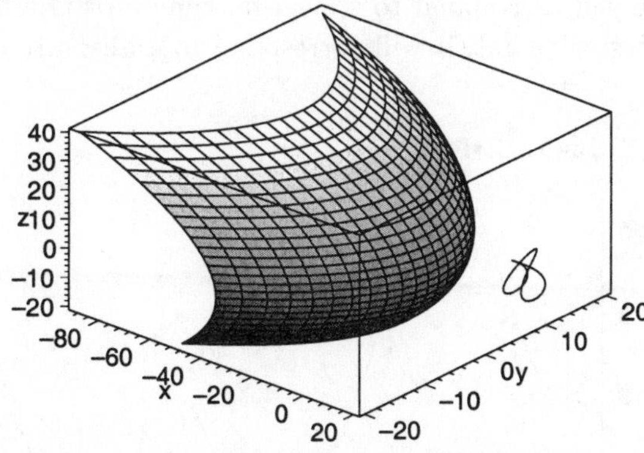


```
> F:=(f(y,z)-p(t))^2+(y-q(t))^2+(z-r(t))^2;
```

$$F := (-12 - \frac{1}{10}\,y^2 - \frac{1}{30}\,z^2 - 4\sin(t))^2 + (y - 9 - 2\cos(3\,t))^2$$
$$+ (z + 5 - 7\cos(2\,t))^2$$

```
> Fy:=diff(F,y); Fz:=diff(F,z); Ft:=diff(F,t);
```

$$Fy := -\frac{2}{5}\left(-12 - \frac{1}{10}y^2 - \frac{1}{30}z^2 - 4\sin(t)\right)y + 2y - 18 - 4\cos(3t)$$

$$Fz := -\frac{2}{15}\left(-12 - \frac{1}{10}y^2 - \frac{1}{30}z^2 - 4\sin(t)\right)z + 2z + 10 - 14\cos(2t)$$

$$Ft := -8\left(-12 - \frac{1}{10}y^2 - \frac{1}{30}z^2 - 4\sin(t)\right)\cos(t)$$
$$+ 12\,(y - 9 - 2\cos(3t))\sin(3t) + 28\,(z + 5 - 7\cos(2t))\sin(2t)$$

```
> s:=fsolve({Fy=0,Fz=0,Ft=0},{y,z,t},t=0..2*Pi);
```

$$s := \{t = 3.608215105,\ y = 2.715421307,\ z = -.4818113946\}$$

```
> assign(s);
> P:=[f(y,z),y,z]; Q:=[p(t),q(t),r(t)];
```

$$P := [3.254910638,\ 2.715421307,\ -.4818113946]$$
$$Q := [14.20051043,\ 8.659804296,\ -.833392376]$$

```
> a:=sqrt(F);
```

$$a := 12.46055589$$

#The points P and Q could correspond to any local maximum or minimum of the function F. We can see by looking at the above graph that there are several such points, and there is no reason to suspect that we have found the desired ones.

How we proceed from here varies from one problem to the next. In our case, we determine the distance between the point P on the surface and an arbitrary point on the curve. This defines a function of t that can be plotted. The value t that we computed above will correspond to a local maximum or minimum point on this graph. If it is the desired point, it will correspond to a absolute minimum point.#

```
> t:='t':
> plot(sqrt(F),t=0..2*Pi);
```

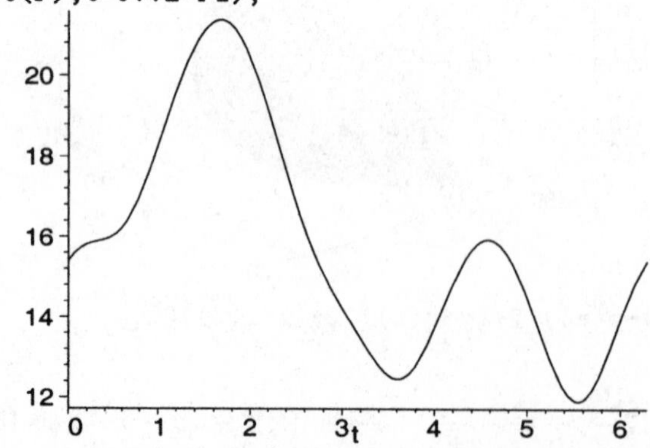

#As expected, this plot has several local maxima and minima. Not surprisingly, the **absolute minimum does not occur at the point t determined above.** It seems reasonable to suspect that the correct t-value is not too far from the one corresponding to the absolute minimum in the above graph. Let's try to solve the system again, using the plot to determine a better range for t. #

```
> s:=fsolve({Fy=0,Fz=0,Ft=0},{y,z,t},t=5..6);
```

$$s := \{t = 5.376196718, y = 2.397490682, z = -4.017401388\}$$

```
> assign(s):
> P:=[f(y,z),y,z]; Q:=[p(t),q(t),r(t)];
```

$$P := [2.887220046, 2.397490682, -4.017401388]$$

$$Q := [12.84939231, 7.174333714, -6.685537677]$$

```
> a:=sqrt(F);
```

$$a := 11.36582848$$

#It looks as if we have found an answer to the problem. If so, then the minimum distance is the value $a = 11.36582848$, between the point P on the surface and Q on the curve.

Have we, however, established that this answer corresponds to the absolute minimum and not just another local minimum? Certainly not! If the point P on the surface is the correct point, then we have indeed established that our answer is correct. However, what if the curve takes another dip closer yet to the surface at some point on the surface far removed from P? This is an important, but very difficult issue to resolve, and there is no swift, clean mathematical way to proceed. We cannot prove, in a pure mathematical sense, that our answer is correct, but on a practical level, we can offer very convincing evidence.

A thorough search for other answers will be conducted. It will help if we restrict our search to as small a window as possible, and so our approach begins with another look at the active plot window showing the curve and surface together. The distance function, F, (actually distance squared) depends on the variables y, z, and t, and we wish to establish small ranges for each of these variables and to restrict our search to these intervals. A range for the variable t is easy to establish. We simply use the interval from 0 to 2π. By turning the plot around, and looking at it carefully from different directions, it is possible to establish the ranges $y = 0..6$, and $z = -10..2$. To see these ranges, **it definitely helps to see the plot in a "true" scale**. Click the plot window, then open the **Projection** menu and click on **Constrained**, or click the [1:1] button.

Now that we have a fairly narrow band of possible solutions, we blanket the entire region with a large number of points, evaluate F at each of the points, and check to see whether there are any values of F smaller than $a^2 = 11.36582848^2$. If we are careful how we do this, it will not take very much computer time.

We partition the interval $0 \leq y \leq 6$ into 60 points, the interval $-10 \leq z \leq 2$ into 120 points, and the interval $0 \leq t \leq 2\pi$ into 60 points, all equally spaced over each interval. This places the partition points 0.1 units apart— only slightly more for the t interval. The function F is then evaluated at each of the corresponding 432,000 points. It is important to speed things up, so we suppress output unless we find a point where F is smaller than $a^2 = 11.36582848^2$. To speed things up even more, we turn down the value of Digits to 3 before running the program, and then return its value to 10 afterwards. We can always be more accurate later if we actually find a point in our search. The following program took 3 minutes and 34 seconds of computer time on a Power Macintosh 7200/90.#

```
> y:='y': z:='z':
> Digits:=3: for j from 1 to 60 do for k from
  1 to 120 do

  for i from 1 to 60 do y:=0+j*.1; z:=-10+k*.1; t:=i*Pi/30;

  if evalf(F)<11.3658^2 then print(y,z,t) else

  break fi; od;od;od; Digits:=10:
```

****

When this program is run, there is no output. This should mean that the answer we obtained above is correct. Can the program be trusted? It could be run again with $a^2 = 11.36582848^2$ replaced by a slightly larger number. The program should (and does) begin to return points y, z, and t close to the values where we expect the minimum to occur.

In a pure mathematical sense, the minimum could still occur at some point in between these values, but, in a practical sense, surely we have enough evidence to declare our answer to be correct.

In the last example, we were able to express the function we wished to minimize—distance—easily as a function of several independent variables. In many optimization problems, it is either difficult or impossible to express the function being optimized in terms of independent variables. In these cases, the method of Lagrange can be used to optimize the function.

Suppose we wish to optimize a real-valued function $f(x, y, z)$, where x, y, and z are not independent variables, but must satisfy the constraint $g(x, y, z) = 0$. We introduce a new variable λ, called a Lagrange multiplier, and define the function

$$F(x, y, z, \lambda) = f(x, y, z) + \lambda g(x, y, z)$$

where x, y, z, and λ are regarded as independent variables. According to the method of Lagrange, the optimum value of the original function f, if it exists, will occur at a critical point of F, that is to say, at a point (x, y, z, λ), where the four partials of F are zero (or fail to exist). Setting the partial of F with respect to λ equal to zero simply returns the constraint $g(x, y, z) = 0$.

More generally, to optimize $f(x_1, x_2, \ldots, x_n)$, subject to the constraints

$$g_1(x_1, x_2, \ldots, x_n) = 0,\ g_2(x_1, x_2, \ldots, x_n) = 0, \ldots, g_p(x_1, x_2, \ldots, x_n) = 0,$$

we introduce the Lagrange multipliers $\lambda_1, \lambda_2, \ldots, \lambda_p$, and define the function

$$F(x_1, x_2, \ldots, x_n, \lambda_1, \lambda_2, \ldots, \lambda_p) = f(x_1, x_2, \ldots, x_n) + \sum_{j=1}^{p} \lambda_j g_j(x_1, x_2, \ldots, x_n),$$

where the variables $x_1, x_2, \ldots, x_n, \lambda_1, \lambda_2, \ldots, \lambda_p$ are regarded as $n + p$ independent variables. Then the optimum value of f, if it exists, will occur at a critical point of F.

Example 11.13 *Find the minimum distance between the curves*

$$8x^2+3y^2+80x-42y+251=0 \text{ and } 108x^2+68xy+57y^2+340x+570y+1225=0$$

and the points on each curve where the minimum occurs.

We begin by plotting the two curves. Each is an ellipse, and the plot can be used to get an approximate location for the two points (one on each ellipse) which are closest to each other. This kind of information is frequently needed in order to help Maple solve more complicated problems. The plot, however, is misleading. Notice the different scales on each axis. Later, this will cause a problem, but it would probably not be anticipated at this time.

Let $P(x, y)$ be an arbitrary point on the first curve, and let $Q(u, v)$ be an arbitrary point on the second. The points P and Q are obviously different points, and so variables other than x and y must be used for one of the points. We minimize the distance between P and Q, using the method of Lagrange. Just as in the previous problem, minor complications can be avoided by minimizing, instead, the square of the distance between P and Q.

****

```
> curve1:=8*x^2+80*x+3*y^2-42*y+251=0:
> curve2:=108*x^2+68*x*y+57*y^2+340*x+570*y+1225=0:
> with(plots):

Warning, the name changecoords has been redefined
> implicitplot({curve1,curve2},x=-15..10,
  y=-10..15);
```

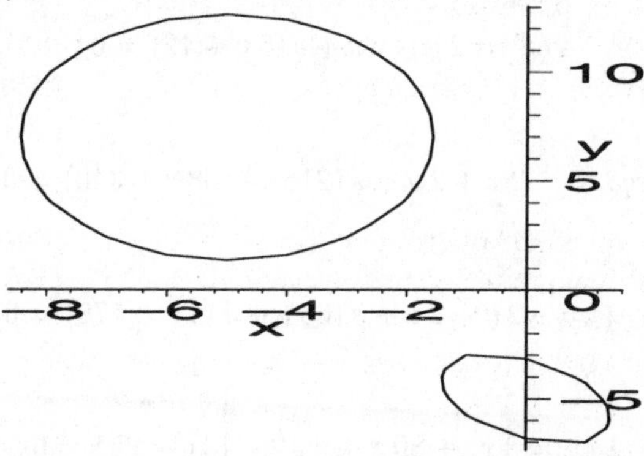

We will need some fairly accurate estimates for the solutions to this complicated system of equations in order to solve it. Maple's first plot is not drawn to scale, and it would not lead to very accurate estimates. The plot above was made by selecting the original plot and then clicking the [1:1] button in the menu-bar. With this plot, we can create a reasonable small window in which the solution is contained.#

```
> g1:=lhs(curve1); g2:=subs(x=u,y=v,lhs(curve2));
```

$$g1 := 8\,x^2+80\,x+3\,y^2-42\,y+251$$

$$g2 := 108\,u^2 + 68\,u\,v + 57\,v^2 + 340\,u + 570\,v + 1225$$

```
> f:=(x-u)^2+(y-v)^2;
```

#This is the square of the distance between P and Q.#

$$f := (x-u)^2 + (y-v)^2$$

#The variables λ_1 and λ_2 can be used in a Maple work session by using the unassigned names "lambda1" and "lambda2" as input variables. They will appear in output statements as Greek letters. We take an easier approach, however, and simply let r and s denote the Lagrange multipliers.#

```
> F:=f+r*g1+s*g2;
```

$$F := (x-u)^2 + (y-v)^2 + r\,(8\,x^2 + 80\,x + 3\,y^2 - 42\,y + 251) + s\,(108\,u^2 + 68\,u\,v + 57\,v^2 + 340\,u + 570\,v + 1225)$$

```
> eq1:=diff(F,x)=0; eq2:=diff(F,y)=0; eq3:=diff(F,u)=0;
  eq4:=diff(F,v)=0; eq5:=g1=0; eq6:=g2=0;
```

$$eq1 := 2\,x - 2\,u + r\,(16\,x + 80) = 0$$

$$eq2 := 2\,y - 2\,v + r\,(6\,y - 42) = 0$$

$$eq3 := -2\,x + 2\,u + s\,(216\,u + 68\,v + 340) = 0$$

$$eq4 := -2\,y + 2\,v + s\,(68\,u + 114\,v + 570) = 0$$

$$eq5 := 8\,x^2 + 80\,x + 3\,y^2 - 42\,y + 251 = 0$$

$$eq6 := 108\,u^2 + 68\,u\,v + 57\,v^2 + 340\,u + 570\,v + 1225 = 0$$

#By looking at the plot, we can see approximate locations for the values of x, y, u, and v that will minimize f. These values are used to create a range that can be used to help solve the system of equations.#

```
> region:=x=-4..-3.4,y=1.5..2,u=-1.6..-0.5,v=-4..-2.8;
```

$$region := x = -4.. - 3.4,\ y = 1.5..2,\ u = -1.6.. - .5,\ v = -4.. - 2.8$$

```
> fsolve({eq1,eq2,eq3,eq4,eq5,eq6},{x,y,u,v,r,s},{region});
```

$\{r = .2947673971,\ s = .06097990488,\ u = -1.043919956,\ v = -3.023912069,$
$x = -3.821942798,\ y = 1.680305465\}$

This solution is an **impressive display of Maple's computational power!** Earlier versions of Maple were not able to solve this system.

```
> assign(%):
> check:=eq1,eq2,eq3,eq4,eq5,eq6;
```

$$check := 0 = 0,\ -.7\,10^{-8} = 0,\ 0 = 0,\ .3\,10^{-8} = 0,\ .1\,10^{-6} = 0,\ 0 = 0$$

#The two points P and Q and the distance between the curves are as follows.#

```
> P:=[x,y]; Q:=[u,v]; CurveSeparation:=sqrt(f);
```

$$P := [-3.821942798,\ 1.680305465]$$

$$Q := [-1.043919956,\ -3.023912069]$$

$$CurveSeparation := 5.463247525$$

#We are done, but what would we have done if Maple had been unable to solve this system? Certainly we can't expect such a remarkable success with every system of equations that we encounter.

Let us discuss a strategy that could have been used if Maple had been unable to solve this (or possibly another) system of equations. **Reducing the number of equations and the number of unknowns** would certainly lead to a computationally less complicated system. We do not have even approximate values for the Lagrange multipliers r and s, so we choose to eliminate these variables. Incidentally, these variables can always be eliminated, because Lagrange multipliers always appear as first-power terms, and so the equations involved can always be solved for them.#

```
> r1:=solve(eq1,r); s1:=solve(eq3,s);
```

$$r1 := \frac{1}{8}\,\frac{-x+u}{x+5}$$

$$s1 := -\frac{1}{2}\,\frac{-x+u}{54\,u+17\,v+85}$$

```
> eq21:=subs(r=r1,eq2);eq41:=subs(s=s1,eq4);
```

$$eq21 := 2\,y - 2\,v + \frac{1}{8}\,\frac{(-x+u)\,(6\,y-42)}{x+5} = 0$$

$$eq41 := -2\,y + 2\,v - \frac{1}{2}\,\frac{(-x+u)\,(68\,u+114\,v+570)}{54\,u+17\,v+85} = 0$$

```
> fsolve({eq21,eq41,eq5,eq6},{x,y,u,v},{region}):
```

There is no need to pursue this alternate approach any further, because we already have a solution. Consequently, the output was omitted.

****

Notice that, once we had values for the six unknowns, we confirmed that the original six equations $eq1, \ldots, eq6$ were satisfied.

If f is a continuous real-valued function in a closed and bounded region D, then f has an absolute maximum and an absolute minimum in D. That is is say, there are points (x_1, y_1) and (x_2, y_2) in D such that

$$f(x_1, y_1) \le f(x, y) \le f(x_2, y_2)$$

for all (x, y) in D. The absolute maximum or minimum could occur at an interior point (x_0, y_0) of D, in which case, (x_0, y_0) would be a critical point of f. On the other hand, the absolute maximum or minimum could occur at a boundary point of D, in which case no special role is played by the partials of f.

To solve such a problem, we collect all the critical points of f interior to the region D. Points on the boundary of D are simply points that satisfy some constraint condition $g(x, y) = 0$. We use either substitution methods or the method of Lagrange to find all of the candidates for maximum or minimum on the boundary of D. **We know the absolute maximum and minimum exist, so once we have a collection of all the candidates interior to D and on its boundary**, all we have to do to finish the problem is evaluate f at each of the candidates. The absolute maximum and minimum values of f will be the largest and smallest of these numbers.

Example 11.14 *Let D be the region in the (x, y) plane bounded by the curves $y = 16 - x^2$ and $y = x^2 - 16$, and let f be defined by*

$$f(x, y) = \frac{8(x-3)^2 + 5(y+1)^2 + 1}{x^2 + 2y^2 + 1}.$$

Find the absolute maximum and absolute minimum of f in D and the points where they occur.

An explanation of our confusing array of names could make our solution more readable. As usual, the letter "p" is used in a name (p for prime) to denote a derivative. The names ubf, and lbf refer to the upper and lower boundary values of f.

****

```
> f:=(x,y)->(8*(x-3)^2+5*(y+1)^2+1)/(x^2+2*y^2+1):
> plot3d(f(x,y),x=-4..4,y=x^2-16..16-x^2);
```

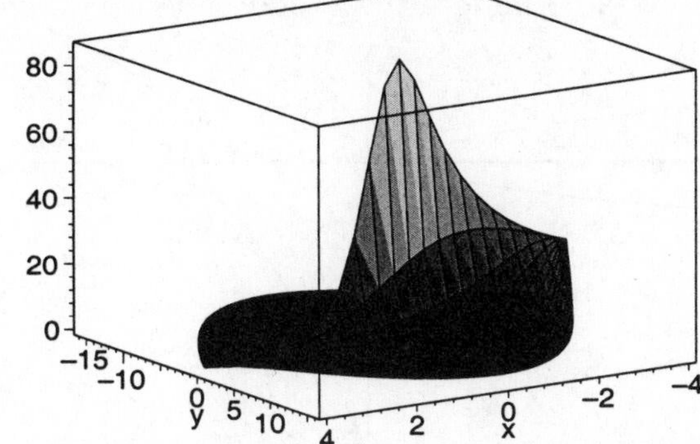

#The absolute maximum appears to occur near the origin.#

```
> fpx:=diff(f(x,y),x);fpy:=diff(f(x,y),y);
```

$$fpx := \frac{16\,x - 48}{x^2 + 2\,y^2 + 1} - 2\,\frac{(8\,(x-3)^2 + 5\,(y+1)^2 + 1)\,x}{(x^2 + 2\,y^2 + 1)^2}$$

$$fpy := \frac{10\,y + 10}{x^2 + 2\,y^2 + 1} - 4\,\frac{(8\,(x-3)^2 + 5\,(y+1)^2 + 1)\,y}{(x^2 + 2\,y^2 + 1)^2}$$

```
> eqx:=numer(normal(fpx))=0; eqy:=numer(normal(fpy))=0;
```

$$eqx := 22\,x\,y^2 - 140\,x + 48\,x^2 - 96\,y^2 - 48 - 20\,x\,y = 0$$

$$eqy := -22\,y\,x^2 - 302\,y + 10\,x^2 - 20\,y^2 + 10 + 192\,x\,y = 0$$

```
> fsolve({eqx,eqy},{x,y});
```

$$\{y = -1.034009312,\ x = 3.031155276\}$$

```
> assign(%):  a1:=x; b1:=y; x:='x':  y:='y':
```

$$a1 := 3.031155276$$

$$b1 := -1.034009312$$

#A good way to find other solutions of this system of equations is to plot the curves *eqx* and *eqy*. Each intersection point is a solution to the system.#

```
> with(plots):
```

Warning, the name changecoords has been redefined

```
> implicitplot({eqx,eqy},x=-4..4,y=-16..16);
```

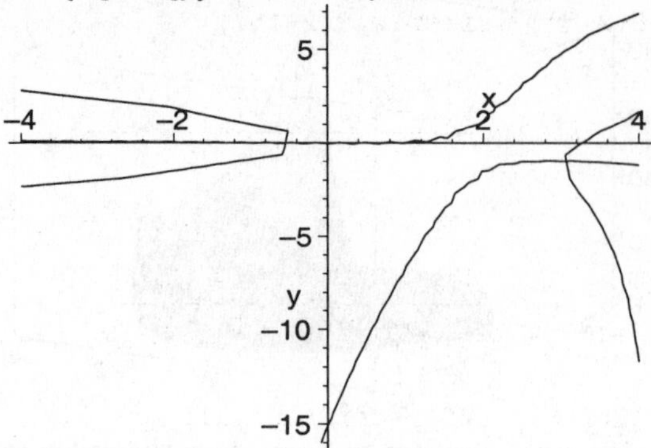

#There appears to be only one point of intersection, but we know from the preceding plot that there should be another solution near the origin. The curve near the origin has a curious feature. It appears to simply end. Click the plot to activate the plot tool bar. **Click the button that calls for no axes.** #

```
> implicitplot({eqx,eqy},x=-4..4,y=-16..16);
```

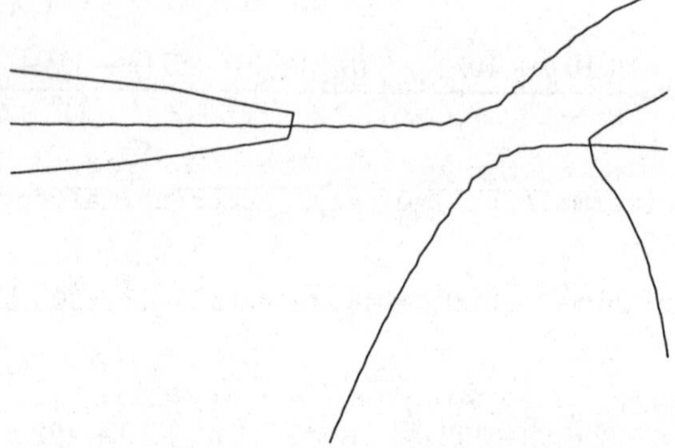

#With the axes removed, a **second point of intersection, hidden by the *x*-axis, is quite visible.**#

```
> fsolve({eqx,eqy},{x,y},x=-.5..0,y=-0.5..0.5);
```

$$\{x = -.3093760629, y = .03009311131\}$$

```
> assign(%): a2:=x; b2:=y; x:='x': y:='y':
```

$$a2 := -.3093760629$$

$$b2 := .03009311131$$

#We have two critical points interior to the region. The absolute maximum or absolute minimum can also occur at a boundary point. We search for these boundary points next.#

```
> ubf:=f(x,16-x^2); lbf:=f(x,x^2-16);
```

$$ubf := \frac{8\,(x-3)^2+5\,(17-x^2)^2+1}{x^2+2\,(16-x^2)^2+1}$$

$$lbf := \frac{8\,(x-3)^2+5\,(x^2-15)^2+1}{x^2+2\,(x^2-16)^2+1}$$

```
> ubfp:=diff(ubf,x); lbfp:=diff(lbf,x);
```

$$ubfp := \frac{16\,x-48-20\,(17-x^2)\,x}{x^2+2\,(16-x^2)^2+1} - \frac{(8\,(x-3)^2+5\,(17-x^2)^2+1)\,(2\,x-8\,(16-x^2)\,x)}{(x^2+2\,(16-x^2)^2+1)^2}$$

$$lbfp := \frac{16\,x-48+20\,(x^2-15)\,x}{x^2+2\,(x^2-16)^2+1} - \frac{(8\,(x-3)^2+5\,(x^2-15)^2+1)\,(2\,x+8\,(x^2-16)\,x)}{(x^2+2\,(x^2-16)^2+1)^2}$$

```
> eq1:=numer(normal(ubfp))=0; eq2:=numer(normal(lbfp))=0;
```

$$eq1 := -1884\,x^3+25056\,x+18\,x^5-3024\,x^2-24624+288\,x^4=0$$

$$eq2 := 676\,x^3+5256\,x-62\,x^5-3024\,x^2-24624+288\,x^4=0$$

```
> s1:=fsolve(eq1,x);
```

$s1 :=$

$$-20.53089992,\ -3.970937003,\ 1.340570278,\ 3.029206841,\ 4.132059807$$

```
> s2:=fsolve(eq2,x);
```

$$s2 := -4.000876796,\ 3.829826669,\ 4.839489667$$

```
> g:=x->16-x^2;
```

$$g := x \to 16 - x^2$$

#The upper and lower boundaries consist of points (x, y) with $y = \pm g(x)$ and $-4 \le x \le 4$. Only 4 of the 6 values of x in the list of candidates are in the right interval. Finally, the end points of the interval [-4, 4] are always candidates.#

```
> a3:=s1[2]:  b3:=g(a3):  a4:=s1[3]:  b4:=g(a4):

  a5:=s1[4]:  b5:=g(a5):  a6:=s2[2]:  b6:=-g(a6):

  a7:=-4.0:  b7:=0:  a8:=4.0:  b8:=0:
```

#We have eight candidates, $(a1, b1), \ldots, (a8, b8)$. The corresponding values of f are listed next. The first two values correspond to interior points of D, the next three values correspond to points on the upper boundary of D, the next value corresponds to a point on the lower boundary of D, and the last two correspond to corner points on the boundary.#

```
> f(a1,b1), f(a2,b2), f(a3,b3);
  f(a4,b4),f(a5,b5), f(a6,b6);
  f(a7,b7), f(a8,b8);
```

$$.08222680411, 85.57549110, 23.54493540$$

$$2.901401766, 2.972430935, .3674330959$$

$$23.41176471, .8235294118$$

****

The absolute maximum value, $M = 85.57549110$, occurs at the interior point $(a2, b2) = (-.3093760629, .03009311131)$, and the absolute minimum value, $m =$ 08222680411, occurs at $(a1, b1) = (3.031155276, -1.034009312)$.

11.6 Exercise Set

New Maple Commands in this Chapter (and a few old commands as a reminder)

animate( )	contourplot( )	D( )	D[]( )	diff( )
display( )	grad( )	gradplot( )	gradplot3d( )	implicitdiff( )
implicitplot( )	implicitplot3d( )	matadd( )	plot3d( )	proc( )
scalarmul( )	seq( )	sum( )	unapply( )	

1. This problem could require more memory than is available, if all of its parts are done in one Maple work session. Keep an eye on the memory bar in the lower

right-hand corner of the work sheet. Consider the function

$$f(x,y) = (83 - 5x - 7y)e^{-0.57x^2 - 0.43y^2 + 0.48xy + 0.24x + 5.82y - 26}.$$

a) Plot f so that all of the significant features of the graph are shown.
b) Display a contour plot by using the same plot window used in part a), making appropriate adjustments, and dragging the plot to a view point that looks down from above.
c) Display another contour plot, over a much smaller range, and use it to estimate (visually) the x and y coordinates of the absolute maximum and minimum points of f. (Use the contourplot( ) command in the plots package.)

Let (x_1, y_1) and (x_2, y_2) denote the approximate coordinates of the maximum and minimum points (respectively) found in this manner. These points will be referred to in subsequent parts of the problem.
d) Display two vertical sections of the graph that pass through the point (x_1, y_1). One of the vertical sections should be parallel to the x-axis, the other to the y-axis.
e) Display a vertical section that passes through the points (x_1, y_1) and (x_2, y_2).
f) Create an animation that plots sections of the graph of f, parallel to the x-axis, that start on one side of the point (x_0, y_0) and end on the other side.
g) This animation is more involved. Create an animation that plots vertical sections of the graph of f that are all perpendicular to the line L passing through the points (x_1, y_1), and (x_2, y_2). Arrange the sections so that they are all centered over the line L. The sections should start on one side of both points and end on the other side of both points.

2. Use Maple's differentiation commands to compute, directly, the partials of $f(x,y) = x^4 + 5x^2y^3 - 8y^3 + 7x^3$. Confirm that the answers are correct by using the definition of a partial derivative as a limit of a certain difference quotient. Simplify the difference quotients so that their limits are obvious.

3. Use Maple's differentiation commands to compute, directly, the partials of

$$f(x,y,z) = \frac{x^2\cos(x+y)}{x^2 + y^3 + z^4}$$

at the point $(x_0, y_0, z_0) = (5, -8, 3)$. Confirm that the answers are correct by using the definition of a partial derivative as a limit of a certain difference quotient.

4. Use Maple's diff( ) command to compute the first-order and second-order partial derivatives of $f(x,y) = \cos(x^2y)e^{3y-5x}$. Compute the partials again using Maple's D( ) command. Sometimes an expression is desired as output, sometimes a function is desired as output. Both commands are useful. Evaluate all of the partials, as decimals, at the point $(x,y) = (8,14)$.

5. The equation $x^3y^2 + 5xz^2y^3 + 4yz^3 - 7x^2 = 17z + 9$ defines z implicitly as a function $z = f(x,y)$ of x and y. Use the diff( ) command and implicit differentiation to determine (as decimal numbers) the partials of z with respect to x and y at the point $(x_0, y_0) = (5, 7)$. The function $z = f(x,y)$ might not be uniquely defined. Evaluate the partials for every such determination $z = f(x,y)$.

Before you can differentiate z, implicitly, with respect to x or y, Maple must be told that z depends on x and y. These issues are discussed in Example 11.4. Substitutions can be tricky is problems like this. In a command such as

```
> subs(eqn1,eqn2,eqn3,expr);
```

Maple first performs the substitution *eqn*1, everywhere in *expr*, before it begins the substitution *eqn*2, etc. After differentiating implicitly, simplify notation, and replace variables by numbers as soon as possible. Check your answers with Maple's implicitdiff( ) command.

6. A general (nice) equation in the variables x, y, and z is modeled by the equation $F(x, y, z) = 0$, where F is any real-valued differentiable function of three variables. It is natural to attempt to solve such an equation for z (for example) in terms of x and y. When such a solution exists, even if it cannot be found, we say that $F(x, y, z) = 0$ defines z implicitly as a function of x and y.

If $F(x_0, y_0, z_0) = 0$ and if the partial $F_z(x_0, y_0, z_0) \neq 0$, then it turns out that there is a differentiable function $g(x, y)$, with $z_0 = g(x_0, y_0)$, such that $z = g(x, y)$ solves the equation $F(x, y, z) = 0$ in some neighborhood of (x_0, y_0). Use the *Chain Rule* (Maple knows it well) to find formulas for the partials of g with respect to x and y at x_0, y_0. In the process, you will find a reason for the condition $F_z(x_0, y_0, z_0) \neq 0$. (Hint: Before you differentiate, replace z by $g(x, y)$. Maple must be told that z depends on x and y.)

7. Building on the last problem, a general system of two (nice) equations in three variables x, y, and z is modeled by the system of equations

$$F(x, y, z) = 0,\ G(x, y, z) = 0$$

where F, and G are differentiable functions of three variables. It is natural to think of this as a system of two equations in two unknowns, (think of one of the variables as a given number, rather than as an unknown), and to attempt to solve such a system for the unknowns, say y and z (for example) in terms of x. When such a solution exists, even if it cannot be found, we say that the system

$$F(x, y, z) = 0,\ G(x, y, z) = 0$$

defines y and z implicitly as functions of x. Assuming that such a solution $y = p(x)$, $z = q(x)$ exists, take advantage of Maple's understanding of the *Chain Rule* to find formulas for $\frac{dy}{dx} = p'(x)$, $\frac{dz}{dx} = q'(x)$ in terms of the partials of F and G. In the process, if

$$F(x_0, y_0, z_0) = 0,\ G(x_0, y_0, z_0) = 0,$$

find a condition on the partials of F and G that must be met in order for the solution $y = p(x)$, $z = q(x)$ with $y_0 = p(x_0)$, $z_0 = q(x_0)$ to exist. Maple must be told that y and z depend on x, so substitute $y = p(x)$, $z = q(x)$ into these equations before you begin to differentiate. After the differentiation, it might help to simplify notation. The symbols *yp* and *zp* are good names for diff($p(x)$, x), and diff($q(x)$, x). As a final step, **check the answers, using the implicitdiff( ) command.**

8. Find the directional derivative of the function

$$f(x, y, z) = \frac{e^{2zx}}{y^2 x^3}$$

at the point $P(4, 9, 2)$ in the direction $\vec{w} = 7\vec{i} - 15\vec{j} + 3\vec{k}$. Use this to approximate (as a decimal) the change in f as the point (x, y, z) moves a distance of 0.02 units from P in the direction of $\vec{w}$. Compare this to the actual change in f (expressed as a decimal).

9. Let $f(x, y, z) = 3x^2 - y^2 + 2z^2 - 8x - 5y + 7z - 6$. Find the point(s) on the level surface $f(x, y, z) = 0$, if they exist, where ∇f has the same direction and sense as the vector $\vec{n} = 2\vec{i} - \vec{j} + 3\vec{k}$.

10. Suppose that the effect of three economic variables, p, r and t on the value, V, of the U.S. dollar (the trading rate of the dollar in yen) is being studied, where the variables p, r, and t are the U.S. prime (annual) interest rate, the U.S monthly index of inflation, both as decimal rates, and the monthly U.S. trade deficit, in billions of dollars, respectively. Suppose that, in the short term, the effects of these three variables on V is thought to be

$$V(p, r, t) = 101 + 3543p^2 + 841r - 0.0174t^3.$$

Suppose that the prime interest rate is currently 7.4%, that the monthly index of inflation is 0.38%, and that the monthly trade deficit is 14 billion dollars, so that $p_0 = 0.074$, $r_0 = 0.0038$, and $t_0 = 14$. By adjusting the values of p, r, and t, economists want to bring the value of the dollar down, but not at too fast a rate, relative to the other variables, for fear of creating an unstable economy. Suppose that a desirable rate of decrease is .23 with respect to any combination of the variables. Use gradient vectors and directional derivatives to describe a range of allowable directions

$$\vec{u} = u_p\vec{i} + u_r\vec{j} + u_t\vec{k}$$

in which to move the variables p, r, and t. Here $\vec{u}$ denotes a unit vector in the (p, r, t) coordinate system. To specify allowable values for the vector $\vec{u}$, write your answer as a system of two equations in the unknowns u_p, u_r, and u_t that the components of the unit vector $\vec{u}$ must satisfy in order to be an allowable direction.

The U.S. monthly trade deficit is difficult to control. Suppose we assume that it remains constant at $t = t_0 = 14$ billion dollars during the adjustment. Determine the corresponding direction vectors $\vec{u}$ (there are two). For one of these two direction vectors, determine the parametric equation of the line through $P_0(p_0, r_0, t_0)$ with direction $\vec{u}$. Specify a few reasonable values of p, r, and t along this line and compute the corresponding values of $V(p, r, t)$.

11. Suppose that a hill is modeled by the graph of the function

$$z = f(x, y) = (8.5 - 0.0763x^2 - 0.274y^2 + 1.4x - 0.98y)\cos(5x + 7y) + 500$$

where the elevation $z = f(x, y)$ is in feet and the coordinates of the point (x, y) are in miles. A projectile is fired from the point on the hill corresponding to $(x, y) =$

(13.423, −37.25438) with an initial speed of 97 ft/sec in a direction perpendicular to the hill. How long does it take for the projectile to fall back down and strike the hill? What are the coordinates of the contact point? Assume that the only force acting on the projectile after it is fired is gravitational acceleration, $\vec{g} = -32\vec{k}$ in ft/sec^2.

12. Show that every tangent plane to the graph of $f(x,y) = \frac{x^2-3y^2}{5x+4y}$ passes through the origin.

13. Let f and g be real-valued differentiable functions of x, y, and z. Use Maple to prove that $\nabla(fg) = f\nabla g + g\nabla f$.

14. Find the critical points of $f(x,y) = x^4 - y^3 - 84x^2 + 22y^2 - 133y + 2431$ and the values of f at the critical points. Use the *Second Derivative Test* (if it applies) at each critical point to decide whether it corresponds to a local maximum, local minimum, or saddle point.

15. Find the critical points of $f(x,y) = (x^4 + y^2)e^{1-x^2-y^2}$ and the values of f at the critical points. Use the *Second Derivative Test* (if it applies) at each critical point to decide whether it corresponds to a local maximum, local minimum, or saddle point. If the *Second Derivative Test* does not apply, try to provide some evidence in support of some kind of conclusion.

16. Find the absolute maximum of the function f in problem 1 and the point where it occurs. Absolute maximums do not always exist. How do we know one exists in this case? Provide convincing evidence that it exists and that it must be the candidate of choice. Does the absolute minimum exist? Explain your answer.

17. Find the absolute maximum and absolute minimum of

$$f(x,y,z) = x^2 + 2x + y^2 + y + z^2 - 6z + 2$$

on and inside the ellipsoid

$$\frac{x^2}{4} + \frac{y^2}{9} + \frac{z^2}{25} = 1.$$

Be sure to collect all possible points where a maximum or minimum could occur. Absolute maxima and minima do not always exist. How do we know they exist in this case? Provide convincing evidence that they exist and that they must be the candidates of choice.

18. A propane storage tank in the shape of a cylinder with hemispherical ends must hold 10,000 cubic meters of gas. What dimensions will minimize the amount of material needed to build the tank?

Project: Implicit Differentiation

The system of equations

$$x = u\ln(yv), \; y = v\ln(5 + xu^2)$$

defines $u = f(x,y)$ and $v = g(x,y)$ implicitly as functions of x and y. Use the diff( ) command and implicit differentiation to compute (as decimals) the values of the partials of u and v with respect to x and y at the point $(x,y) = (4,8)$. Show that the values of u and v corresponding to $x = 4$ and $y = 8$ are uniquely determined. (This does not imply that the functions f and g are uniquely determined near the point $(x,y) = (4,8)$, but it makes it seem likely.). The implicitdiff( ) command can be used to compute these partials directly. **Check your answers, using this very impressive command.**

Intuitively, this problem should seem reasonable. Think of x and y as given (rather than as unknowns), and the system of two equations in the two unknowns u and v, should be solvable for u and v. With a simpler set of equations, one might actually be able to solve for u and v in terms of x and y. In that case, the partials could be computed directly. In the present case, the equations are too complicated to actually solve for u and v in terms of x and y, and so the partials must be computed by implicit differentiation.

The remarks concerning implicit differentiation and substitution in problem 11.6 and the issues discussed in Example 11.4 are relevant here as well. Maple can be used in a very elegant way to solve this problem, but you must be careful in assigning names and in using the substitution and differentiation commands.

Project: Least-Squares Regression Analysis

1. Suppose that experimental, or statistical data is collected in the form of points (data points), (x_1, y_1), (x_2, y_2),..., (x_n, y_n). Suppose that plotting the points shows that they tend to cluster near some straight line in the (x,y) plane. The so-called "least-squares regression line" is the line $y = f(x) = ax + b$ that "best fits" the data points in the sense that

$$S = \sum_{j=1}^{n}(f(x_j) - y_j)^2 = \sum_{j=1}^{n}(ax_j + b - y_j)^2$$

is as small as possible. In other words, it is the line for which the sum of the squares of the error terms is as small as possible. Individual error terms are squared so that they are always positive. Consequently, they always accumulate in the sum. Simply summing the error terms (without squaring) could lead to a small sum (or zero) even if the individual error terms are quite large.

Determine formulas for a and b in terms of the data points. These are classic formulas, which you will probably find in your main text book. (Hint: Think of S as a function of a and b, and minimize S with respect to a and b.) Maple's bracket notation—$x[j]$, $y[j]$, $(j = 1, \ldots, n)$—can be used to label the data points, and Maple's sum( ) command can be used to form the sums.

2. Suppose that the data points (x_1, y_1), (x_2, y_2),..., (x_n, y_n), appear to almost fit a quadratic $y = ax^2 + bx + c$. Find formulas for a, b, and c in the same "least-squares" way that was used in the previous problem. Maple's sum( ) command should be used to form the sums. Test your results on a data set consisting of 10

points. To generate a meaningful data set in a convenient way, start with a specific function of the form $f(x) = Ax^2 + Bx + C + 0.2\sin(x)$, which is a quadratic together with a small alternating "chaotic" term. Use it to generate the y-coordinates of the 10 data points. If x and y are names of Maple sequences that represent the data points, then the jth point of x or y can be accessed conveniently via Maple's bracket notation $x[j]$, or $y[j]$. Check to see whether the "least squares" values of a, b, and c are close to the values of A, B, and C.

3. Suppose that the data points (x_1, y_1, z_1), (x_2, y_2, z_2),..., (x_n, y_n, z_n) appear to lie close to a plane $z = f(x, y) = ax + by + c$. The corresponding "mean-square" error term would take the form

$$S = \sum_{j=1}^{n}(f(x_j, y_j) - z_j)^2 = \sum_{j=1}^{n}(ax_j + by_j + c - z_j)^2.$$

Find formulas for a, b, c in terms of the data points that would minimize S.

Test your results on a data set consisting of 10 points. Generate a data set in much the same way that a data set was generated in the preceding problem, by using a function of the form $f(x, y) = Ax + By + Cz + 0.2\sin(xyz)$. Enter it as a Maple sequence, so that computations can be performed with Maple's sum( ) command. Check to see whether the "least squares" values of a, b, and c, are close to the values of A, B, and C.

Project: The Gradient in Cylindrical Coordinates

Let $f(r, \theta, z)$ be a real-valued differentiable function of the variables r, θ, and z in a cylindrical coordinate system. Prove (using Maple) that the gradient vector

$$\nabla f = \frac{\partial f}{\partial x}\vec{i} + \frac{\partial f}{\partial y}\vec{j} + \frac{\partial f}{\partial z}\vec{k}$$

can be expressed in the cylindrical form

$$\nabla f = \frac{\partial f}{\partial r}\vec{e_r} + \frac{1}{r}\frac{\partial f}{\partial \theta}\vec{e_\theta} + \frac{\partial f}{\partial z}\vec{e_z}$$

where $\vec{e_r}$, $\vec{e_\theta}$, $\vec{e_z}$ are mutually orthogonal unit vectors in the r, θ, and z directions. More specifically, if $P(r, \theta, z)$ is a point in the cylindrical coordinate system, then $\vec{e_r}$ is the horizontal unit vector based at P projecting radially away from the z-axis, $\vec{e_\theta}$ is the horizontal unit vector based at P tangent to the cylinder $x^2 + y^2 = r^2$ at P and pointing in the counterclockwise direction (towards increasing θ values), and $\vec{e_z}$ is the vertical unit vector $\vec{k}$ based at P.

Project: Finding Local Extreme Values

It can be very difficult to find a local or absolute maximum or minimum of a differentiable function $f(x, y, z)$ of three variables, and the complexity of the problem grows enormously as the number of variables increases beyond three. It could well be a simple matter to get Maple to compute partial derivatives and set them equal to zero, but getting Maple to solve the system of equations is another matter.

The gradient vector always points in the direction in which f increases at its greatest rate, so ∇f can be used to find the approximate locations of the candidates for local maxima, and minima.

To search for a local maximum, we start with a point (x_0, y_0, z_0) that serves as our initial guess. We can then use Maple's gradient and plotting tools to move from (x_0, y_0, z_0) in the direction of ∇f as far as possible towards a local maximum in this direction. (Maple's elementary plot( ) command should be used here to plot a function of one variable, regardless of how many variables appear in f.) This will determine a point (x_1, y_1, z_1) which is closer than (x_0, y_0, z_0) to a local maximum. The procedure can be repeated as often as necessary until a point sufficiently close to a local maximum is determined. A similar procedure can be used to find approximate locations of local minimum points.

In this problem, find (approximately) the coordinates of a point that is located in the first octant of three-dimensional space, and is a local minimum of

$$f(x, y, z) = 48xy + 17yz + x^2 z + \frac{83000}{xy^2 z}.$$

Compute the partial derivatives and set them equal to zero. Can Maple solve this system of equations? Probably not!

Use a gradient search method to find an approximate location of the local minimum. Assign names to each successive approximation (x_j, y_j, z_j) determined by this method. Maple will treat these points as points in a sequence, if you use names like $X[j] := vector([x_j, y_j, z_j])$ (notice the brackets on $X[j]$). You can then use Maple's seq( ) command to review how the points $X[1], X[2], \ldots,$ seem to be converging, and how the values of f at these points seem to be decreasing and leveling off to some value.

Now return to the system of equations obtained by setting all three partial derivatives equal to zero. Use the approximation obtained with this gradient approach to create a very small window in space where the solution is now known to exist. Insert this window, in the form of three ranges, inside Maple's fsolve( ) command. See whether Maple with this very big hint, can now solve the system of equations to produce a solution inside this small window. Maple might still balk, but, if it produces a solution, check to see whether it improves the accuracy of the solution. It should, given the strength of the hint.

If f is a real-valued differentiable function of $n > 3$ variables, then ∇f is defined in the same expected way. It turns out that ∇f always points in the direction in which f increases at the greatest rate. This important feature, which holds regardless of how many variables are involved, can be used to create a gradient search method for local maxima and minima that is very similar to the method used in this problem.

Project: A Summer Job

1. It's summer, you have three months to make money before school starts, and you have just been offered \$7,000 to deliver 10,000 tropical birds from a warehouse

on the waterfront to a dealer in a town 40 miles away. Rather than let someone else strike first at the opportunity, you readily accept the job—\$7,000 seems like so much money for such a small job. But then, after you accept, your enthusiasm is tempered by rising doubts about costs. You have agreed to pay for all of the expenses involved.

Arrangements are made to rent a small flatbed truck that will cost \$20 for each run from the warehouse to the dealer and back. In order to transport the birds, you have agreed to build (and pay for) a rectangular cage meeting certain specifications. For the welfare of the birds, no more than 10 birds per cubic foot will be allowed in the cage. In order to fit safely on the truck bed, the cage cannot exceed 6 feet in width, 12 feet in length, or 3 feet in height. All that's left to do now is to arrange for the construction of the rectangular cage. The material for the sides and top will cost \$10 per square foot, the material for the bottom will cost \$14 per square foot. The two ends, which must contain special capture mechanisms, will cost \$24 per square foot. To maintain a constant temperature in the box, two special heating strips must be mounted on the bottom of the box running its entire length. The heating material costs \$8 per linear foot. Finally, perches for the birds will be installed at a cost of \$4 per cubic foot. When the delivery service has been completed, the dealer has agreed to buy the cage for \$400, regardless of what it costs you to build.

A bigger cage means fewer trips, but a bigger cage is much more expensive to build. What should the dimensions of the cage be in order to minimize total costs? What is the cost of the cage? How many trips must be made? How much money is left for you?

The system of equations obtained by setting all of the partial derivatives equal to zero is complicated, but Maple is up to the task of supplying at least one solution. Are there other solutions to this system of equations? This issue, difficult to resolve, will be taken up in the next exercise. In this exercise, assume that there are no other critical points (interior to the region under consideration). This makes the problem easier. Don't forget, however, that the absolute minimum could occur at a boundary point rather than at a critical point.

2. It seems probable that the point we found in the first part of this problem is the minimum point, but, if you are skeptical (as a good scientist should be), you will not feel comfortable assuming that there are no other critical points for this problem. Maple is ready and willing to help supply you with more evidence. Write a simple program that evaluates the cost function at many points, searching (probably without success) for smaller values of the cost function. Write the program thoughtfully, so that you don't display output that you don't need to see. This will speed up the program enormously when it is run (see Example 11.12). Evidence generated in this way will not prove that you already have the absolute minimum (if, in fact, you do), but by spreading a large number of points around the region, it certainly can be used to supply very convincing evidence to support such a conclusion.

Project: Managing a Large Apartment Building

Suppose you are the manager of a large apartment building with 1450 units. Currently, the rent is p_0 = 758 dollars per month for each unit, and there are

$x_0 = 1288$ units occupied. Fixed overhead costs are \$114,000 per month (mostly for salaries, property taxes, and mortgage payments), monthly utility costs average \$47 per occupied unit, and monthly (essential) repair costs average \$11 per occupied unit. In addition, you have a special monthly budget to pay for advertising and optional remodeling and repair projects. This budget amounts to \$50 for each occupied unit, and it is up to you to decide how much you wish to spend on each of these two items, but you are expected to spend the full amount each month. Currently, these costs are $a_0 = 23000$ dollars per month for advertising and $r_0 =$ 41400 dollars per month per unit for optional remodeling and repair projects. It is also up to you to decide how much rent to charge for each unit.

Let p denote the monthly rental charge for each unit, let a denote the total monthly advertising cost, and let r denote the monthly cost for optional remodeling and repair. Suppose these variables are currently

$$(p, a, r) = (p_0, a_0, r_0) = (758, 23000, 41400).$$

You are considering some changes in these variables. Naturally, the building owners are expecting your choices to generate as large a profit as possible under the circumstances.

The results of a market analysis are available. The number of units occupied, denoted by x, will naturally decrease (increase) from $x_0 = 1288$ if the monthly rent p increases (decreases) from p_0. By the same token, x will change as a changes and as r changes . It is **reasonable to assume that the relation between x and the variables p, a, and r is linear**, as long as p, a, and r do not move too far from p_0, a_0, and r_0, respectively. A market analysis suggests that these rates of change are (in units per dollar)

$$\frac{\partial x}{\partial p} = -1.6, \quad \frac{\partial x}{\partial a} = 0.005, \quad \frac{\partial x}{\partial r} = 0.011.$$

Furthermore, this linear relation, with these rates of change, is expected to be valid for all p, a, and r in the intervals

$$|p - p_0| \leq 100, \ |a - a_0| \leq 10000, \ |r - r_0| \leq 10000.$$

The three variables, p, a, and r are not independent variables. It turns out that total monthly profit, P, can be expressed in terms of just two of the three variables. Find an expression for P in terms of p and r. Determine the values of p and r, within these allowable intervals, that will generate the largest profit. For these values of p and r, give values for all of the economic terms involved in this problem.

A drastic change in the values of p, a, and r might be regarded as risky, in spite of the higher profits that should be generated by this model. A more conservative approach should also be considered. Suppose that you wish to move the variables p and r only slightly from their current values p_0 and r_0, perhaps to test the waters before you take a more dramatic plunge. Determine a direction to move the variables p and r from p_0 and r_0 that will increase P at the greatest rate.

Chapter 12

Multiple Integration

Double, triple, and even higher-order integrals play an important role in mathematics and in its applications. In this chapter, we will see that Maple's familiar integration command, int( ), can be used in a very straightforward way to compute all of these integrals.

12.1 Double Integrals

If $f(x, y)$ is a bounded real-valued function of x and y that is defined on a rectangle $R = [a, b] \times [c, d]$, then the double integral of f over R is defined as the limit

$$\int\int_R f(x, y)\, dxdy = \lim_{\|\mathcal{P}\| \to 0} S(\mathcal{P}, f)$$

where $S(\mathcal{P}, f)$ is the Riemann sum

$$S(\mathcal{P}, f) = \sum_{j=1}^{n} f(x_j, y_j)\Delta x_j \Delta y_j.$$

This definition is a bit abbreviated, and other details should be mentioned. In particular, $\mathcal{P} = \{R_1, R_2, \ldots, R_n\}$ is a partition of R into n subrectangles, Δx_j and Δy_j denote the dimensions of R_j, and (x_j, y_j) represents an arbitrary point in R_j. The limit is taken as $\|\mathcal{P}\|$; the maximum of all of the dimensions, $\Delta x_j, \Delta y_j$ in $\mathcal{P}$, tends to zero; and the limit must be independent of the choices of the points (x_j, y_j) $(j = 1, \ldots, n)$. Taking the limit as $\|\mathcal{P}\| \to 0$ is simply a convenient way of saying that all of the dimensions $\Delta x_j, \Delta y_j$ tend to 0. When Riemann sums are actually constructed, a double-indexing scheme for the subrectangles and points is usually more convenient. The details can be found in any standard calculus text, along with a theorem stating that the double integral of a continuous function over a rectangle always exists.

The definition of a double integral could be used to create a numerical technique for approximating its value. Rather than move in this direction, however, we shall simply take advantage of Maple's own numerical techniques. Still, this definition is important to understand for other reasons. It gives intuitive meaning to the individual symbols in $\int\int_R f(x, y)\, dxdy$. Think of $dx\, dy$ as the area of a infinitesimally small rectangle of dimensions dx by dy located at the point (x, y) in R. Multiply this area by the value of f at that same point (x, y) to get $f(x, y)dxdy$. The symbol $\int\int_R f(x, y)\,dxdy$ represents, intuitively, the sum of all the terms $f(x, y)dxdy$ as (x, y) ranges over all of the points in R. This interpretation helps immensely in setting up integrals, and also in setting up applications.

Maple is an excellent environment to help us gain some insight into this definition, and so we begin our study of double integrals with an example involving approximation by Riemann sums.

Example 12.1 *Let $f(x, y) = xe^{xy}$, and let R be the rectangle $R = [-1, 2] \times [1, 3]$. Approximate the value of $\int\int_R f(x, y)dxdy$ with a Riemann sum obtained by partitioning R into 20^2 subrectangles, using a grid of 20 equally spaced horizontal lines and 20 equally spaced vertical lines, and by letting (x_j, y_j) be the midpoint of the j-th subrectangle for each $j = 1, 2, \ldots, 400$.*

Setting up the partition points and the corresponding midpoints is easily accomplished with Maple's seq( ) command. The coordinates of the midpoints of the subrectangles are easier to specify with a double-indexing scheme.

****

```
> f:=(x,y)->x*exp(x*y):
> lx:=(2-(-1))/20; ly:=(3-1)/20;
```

$$lx := \frac{3}{20}$$

$$ly := \frac{1}{10}$$

```
> X:=seq(-1+j*lx,j=0..20);
```

$$X := -1, \frac{-17}{20}, \frac{-7}{10}, \frac{-11}{20}, \frac{-2}{5}, \frac{-1}{4}, \frac{-1}{10}, \frac{1}{20}, \frac{1}{5}, \frac{7}{20}, \frac{1}{2}, \frac{13}{20}, \frac{4}{5}, \frac{19}{20}, \frac{11}{10}, \frac{5}{4}, \frac{7}{5}, \frac{31}{20}, \frac{17}{10}, \frac{37}{20}, 2$$

```
> Y:=seq(1+j*ly,j=0..20);
```

$$Y := 1, \frac{11}{10}, \frac{6}{5}, \frac{13}{10}, \frac{7}{5}, \frac{3}{2}, \frac{8}{5}, \frac{17}{10}, \frac{9}{5}, \frac{19}{10}, 2, \frac{21}{10}, \frac{11}{5}, \frac{23}{10}, \frac{12}{5}, \frac{5}{2}, \frac{13}{5}, \frac{27}{10}, \frac{14}{5}, \frac{29}{10}, 3$$

```
> mx:=seq((X[j]+X[j+1])/2,j=1..20);
```

#This gives us the x-coordinates of the midpoints.#

$$mx := \frac{-37}{40}, \frac{-31}{40}, \frac{-5}{8}, \frac{-19}{40}, \frac{-13}{40}, \frac{-7}{40}, \frac{-1}{40}, \frac{1}{8}, \frac{11}{40}, \frac{17}{40}, \frac{23}{40}, \frac{29}{40}, \frac{7}{8}, \frac{41}{40}, \frac{47}{40}, \frac{53}{40}, \frac{59}{40}, \frac{13}{8}, \frac{71}{40}, \frac{77}{40}$$

```
> my:=seq((Y[j]+Y[j+1])/2,j=1..20);
```

#This gives us the y-coordinates of the midpoints.#

$$my := \frac{21}{20}, \frac{23}{20}, \frac{5}{4}, \frac{27}{20}, \frac{29}{20}, \frac{31}{20}, \frac{33}{20}, \frac{7}{4}, \frac{37}{20}, \frac{39}{20}, \frac{41}{20}, \frac{43}{20}, \frac{9}{4}, \frac{47}{20}, \frac{49}{20}, \frac{51}{20}, \frac{53}{20}, \frac{11}{4}, \frac{57}{20}, \frac{59}{20}$$

```
> evalf(sum(sum(f(mx[i],my[j])*lx*ly,i=1..20),j
  =1..20));
```

$$126.1633834$$

****

By way of comparison, Maple's decimal value for this integral turns out to be 127.4384921, as we shall see shortly.

The definition of a double integral can be extended to planar regions other than rectangles, as follows. Let $f(x, y)$ be a bounded real-valued function of x and y defined over an arbitrary bounded planar region D. We enclose D in a rectangle R ($D \subseteq R$), and extend the definition of f to all of R by defining $f(x, y)$ to be 0 for all points (x, y) in R outside of D. Then define

$$\int\int_D f(x, y)\, dxdy = \int\int_R f(x, y)\, dxdy.$$

It can be shown that the double integral of a continuous function f over a planar region D always exists, as long as D is not "unreasonably complicated." Conditions on D which would imply the existence of the integral are discussed in any standard calculus text.

Example 12.2 *Let $f(x, y) = 2x^2 + 5y^2$, and let D be the closed disk of radius 2 centered at the origin. Approximate the value of $\int\int_D f(x, y)dxdy$ with a Riemann sum obtained by partitioning $R = [-2, 2] \times [-2, 2]$ into 20^2 subrectangles, using a grid of 20 equally spaced horizontal lines and 20 equally spaced vertical lines, and by letting (x_j, y_j) be the midpoint of the j-th subrectangle for each $j = 1, 2, \ldots, 400$.*

The rectangle $R = [-2, 2] \times [-2, 2]$ is defined by the same interval in both the x and y directions, so we need to partition only one of these intervals to create the rectangular partition of R.

We have used the very convenient piecewise( ) function infrequently, so it could help to look it up in the Index or the Help File. When the command is used with three arguments, the first argument is a condition $cond_1$, the second argument is the value under $cond_1$, and the third argument is the value otherwise. Also, as a reminder, notice that the keyboard symbols <= are used to denote the "less than or equal to" relation ($\leq$) in a Maple input statement .

****

```
> f:=(x,y)->piecewise(x^2+y^2<=4,2*x^2+5*y^2,0);
```

$$f := (x, y) \to \text{piecewise}(x^2 + y^2 \leq 4,\ 2\,x^2 + 5\,y^2,\ 0)$$

\# There are other ways to extend $f(x, y)$ to be 0 outside of D, but it is **much more efficient to set it up with the piecewise( ) command.** We could have, for example, extended $f(x, y)$ with an input statement of the form:

```
>h:=(x,y)->if x^2+y^2<=4 then 2*x^2+5*y^2 else 0 fi:
```

If we had used this approach, **Maple would not let us use the expression $h(x, y)$ until after we had evaluated (x, y).** This would cause major complications in setting up the Riemann sum we wish to evaluate. Notice, however, **how easy it is to use the expression $f(x, y)$ in an input statement.**#

```
> f(x,y);
```

$$\begin{cases} 2\,x^2 + 5\,y^2 & x^2 + y^2 \leq 4 \\ 0 & otherwise \end{cases}$$

```
> l:=4/20;
```

$$l := \frac{1}{5}$$

```
> P:=seq(-2+j*l,j=0..20);
```

$$P := -2, \frac{-9}{5}, \frac{-8}{5}, \frac{-7}{5}, \frac{-6}{5}, -1, \frac{-4}{5}, \frac{-3}{5}, \frac{-2}{5}, \frac{-1}{5}, 0, \frac{1}{5}, \frac{2}{5}, \frac{3}{5}, \frac{4}{5}, 1, \frac{6}{5}, \frac{7}{5}, \frac{8}{5}, \frac{9}{5}, 2$$

```
> m:=seq((P[j]+P[j+1])/2,j=1..20);
```

$$m := \frac{-19}{10}, \frac{-17}{10}, \frac{-3}{2}, \frac{-13}{10}, \frac{-11}{10}, \frac{-9}{10}, \frac{-7}{10}, \frac{-1}{2}, \frac{-3}{10},$$
$$\frac{-1}{10}, \frac{1}{10}, \frac{3}{10}, \frac{1}{2}, \frac{7}{10}, \frac{9}{10}, \frac{11}{10}, \frac{13}{10}, \frac{3}{2}, \frac{17}{10}, \frac{19}{10}$$

```
> S:=evalf(sum(sum(f(m[i],m[j])*l^2,j=1..20),i=1..20));
```

$$S := 88.96160000$$

\#Maple's decimal value for this integral turns out to be 87.96459431, as we shall see next. #

****

Double integrals are computed by Maple via the familiar int( ) command in an expected way. If D is the domain defined by

$$\{(x,y)|p(x)\leq y\leq q(x), a\leq x\leq b\},$$

then

```
> A:=int(f(x,y),y=p(x)..q(x));
```

defines the integral

$$A:=\int_{p(x)}^{q(x)} f(x,y)dy,$$

and so

```
> B:=int(A,x=a..b);
```

defines the integral

$$B:=\int_a^b\int_{p(x)}^{q(x)} f(x,y)dydx.$$

Of course, both integrals can be brought together in one input statement as

```
> B:=int(int(f(x,y),y=p(x)..q(x)),x=a..b);
```

Maple's decimal value for the integral in Example 12.1, and for the integral in Example 12.2, expressed as the iterated integral

$$\int_{-2}^{2}\int_{-\sqrt{4-x^2}}^{\sqrt{4-x^2}}(2x^2+5y^2)dxdy,$$

are computed as follows.

****

```
> f:=x*exp(x*y):
> answer:=evalf(int(int(f,x=-1..2),y=1..3));
```

$$answer := 127.4384921$$

```
> f:=2*x^2+5*y^2:
> evalf(int(int(f,y=-sqrt(4-x^2)..sqrt(4-x^2)),x=-2..2));
```

$$87.96459431$$

****

The integrals $A=\int_{p(x)}^{q(x)} f(x,y)dy$, and $B=\int_a^b A\,dx$, which together determine the double integral

$$B=\int_a^b\int_{p(x)}^{q(x)} f(x,y)dydx,$$

can also be evaluated separately. Evaluating the integrals separately allows the intermediate antidifferentiation processes to be observed.

Example 12.3 *Evaluate the iterated integral* $\int_{-3}^{3}\int_{3x+7}^{9-x^2} x\sin(y)\,dydx$ *directly, using a single Maple input statement. Evaluate the integral again as a sequence of two separate single integrals. Use indefinite integration, so that the individual antidifferentiation processes can be observed in greater detail.*

****

```
> answer:=int(int(x*sin(y),y=3*x+7..9-x^2),x=-3..3);
```

$$answer := \frac{1}{9}\cos(16)+\sin(16)-\frac{1}{9}\cos(2)-\sin(2)$$

```
> A1:=int(x*sin(y),y);
```

$$A1 := -x\cos(y)$$

```
> A2:=subs(y=9-x^2,A1)-subs(y=3*x+7,A1);
```

$$A2 := -x\cos(9-x^2)+x\cos(3x+7)$$

```
> B1:=int(A2,x);
```

$$B1 := -\frac{1}{2}\sin(-9+x^2)+\frac{1}{9}\cos(3x+7)+\frac{1}{9}(3x+7)\sin(3x+7)-\frac{7}{9}\sin(3x+7)$$

```
> B2:=subs(x=3,B1)-subs(x=-3,B1);
```

$$B2 := \frac{1}{9}\cos(16)+\sin(16)-\frac{1}{9}\cos(-2)+\sin(-2)$$

#The expression $B2$ differs only slightly from the answer we obtained by a direct evaluation of the double integral. Using elementary trigonometric identities, we can easily see that they are the same. On another problem, two answers might look very different. One way to compare answers to see whether they are equivalent is to subtract one from the other and attempt to show that this difference is 0. Maple is particularly good at producing a zero, if an expression is equivalent to zero. #

```
> compare:=simplify(answer-B2,trig);
```

$$compare := 0$$

What else could we have done if Maple had not been so cooperative? Try using the Help File. Initiate a **Full Text Search** by entering **"trig identities"**. This would lead to a host of items worth investigating.#

****

Maple's help file is unusually silent about numerical approximation techniques for double integrals. Certainly, it has such a technique, and, from our knowledge of Maple's single integral, we can almost guess how to write the input statement. Recall that with the statement

```
> evalf(int(f(x),x=a..b));
```

Maple first tries to find an antiderivative $F(x)$ of $f(x)$, evaluates $F(b)-F(a)$ exactly, and then finishes with a decimal approximation of $F(b) - F(a)$. If it cannot find an antiderivative of $f(x)$, then a numerical approximation technique (similar to Simpson's Rule) is used to evaluate the integral directly. Change the active form, int( ), of the integration command to the inert form, Int( ),and the input statement

```
> evalf(Int(f(x),x=a..b));
```

immediately activates the numerical approximation technique, regardless of whether $f(x)$ has an elementary antiderivative. In the same way,

```
> evalf(int(int(f(x,y),x=a..b),y=c..d));
```

first attempts an exact evaluation by way of antidifferentiation, before it gives a decimal approximation, and it uses a numerical approximation technique only if an antiderivative cannot be found. In contrast, the input statement

```
> evalf(Int(Int(f(x,y),x=a..b),y=c..d));
```

always evaluates by way of a numerical integration technique regardless of whether the integral can be evaluated by antiderivative techniques. **The performance differences can be quite dramatic**! As you can imagine, numerical approximation techniques for double integrals take considerable time even for simple integrals. It often takes less time for Maple to evaluate a double integral by antidifferentiation techniques. Even simple integrals that might take Maple 10 seconds or less to evaluate with the command evalf(int(int( ))) might take 6 minutes or more to evaluate with the command evalf(Int(Int( ))).

If Maple cannot evaluate a double integral exactly, it might still be possible to avoid using Maple's slow numerical techniques for double integrals. Try setting up the integral as a sequence of two separate single integrals. If the first integral can be evaluated exactly, then numerical techniques need to be applied only to the second single integral. As we have noticed throughout this manual, numerical techniques on single integrals are usually quite fast.

Example 12.4 *Let D be the region in the (x, y) plane bounded by the curves $y = 2e^{3x}$ and $y = 7x^3+28$. Evaluate both iterated integrals associated with $\int\int_D x^2y^3dxdy$, and verify their equivalence.*

To set up the first iterated integral, fix x, then collect the terms $f(x, y)dxdy$ with respect to y as y goes from its value on the bottom curve to its value on the top curve. This determines (in a manner of speaking) a vertical "line of values" at x. To finish the integral, we collect these "lines" with respect to x as x goes from the left-hand "corner" of the region D to the right-hand "corner" of D.

****

```
> f:=x->2*exp(3*x):  g:=x->7*x^3+28:
> plot({f(x),g(x)},x=-2..1);
```

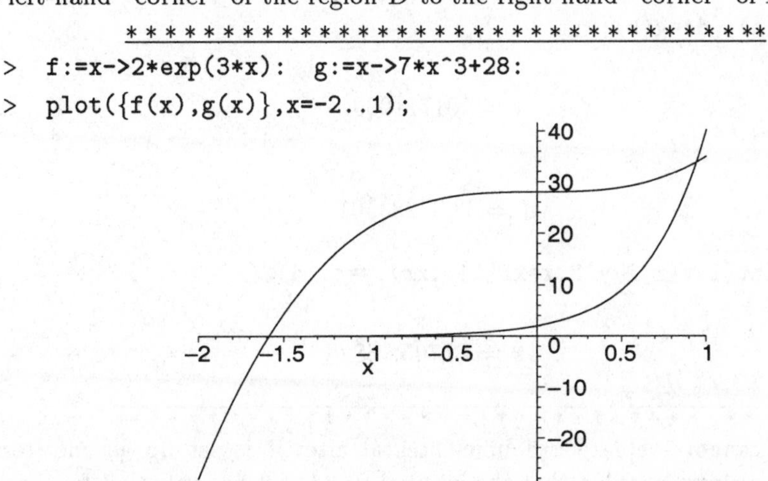


```
> b:=fsolve(f(x)=g(x),x);
```

$$b := .9431553952$$

```
> a:=fsolve(f(x)=g(x),x,x=-2.0..-1.5);
```

$$a := -1.587077646$$

```
> A1:=int(int(x^2*y^3,y=f(x)..g(x)),x=a..b);
```

$$A1 := 85705.16829$$

To set up the other iterated integral, fix y, then collect the terms $f(x,y)dxdy$ with respect to x as x goes from its value on the left-hand curve to its value on the right-hand curve. To find these x-values, we must solve the equations $y = g(x)$ and $y = f(x)$ for x in terms of y. We denote their values by xl (for x on the left) and xr (for x on the right). This collection gives us (in a manner of speaking) a horizontal "line of values" at y. To finish the double integral, we collect these *lines* with respect to y as y goes from the bottom "corner" of D to the upper "corner" of D.

```
> xl:=solve(g(x)=y,x); xr:=solve(f(x)=y,x);
```

$$xl := \frac{1}{7}(-1372+49\,y)^{1/3},\ -\frac{1}{14}(-1372+49\,y)^{1/3}+\frac{1}{14}I\sqrt{3}(-1372+49\,y)^{1/3},$$
$$-\frac{1}{14}(-1372+49\,y)^{1/3}-\frac{1}{14}I\sqrt{3}(-1372+49\,y)^{1/3}$$

$$xr := \frac{1}{3}\ln(\frac{1}{2}y)$$

```
> c:=g(a); d:=g(b);
```

$$c := .01711009$$

$$d := 33.87283501$$

```
> A2:=int(int(x^2*y^3,x=xl[1]..xr),y=c..d);
```

$$A2 := 85705.16828$$

****

If Maple cannot evaluate a double integral after it is set up as one iterated integral, it is at least possible that the integral could still be evaluated by changing the order of integration and evaluating the alternate iterated integral instead.

Example 12.5 *Find the exact symbolic value for the double integral*

$$\int_0^5 \int_{y^2}^{25} e^{-x\sqrt{x}/10} y^3 \, dx dy$$

We first try to evaluate the integral directly. A glance at the integral would suggest that Maple probably cannot find the antiderivative with respect to x of $e^{x\sqrt{x}/10}$. To set up the alternate iterated integral, we must determine D—the region of integration. There is no need to use Maple here. In the integral, as it is expressed in the example, for y fixed, x goes from $x = y^2$ to $x = 25$, and then y goes from $y = 0$ to $y = 5$. It follows that D is the region bounded by $y = \sqrt{x}$, $x = 25$ and the x-axis.

****

```
> int(int(exp(-x*sqrt(x)/10)*y^3,x=y^2..25),y=0..5);
```

$$\int_0^5 \int_{y^2}^{25} e^{(-1/10\, x^{3/2})}\, y^3 \, dx \, dy$$

```
> int(int(exp(-x*sqrt(x)/10)*y^3,y=0..sqrt(x)),x=0..25);
```

$$-225\, e^{(-25/2)} + \frac{50}{3}$$

****

When a double integral is transformed into a polar-coordinate integral, the variables x and y are replaced by

$$x = r\cos(\theta), \;\; y = r\sin(\theta),$$

and the area differential $dxdy$ is replaced by $rdrd\theta$. Of course, the differentials are not inserted inside Maple's integration command, but, when polar-coordinated integrals are set up, the extra factor of r must be inserted in the integrand.

Example 12.6 *Let D be the region bounded by $x+y \geq 0$ and $x^2+y^2 \leq 4$. Evaluate the double integral*

$$\int\int_D xye^y dxdy$$

first as a cartesian integral, and then as a polar coordinate integral.

The boundary curves are plotted below. The region D lies inside both the circular disk and the "upper", or "right" half-plane formed by the line. The points of intersection between the two curves are $(x,y) = (\mp\sqrt{2}, \pm\sqrt{2})$. From the plot, we can see that the cartesian integral will have to be split into two integrals, depending on whether $x \leq \sqrt{2}$, or $x \geq \sqrt{2}$.

****

```
> with(plots):
Warning, the name changecoords has been redefined
```

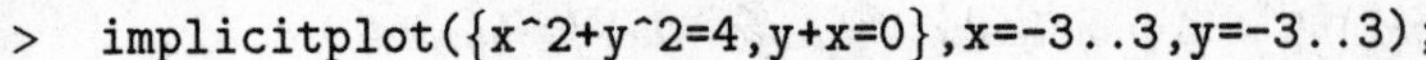

```
> implicitplot({x^2+y^2=4,y+x=0},x=-3..3,y=-3..3);
```

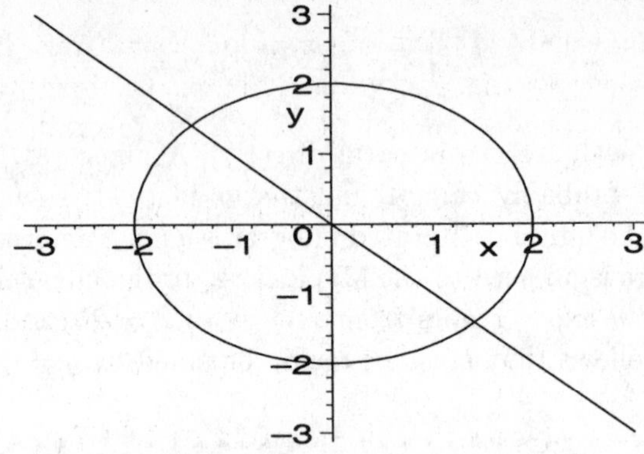


```
> f:=(x,y)->x*y*exp(y):
> cart:=int(int(f(x,y),y=-sqrt(4-x^2)..sqrt(4-x^2)),x=sqrt(2)..2)

  +int(int(f(x,y),y=-x..sqrt(4-x^2)),x=-sqrt(2)..sqrt(2));
```

$$cart := (5\,e^{(2\sqrt{2})} - 3\sqrt{2}\,e^{(2\sqrt{2})} - 3\sqrt{2} - 5)\,e^{(-\sqrt{2})}$$
$$+ (-5\,e^{(2\sqrt{2})} + 3\sqrt{2}\,e^{(2\sqrt{2})} - 3\sqrt{2} - 5)\,e^{(-\sqrt{2})} + 10\,e^{(\sqrt{2})} - 6\sqrt{2}\,e^{(\sqrt{2})}$$

#The letter θ can be used in a Maple work session by using the name "theta". We used the letter t instead, just for the sake of convenience. To set up the limits of integration, think of holding $\theta = t$ fixed, and "collect" differentials terms with respect to r as r goes from $r = 0$ to $r = 2$. This sweeps out a "thin circular sector" at $\theta = t$ of angle opening $d\theta = dt$. Then we collect these "thin sectors" with respect to $\theta = t$ as t goes from $t = -\pi/4$ to $t = 3\pi/4$.#

```
> pol:=int(int(r*f(r*cos(t),r*sin(t)),r=0..2),t=-Pi/4..3*Pi/4);
```

$$pol := 10\,e^{(\sqrt{2})} - 6\sqrt{2}\,e^{(\sqrt{2})} - 6 + (-10 - 6\sqrt{2} + 6\,e^{(\sqrt{2})})\,e^{(-\sqrt{2})}$$

```
> cart2:=simplify(cart);
```

$$cart2 := -2\,(3\,e^{(-2\sqrt{2})}\sqrt{2} + 5\,e^{(-2\sqrt{2})} - 5 + 3\sqrt{2})\,e^{(\sqrt{2})}$$

```
> simplify(cart2-pol);
```

$$0$$

****

Maple computed both integrals with relative ease, but the polar integral was clearly easier to set up.

12.2 Applications of Double Integrals

Area is the most immediate application of a double integral. If D is a region in the (x, y) plane, then $\int\int_D dxdy$ is the area of the region D. Obviously, area is also an application of the more familiar single variable integral, and there is no real advantage gained by treating area as a double integral. Volume, on the other hand, is an application that depends, more fundamentally, on double integration.

Example 12.7 *A scenic building in the shape of a circular cylinder with a hemispherical top is built into the side of a mountain. Determine the volume of the building. The mountain is the surface defined by the graph of*

$$z = m(x, y) = 4850 - 0.0019x^2 - 0.0037y^2,$$

where x, y, and z are expressed in feet. The radius of the cylinder is 48 feet. and the central axis of the cylinder (which is vertical, of course) is 83 feet tall from where it strikes the side of the mountain to the top of the cylinder. The axis passes through the point on the mountain corresponding to $(x_0, y_0) = (79, 113)$. This defines the cylinder, and the hemispherical top sits on top of this cylinder.

The hemisphere is the top half of a sphere having radius 48, centered at the point $(79, 113, z_0)$, where $z_0 = m(79, 113) + 83$. Consequently, it is the graph of the function

$$s(x, y) = z_0 + \sqrt{48^2 - (x - 79)^2 + (y - 113)^2}.$$

If D is the closed disk having radius 48 in the (x, y) plane and centered at $(x_0, y_0) = (79, 113)$, then the volume of the scenic building is the volume of the region R defined by

$$R = \{(x, y, z) | (x, y) \text{ is in } D, \text{ and } m(x, y) \le z \le s(x, y)\}$$

Think of a thin, vertical, rectangular tube located over the point (x, y) in D and extending from the mountain's surface to the hemisphere. The area of the tube's base is $dxdy$, its height is $(s(x, y) - m(x, y))$, and so its volume is $(s(x, y) - m(x, y))dxdy$. We simply add, in the sense of a double integral, the volumes of all of these thin tubes, as (x, y) ranges over D.

There is no need to plot these surfaces to evaluate the volume of R, but it is nice to see (in a manner of speaking) the building and how it sits on the mountain. This is also a good opportunity to bring several somewhat incompatible plots together by using the display( ) command. We use the option **scaling=constrained** so that all of the plots are drawn to scale.

****

```
> m:=(x,y)->4850-0.0019*x^2-0.0037*y^2:
> z0:=m(79,113)+83;
```

$$z0 := 4873.8968$$

```
> s:=(x,y)->z0+sqrt(48^2-(x-79)^2-(y-113)^2):
```

```
> Y:=solve((x-79)^2+(y-113)^2=2304,y);
```

$$Y := 113 + \sqrt{-3937 + 158\,x - x^2},\ 113 - \sqrt{-3937 + 158\,x - x^2}$$

```
> a:=79-48; b:=79+48; c:=113-48; d:=113+48;
```

$$a := 31$$

$$b := 127$$

$$c := 65$$

$$d := 161$$

```
> with(plots):
Warning, the name changecoords has been redefined
```

#The output is suppressed on the next three plot input statements to avoid displaying pages of essentially unreadable computer code.#

```
> p1:=plot3d(m(x,y),x=-200..200,y=-200..200):
> p2:=plot3d(s(x,y),x=a..b,y=Y[2]..Y[1]):
> p3:=implicitplot3d((x-79)^2+(y-113)^2=48^2,x=a..b,y=c..d,
  z=m(79,113)-10..z0):
> display({p1,p2,p3},scaling=constrained);
```

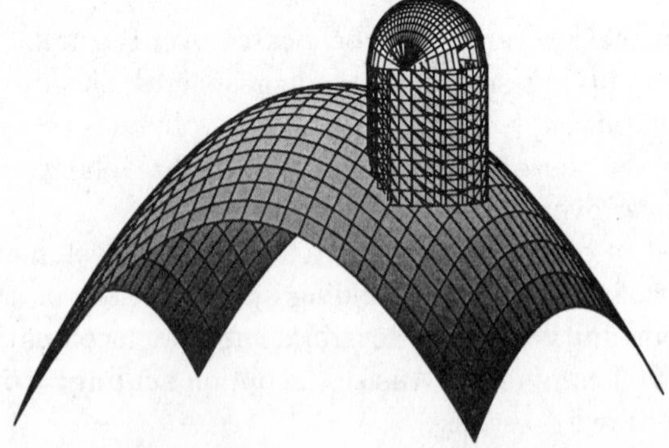

```
> Vol:=evalf(int(int(s(x,y)-m(x,y),y=Y[2]..Y[1]),x=a..b));
```

$$Vol := 855744.0229$$

This is volume in cubic feet.#

****

Let D be a region (a flat plate) in the (x, y) plane. Let $\rho = \rho(x, y)$ denote the mass density of the plate at the point (x, y) in D. Then double integrals can be used to determine the total mass of the plate and its center of mass.

Example 12.8 *Determine the mass and the center of mass of the flat plate bounded by the curves*

$$y = x^4 - 14x^3 + 68x^2 - 136x + 86 \text{ and } y = 23 + 12x - x^2,$$

if the mass density $\rho = \rho(x, y)$ at the point (x, y) on the plate is

$$\rho(x, y) = 0.03(8 - x - y)^2.$$

Let D denote the region bounded by the two polynomial curves. Think of a small square of dimensions $dx \times dy$ located at the point (x, y) in D. Its mass (density times area) is $\rho(x, y)dxdy$, its moment with respect to the y-axis (directed distance to the y-axis times mass) is $x\rho(x, y)dxdy$, and its moment with respect to the x-axis (directed distance to the x-axis times mass) is $y\rho(x, y)dxdy$. To find the mass of the whole plate, we add the individual masses, in the sense of a double integral, as (x, y) ranges over the region D. The total moments are found in the same way.

****

```
> f:=x^4-14*x^3+68*x^2-136*x+86:  g:=23+12*x-x^2:
> plot({f,g},x=0..7);
```

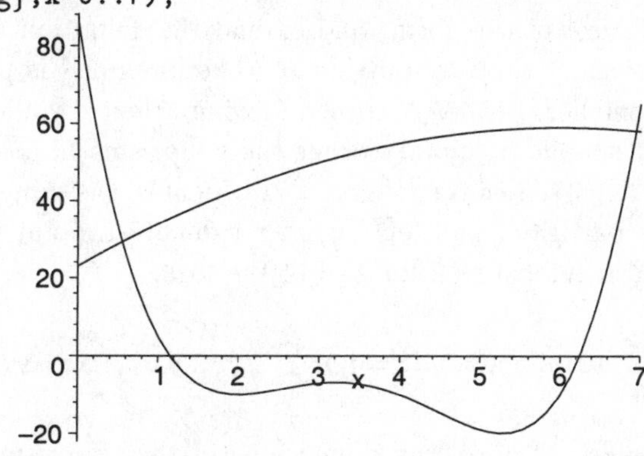


```
> X:=fsolve(f=g,x,x=0..8);
```

$$X := .5527945093, 6.945430258$$

```
> rho:=0.03*(8-x-y)^2:
> M:=evalf(int(int(rho,y=f..g),x=X[1]..X[2]));
```

$$M := 7249.216092$$

```
> Mx:=evalf(int(int(y*rho,y=f..g),x=X[1]..X[2]));
```

$$Mx := 295991.1683$$

```
> My:=evalf(int(int(x*rho,y=f..g),x=X[1]..X[2]));
```

$$My := 33636.69866$$

```
> CenterMass:=[My/M,Mx/M];
```

$$CenterMass := [4.640046349, 40.83078288]$$

****

12.3 Triple Integrals

The definition of a triple integral is very similar to the definition of a double integral. Here, briefly, are some of the highlights.

Suppose that $f(x, y, z)$ is a bounded real-valued function of x, y,and z defined on a solid region R. Enclose R inside a rectangular prism R^*, of the form

$$R \subseteq R^* = [a_1, a_2] \times [b_1, b_2] \times [c_1, c_2],$$

(unless it is already a set of this form) and extend the definition of f by defining f to be zero at points of R^* that are outside R. The prism R^* is partitioned, in the expected way, into smaller prisms. A triple-indexing scheme would usually be used, but, for the sake of simplicity, simply order the subprisms in some way to form a partition $\{P_1, P_2, \ldots, P_n\}$. Let $\Delta x_j \times \Delta y_j \times \Delta z_j$ denote the dimensions of the j-th subprism P_j in the partition, and let (x_j, y_j, z_j) denote a point in P_j. The triple integral of f over R is defined as a limit of of the form

$$\int\int\int_R f(x, y, z)\, dxdydz = \lim \sum f(x_j, y_j, z_j)\Delta x_j \Delta y_j \Delta z_j,$$

where the limit is taken as all of the dimensions of the subprisms in the partition go to zero. The details can be found in any standard calculus text, along with a theorem stating that the triple integral of a continuous function over a "reasonable" solid R always exists.

Think of $dx\, dy\, dz$ as the volume of a infinitesimally small prism ("box") of dimensions dx by dy by dz located at the point (x, y, z) in R. Multiply this volume by the value of f at that same point (x, y, z) to form the term $f(x, y, z)dxdydz$. The symbol $\int\int\int_R f(x, y, z)\, dxdydz$ represents, intuitively, the sum of all the terms $f(x, y, z)dxdydz$ as (x, y, z) ranges over all of the points in R. This interpretation helps immensely in setting up integrals and also in setting up applications.

Maple is an excellent environment for constructing and evaluating Riemann sums, and effort in this direction could help us gain some insight into the definition of a triple integral. Examples 12.1 and 12.2 can be used as models to form similar sums for triple integrals, and a few problems of this sort appear in the exercise set.

Example 12.9 *Evaluate the triple integral* $\int\int\int_R 4x^2y^3\,dxdydz$, *where* R *is the region bounded by the planes* $y = 0$, $z = 0$, $z = 4 - x - y$, *and* $z = 8 + 2x - y$. *Evaluate the integral a second time as a sequence of three separate single integrals. Use indefinite integration so that the individual antidifferentiation processes can be observed in greater detail.*

Some thought must be given to how the plot3d or implicitplot3d( ) commands are to be used, in order to get Maple to present a useful picture of the region R . If too much of each plane is shown, the region will be lost inside all of the drawings. In the present case, a simple device works well. **Notice the use of the option, view=** $z_0..z_1$**, inside the plot3d( ) command.** A coordinate **box is used to show some of the planar faces** (and some numerical values). As an alternative, this box could have been selected from the options menu inside the plot window. Finally, in a key step, **the plot is turned around to see the planar faces** $y = 0$ **and** $z = 0$**.**

The plot shows that we must integrate first with respect to x if we wish to set up the problem in terms of just one (triple) integral.

****

```
> plot3d({0,4-x-y,8+2*x-y},y=0..16/3,x=-4..4,view=0..16/3);
```

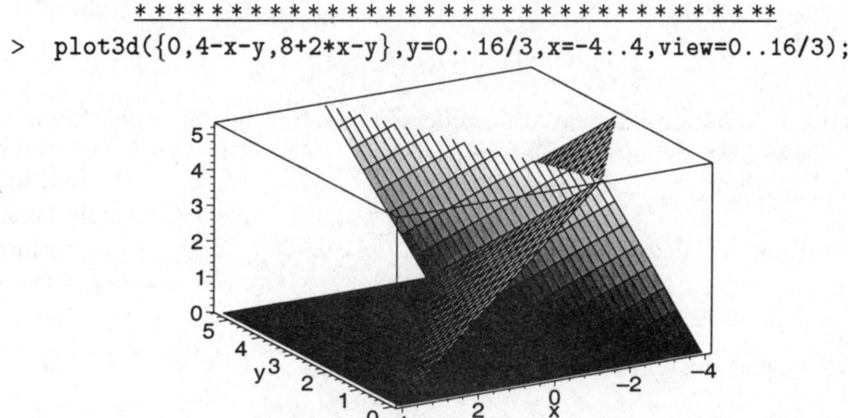

#Look at this plot. For an arbitrary y and z, think of collecting terms first with respect to x , as x goes from one oblique plane to the next (from smaller to larger values of x). This forms, at each y, and z, a horizontal "line of values" (for lack of a better phrase) parallel to the $y = 0$ plane.#

```
> x1:=solve(z=4-x-y,x);
  x2:=solve(z=8+2*x-y,x);
```

$$x1 := -z + 4 - y$$

$$x2 := \frac{1}{2}z - 4 + \frac{1}{2}y$$

Then, collect these "lines" with respect to y, as y goes from $y = 0$ to where it intersects the two oblique planes. This y-coordinate, denoted by $y1$ in the next

input statement, is found by setting $x1 = x2$ and solving for y. We now have—to coin another suggestive phrase—a horizontal "plane of values" for each z.#

```
> y1:=solve(x1=x2,y);
```

$$y1 := -z + \frac{16}{3}$$

Finally, we collect these "planes" as z goes from $z = 0$ to $z = 16/3$, this last value being quite evident from the formula for $y1$.#

```
> Int(Int(Int(4*x^2*y^3,x=x2..x1),y=0..y1),z=0..16/3)
  =int(int(int(4*x^2*y^3,x=x2..x1),y=0..y1),z=0..16/3);
```

$$\int_0^{16/3}\int_0^{-z+16/3}\int_{1/2\,z-4+1/2\,y}^{-z+4-y} 4\,x^2\,y^3\,dx\,dy\,dz = \frac{134217728}{76545}$$

```
> evalf(rhs(%));
```

$$1753.448664$$

#The individual antidifferentiation processes involved in the evaluation of this integral can be observed in the following way.#

```
> A1:=int(4*x^2*y^3,x);
```

$$A1 := \frac{4}{3}\,x^3\,y^3$$

```
> A2:=subs(x=x1,A1)-subs(x=x2,A1);
```

$$A2 := \frac{4}{3}\,(-z+4-y)^3\,y^3 - \frac{4}{3}\,(\frac{1}{2}\,z-4+\frac{1}{2}\,y)^3\,y^3$$

```
> A2:=simplify(A2);
```

$$A2 := -\frac{3}{2}\,y^3\,z^3 + 20\,y^3\,z^2 - \frac{9}{2}\,y^4\,z^2 - 96\,y^3\,z + 40\,y^4\,z - \frac{9}{2}\,y^5\,z + \frac{512}{3}\,y^3 - 96\,y^4 + 20\,y^5 - \frac{3}{2}\,y^6$$

```
> B1:=int(A2,y);
```

$$B1 := -\frac{3}{8}\,y^4\,z^3 + 5\,y^4\,z^2 - \frac{9}{10}\,y^5\,z^2 - 24\,y^4\,z + 8\,y^5\,z - \frac{3}{4}\,y^6\,z + \frac{128}{3}\,y^4 - \frac{96}{5}\,y^5 + \frac{10}{3}\,y^6 - \frac{3}{14}\,y^7$$

```
> B2:=subs(y=y1,B1)-subs(y=0,B1);
```

$$B2 := -\frac{3}{8}(-z+\frac{16}{3})^4 z^3 + 5(-z+\frac{16}{3})^4 z^2 - \frac{9}{10}(-z+\frac{16}{3})^5 z^2 - 24(-z+\frac{16}{3})^4 z + 8(-z+\frac{16}{3})^5 z - \frac{3}{4}(-z+\frac{16}{3})^6 z + \frac{128}{3}(-z+\frac{16}{3})^4 - \frac{96}{5}(-z+\frac{16}{3})^5 + \frac{10}{3}(-z+\frac{16}{3})^6 - \frac{3}{14}(-z+\frac{16}{3})^7$$

```
> B2:=simplify(B2);
```

$$B2 := -\frac{3}{280}z^7 + \frac{1}{3}z^6 - \frac{24}{5}z^5 - \frac{524288}{243}z + \frac{131072}{135}z^2 + \frac{159383552}{76545} - \frac{20480}{81}z^3 + \frac{128}{3}z^4$$

```
> C1:=int(B2,z);
```

$$C1 := -\frac{3}{2240}z^8 + \frac{1}{21}z^7 - \frac{4}{5}z^6 - \frac{262144}{243}z^2 + \frac{131072}{405}z^3 + \frac{159383552}{76545}z - \frac{5120}{81}z^4 + \frac{128}{15}z^5$$

```
> C2:=subs(z=16/3,C1)-subs(z=0,C1);
```

$$C2 := \frac{134217728}{76545}$$

****

Example 12.10 *Find the volume of the region bounded by* $x = 25-y^2$, $x+8y+2z = 56$, $z = 0$, *and* $x = 0$. *Confirm with an alternative calculation.*

It helps to plot the region involved with some care. Notice the method used in this example. **The frame box is used to show the boundaries $x = 0$ and $z = 0$.**

It is necessary to integrate first with respect to z, otherwise two integrals would be required.

****

```
> with(plots):

Warning, the name changecoords has been redefined
```

#Notice that the next two plot input statements end with colons. Output display is definitely undesirable here.#

```
> p1:=plot3d([25-y^2,y,z],y=-5..5,z=0..60):
> p2:=plot3d(28-4*y-x/2,x=0..25,y=-sqrt(25-x)..sqrt(25-x)):
> display({p1,p2});
```

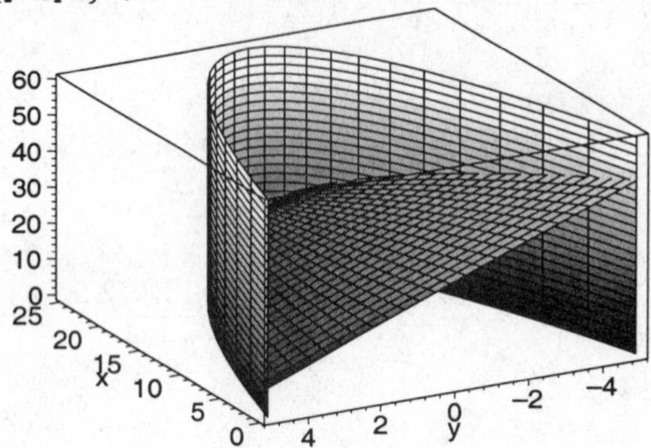

#Several options were added from the plot window to obtain this figure, and the plot was **turned around to show the faces $x = 0$ and $z = 0$.**#

```
> int(int(int(1,z=0..28-4*y-x/2),x=0..25-y^2),y=-5..5);
```

$$\frac{11500}{3}$$

#To confirm this value, we use a different order of integration. There is only one other convenient order.#

```
> int(int(int(1,z=0..28-4*y-x/2),

  y=-sqrt(25-x)..sqrt(25-x)),x=0..25);
```

$$\frac{11500}{3}$$

****

There is an alternative way to plot the region under consideration in the last example. Taking a cue from the way in which the plot $p1$ was formed in the previous work session, we could have plotted the plane $x + 8y + 2z = 56$ in the same fashion. In the plot $p1$ the independent variables were necessarily y and z. In order to plot both surfaces inside one plot command, the same two variables would have to be used for both surfaces. The following single plot command could be used instead of the combination of $p1$, $p2$, and display({p1,p2}).

```
> plot3d([25-y^2,y,z],[56-8*y-2*z,y,z],y=-5..5,z=0..60,
  view=[0..25,-5..5,0..60]);
```

Without the optional argument view=[0..25,-5..5,0..60], control over the plotting range for the x variable would be lost, and Maple would chose a large and fairly undesirable range of values. The plane $x = 0$ will not show in this plot, and so its location will have to be imagined. Of course, the plane $x = 0$ could have been included in the same way with the following plot command.

```
> plot3d([25-y^2,y,z],[56-8*y-2*z,y,z],[0,y,z],y=-5..5,z=0..60);
```

Plotting too many surfaces, however, tends to create a visual mess. This plot command does not lead to a nice picture, although the large range of x values is, for the most part, to blame.

These alternate strategies might work well on other plots, although it appears that our first approach, the one in the previous Maple work session, produces the best picture in this particular case.

Example 12.11 *Determine the volume of the region bounded above by* $x^2+y^2+z^2 = 25$ *and below by* $z = \sqrt{x^2+y^2}/2$ *as a triple integral in rectangular, cylindrical and spherical coordinates.*

****

```
> f:=sqrt(25-x^2-y^2):  g:=sqrt(x^2+y^2)/2:
```

#We begin with some preparations for the plot and for setting up the integrals. The equation, $f = g$, is the curve of intersection between the surfaces. We enter it without square roots.#

```
> eq:=g^2-f^2=0;
```

$$eq := \frac{5}{4}x^2 + \frac{5}{4}y^2 - 25 = 0$$

```
> 4*eq/5;
```

$$x^2 + y^2 - 20 = 0$$

```
> Y:=solve(eq,y);
```

$$Y := \sqrt{20 - x^2},\ -\sqrt{20 - x^2}$$

```
> plot3d({f,g},x=-sqrt(20)..sqrt(20),y=Y[2]..Y[1]);
```

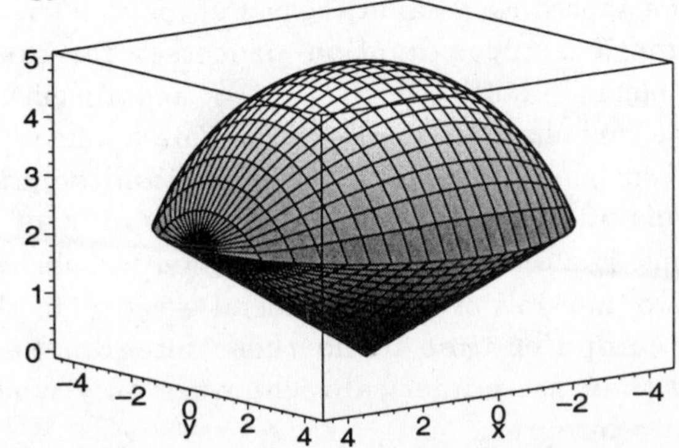

Notice here that the plotting tool bar was activated and the [1:1] button was clicked. This provides a true-scale representation of the region. Other adjustments were also made to the plot to get an acceptable view.#

```
> int(int(int(1,z=g..f),y=Y[2]..Y[1]),x=-sqrt(20)..sqrt(20));
```

#The output—an incomplete answer—is quite complicated, and so it is omitted.

The integral is much easier to compute in spherical coordinates. The ranges on ρ, and θ are straightforward, and the range on ϕ is determined by the angle at the vertex (from the positive z-axis). Elementary right-triangle trigonometry shows that this vertex angle, from the positive z-axis, is $\arctan(\sqrt{20}/\sqrt{5}) = \arctan(2)$.#

```
> alpha:=arctan(2);
```

$$\alpha := \arctan(2)$$

```
> int(int(int(rho^2*sin(phe),rho=0..5),phe=0..alpha),theta=0..2*Pi);
```

$$-\frac{50}{3}\pi\sqrt{5} + \frac{250}{3}\pi$$

```
> evalf(%);
```

$$144.7191422$$

#Finally, we compute the integral in cylindrical coordinates.#

```
> int(int(int(r,z=r/2..sqrt(25-r^2)),r=0..sqrt(20)),theta=0..2*Pi);
```

$$-\frac{50}{3}\pi\sqrt{5} + \frac{250}{3}\pi$$

****

Before we begin the applications in the next section, and before we get too involved in the more applied exercises at the end of the chapter, something should be said about **numerical approximation processes for double and triple integrals.** The command evalf(Int(Int(Int(...)))), and a similar form for double integrals (notice the capitalized I's), is used to initiate a numerical approximation process. As you can imagine, **numerical approximation techniques for double and triple integrals often require a fairly large amount of computer time.**

Frequently, Maple is able to use antidifferentiation techniques to compute the first, or the first two, integrals of a triple integral exactly. In this case, it **often takes much less computer time to do these integrals separately as antiderivatives and then use numerical techniques only when necessary on the last one or two integrals.**

12.4 Applications of Triple Integrals

Centers of mass are very important in our physical world. They are needed to determine, for example, motion stability of a wide range of objects. Here is a classic type of an application of triple integrals.

Example 12.12 *The nose section of a high-speed locomotive has the shape of the region bounded by the planes $z = 0$, $x = 0$, and the surface $z = 3.248 - 1.657y^2 - 0.1299x^2$, where x, y, and z are in meters. Naturally, a model for density can be given only approximately, but suppose that a reasonable value for the density, in $^{kg}/_{m^3}$, at the point (x, y, z) in the nose section is $\delta(x, y, z) = 2400 - 2(x-2)^2 - 3y^2 - 425(z-2)^2$. Determine the mass and the center of mass of the nose section.*

When the region is plotted, the plot options menu is opened, and **Constrained** is chosen from the perspective menu. This allows us to see the nose section in a true scale. As you can see, other adjustments of the plot are also made to get a good view.

Because of the symmetry both of the locomotive and of the density function, it is obvious that the locomotive is balanced on the $y = 0$ coordinate plane, so that the moment Mxz with respect to this coordinate plane is zero, as is the y-coordinate of the center of mass. We compute this moment anyway; it is hardly any extra work.

****

```
> f:=3.248-1.657*y^2-0.1299*x^2:
> plot3d(f,x=0..5,y=-1.5..1.5,view=0..4);
```

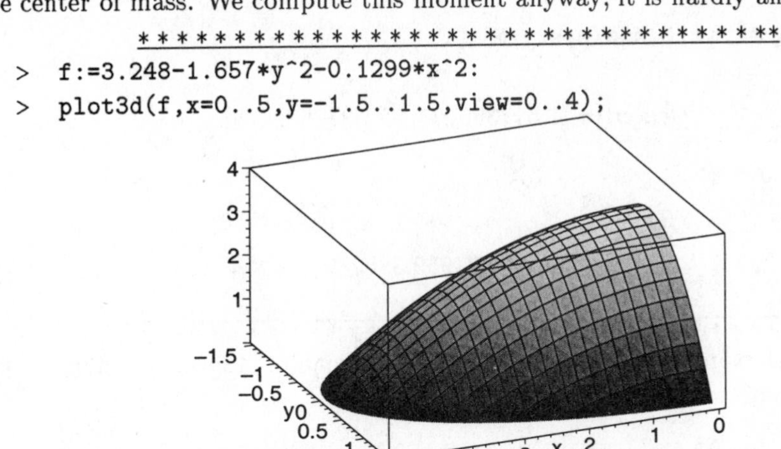


```
> delta:=2400-2*(x-2)^2-3*y^2-425*(z-2)^2:
> Y:=solve(f=0,y);
```

$$Y := \frac{1}{16570}\sqrt{538193600 - 21524430\,x^2},\ -\frac{1}{16570}\sqrt{538193600 - 21524430\,x^2}$$

```
> X:=solve(subs(y=0,f)=0,x);
```

$$X := -5.000384897,\ 5.000384897$$

```
> M:=evalf(int(int(int(delta,z=0..f),y=Y[2]..Y[1]),x=0..X[2]));
```

$$M := 31959.39597 - .1206510996\,10^{-20}\,I$$

```
> Mxy:=evalf(int(int(int(z*delta,z=0..f),y=Y[2]..Y[1]),x=0..X[2]));
```

$$Mxy := 40843.76950 + .4965998632\,10^{-31}\,I$$

```
> Mxz:=evalf(int(int(int(y*delta,z=0..f),y=Y[2]..Y[1]),x=0..X[2]));
```

$$Mxz := 0.$$

```
> Myz:=evalf(int(int(int(x*delta,z=0..f),y=Y[2]..Y[1]),x=0..X[2]));
```

$$Myz := 51689.72427 - .1475591240\,10^{-20}\,I$$

#Because of Maple's methods of integration, several of the above terms have exceedingly small imaginary components. Clearly these imaginary components can be discarded. The easiest way of doing this is to simply **copy and paste the above output values** into new input definitions for these terms. Once they are pasted into an input statement, the offending imaginary parts can then be removed with the delete key. #

```
> M:=31959.39597; Mxy:=40843.76950; Myz:=51689.72427;
```

$$M := 31959.39597$$
$$Mxy := 40843.76950$$
$$Myz := 51689.72427$$

```
> Center:=[Myz/M,Mxz/M,Mxy/M];
```

$$Center := [1.617356107, 0, 1.277989404]$$

```
> mass:=M;
```

$$mass := 31959.39598$$

****

In our final example, we solve a problem that requires the use of many of the ideas of multivariate calculus.

Example 12.13 *A new uncharted planet has been detected, which has the typical circular ellipsoid shape of a planet that rotates about a central axis. The distance between the North and South Poles (the central axis length) is 2800 kilometers, and the planet's equator is circular, with a diameter of 3600 kilometers. Find the volume of the atmosphere, which is 30 kilometers deep.*

Place the origin of an (x, y, z) coordinate system at the center of the planet, with the North Pole on the positive z-axis and the South Pole on the negative z-axis.

The volume of the atmosphere is more difficult to compute than it might appear to be at first glance. The ellipsoid is almost spherical. If it were, then the spherical coordinate variable ρ could be used to specify the altitude of a point above the planet's surface. As it stands, radial lines through the origin are not quite orthogonal to the surface. Consequently, if $(\rho_0, \theta_0, \phi_0)$ represents a point on the surface of the planet, then $(\rho_0+30, \theta_0, \phi_0)$ represents a point in the atmosphere that is (except for a few special points) not quite 30 kilometers in altitude. Ignoring this complication does not lead to a good approximation, and ultimately it dooms the problem and prevents a direct decimal solution.

This problem lends itself to an interesting approximate solution, but, being somewhat lengthy, it was omitted from this manual to save space. **Please look at our web site to find a complete solution.**

12.5 Exercise Set

New Maple Commands in this Chapter (and a few old commands as a reminder)

additionally( )	assume( )	assuming
display( )	evalf( )	evalf(int(int( )))
evalf(Int(Int( )))	evalf(int(int(int( ))))	evalf(Int(Int(Int( ))))
implicitplot3d( )	int(int( ))	int(int(int)))

1. Let R be the rectangle $R = [2,4] \times [-2,3]$, let n be a positive integer, and let $\mathcal{P}_n$ be the partition obtained by dividing R into n^2 subrectangles by using a grid of n equally spaced horizontal lines and n equally spaced vertical lines. Approximate the value of the integral $\int\int_R x^2y^3\,dxdy$ with a Riemann sum over the partition $\mathcal{P}_n$ for the following values of n and the following choices of the points (x_j, y_j), $(j = 1, 2, \ldots, n^2)$. Compare the approximations with the decimal value of the integral obtained via Maple's integration command. Notice how the approximations get better as n gets larger.

a) Let $n = 10$, and let (x_j, y_j), be the midpoint of the j-th subrectangle $(j = 1, 2, \ldots, 100)$.

b) Let $n = 20$, and let (x_j, y_j), be the midpoint of the j-th subrectangle $(j = 1, 2, \ldots, 400)$.

c) Let $n = 20$, and let (x_j, y_j), be the lower left-hand corner of the j-th subrectangle $(j = 1, 2, \ldots, 400)$.

d) Let $n = 50$, and let (x_j, y_j), be the midpoint of the j-th subrectangle $(j = 1, 2, \ldots, 2500)$.

2. Let D be the region bounded by the curve $y = 9 - x^2$ and the x-axis. Enclose D in the rectangle $R = [-3,3] \times [0,9]$, and let $\mathcal{P}_n$ be the partition of D obtained by dividing R into n^2 subrectangles by using a grid of n equally spaced horizontal lines and n equally spaced vertical lines. Approximate the value of the integral $\int\int_D ye^x\,dxdy$ with a Riemann sum over the partition $\mathcal{P}_n$ for the following values of n and the following choices of the points (x_j, y_j), $(j = 1, 2, \ldots, n^2)$. Compare the approximations with the decimal value of the integral obtained via Maple's integration command. Notice how the approximations get better as n gets larger. Use the piecewise( ) command, as we did in Example 12.2, to avoid **evaluation problems.**

a) Let $n = 10$, and let (x_j, y_j), be the midpoint of the j-th subrectangle $(j = 1, 2, \ldots, 100)$.

b) Let $n = 20$, and let (x_j, y_j), be the midpoint of the j-th subrectangle ($j = 1, 2, \ldots, 400$).

c) Let $n = 20$, and let (x_j, y_j), be the lower left-hand corner of the j-th subrectangle ($j = 1, 2, \ldots, 400$).

d) Let $n = 50$, and let (x_j, y_j), be the midpoint of the j-th subrectangle ($j = 1, 2, \ldots, 2500$).

3. Use a single Maple input statement to evaluate the double integral

$$\int\int_D \frac{y^3}{x^2}\, dx dy$$

where D is the region bounded by the curves $xy = 1$ and $10x + 7y = 50$. Evaluate the integral again, as a sequence of two separate single integrals. Use indefinite integration so that the individual antidifferentiation processes can be observed in greater detail.

4. For each of the following functions $f(x, y)$, and regions D, set up and evaluate (if possible) both iterated double integrals for $\int\int_D f(x, y)\, dx dy$. Confirm that the two values are the same. Evaluate each iterated integral a second time, using a sequence of two separate single integrals. Use indefinite integration so that the individual antidifferentiation processes can be observed in greater detail. Notice the advantage of integrating in one order over the other. In some cases, Maple is able to evaluate one iterated integral, but not the other.

a) $f(x, y) = \frac{x^2 y^3}{x^2 + y^2}$, D is the region bounded by the x-axis, $y = x$, and $x = 1$.

b) $f(x, y) = y^2 e^{xy}$, D is the region bounded by $y = x$, $y = 1$, and the y-axis.

c) $f(x, y) = \sqrt{x^7}\cos(x^2 y)$, D is the region bounded by $x = y^2$ and $x = 1$.

5. For each of the following functions $f(x, y)$, and regions D, evaluate, as an approximate decimal, the double integral $\int\int_D f(x, y)\, dx dy$.

a) $f(x, y) = \sqrt{5x^2 + 3y^2}$, D is the closed disk of radius 2 centered at the origin.

b) $f(x, y) = \frac{2x^2}{\sqrt{y+9}}$, D is the triangular region with vertices at the points (3,5), (3,-2), and (8,14).

c) $f(x, y) = \frac{x^2 - y^2}{x^2 + y^2}$, D is the region bounded by the curves $y = x^2 - 6x + 5$ and $y = 2x + 4$.

6. Evaluate, as an approximate decimal, the double integral

$$\int\int_D (x^3 - y^3)\, dx dy$$

where D is the right-hand portion of the two regions bounded by the curves

$$4x^2 + 25y^2 + 32x - 150y + 200 = 0 \text{ and } y - x^2 + 2x + 7 = 0.$$

7. Find the area of the region bounded by

$$x = y^4 - 5y^3 + 2y^2 - 13 \text{ and } x = 57 + 10y - 4y^2.$$

8. Find the mass and center of mass of a thin plate bounded by

$$x = y^4 - 5y^3 + 2y^2 - 13 \text{ and } x = 57 + 10y - 4y^2,$$

where x and y are expressed in centimeters, if the mass density at the point (x, y), in grams per square centimeter, is $\delta(x, y) = 4 + 2x + y/2$.

9. Use a single input statement to compute the double integral, in rectangular coordinates, of $f(x, y) = x^2y^2$ over the closed disk of radius 5 centered at the origin. Evaluate the integral a second time, using a sequence of two separate single integrals. Use indefinite integration so that the individual antidifferentiation processes can be observed in greater detail.

Compute the integral again as a double integral in polar coordinates, and again as a sequence of separate single polar integrals.

10. Compute the double integral of $f(x, y) = \sqrt{16 + x^2 + y^2}$ over the region D bounded by $y = x$, $y = -x$, and the semicircle $y = \sqrt{4 - x^2}$. Try to evaluate the integral in both rectangular and polar coordinates.

11. Let R be the rectangular box (prism) $R = [1, 3] \times [-2, 1] \times [-1, 4]$, let n be a positive integer, and let $\mathcal{P}_n$ be the partition obtained by dividing R into n^3 subprisms by using a gridwork of n equally spaced planes parallel to each of the coordinate planes. Let (x_j, y_j, z_j), be the midpoint of the j-th subprism ($j = 1, 2, \ldots, n^3$). Using this scheme of choosing the n^3 points (x_j, y_j, z_j), approximate the value of the integral $\int\int\int_R xye^{zx}\, dxdydz$ with a Riemann sum over the partition $\mathcal{P}_n$ for the following values of n. Compare the approximations with the decimal value of the integral obtained via Maple's integration command. Notice how the approximations get better as n gets larger. This problem should be accessible on most machines, but its computational complexity increases dramatically as n gets large. The partition $\mathcal{P}_{50}$, for example, has 125,000 subprisms.

a) $n = 10$ b) $n = 20$ c) $n = 50$

12. Let D be the closed sphere of radius 2 centered at the origin. Enclose D in the rectangular box (prism) $R = [-2, 2] \times [-2, 2] \times [-2, 2]$, let n be a positive integer, and let $\mathcal{P}_n$ be the partition obtained by dividing R into n^3 subprisms by using a gridwork of n equally spaced planes parallel to each of the coordinate planes. Let (x_j, y_j, z_j), be the midpoint of the j-th subprism ($j = 1, 2, \ldots, n^3$). Using this scheme of choosing the n^3 points (x_j, y_j, z_j), approximate the value of the integral $\int\int\int_R (xyz - 4)^2\, dxdydz$ with a Riemann sum over the partition $\mathcal{P}_n$ for the following values of n. Use the piecewise( ) command, as we did Example 12.2, to avoid **evaluation problems.** Compare the approximations with the decimal value of the integral obtained via Maple's integration command. Notice how the approximations get better as n gets larger.

a) $n = 10$ b) $n = 20$ c) $n = 50$

Computing the Riemann sum corresponding to $\mathcal{P}_{50}$ will require some patience or a machine with a fast processing time. This partition has 125,000 subprisms.

13. Use a single Maple input statement to evaluate the triple integral

$$\int_0^2 \int_0^{\sqrt{16-4z^2}} \int_0^{8-4z} xy\, dydxdz.$$

Evaluate the integral again, as a sequence of three separate single integrals. Use indefinite integration so that the individual antidifferentiation processes can be observed in greater detail.

14. Set up and evaluate all six iterated integrals for the volume of the region bounded below by $z = 0$, above by $z + 2y = 4$, and on the side by the cylinder $y = x^2$.

15. Compute the triple integral of $f(x, y, z) = x^3(xz - y)^2$ over the region D, which is bounded by the planes $x = 0$ and $3x+7z = 85$ and by the cylinder $3y^2+8z^2 = 28$.

16. Find the volume of the region D, which is bounded by the planes $2x+3y+z = 8$ and $2x + 3y - z = 8$ and by the cylinder $y = x^2 - 4$.

17. Find the volume of the region D, which is bounded by $y = 8x^2 + 3z^2 - 10$ and $y = 10 - 2x^2 - 5z^2$.

18. Find the volume of the region that is bounded above and below by $z = 900 - 25x^2 + 24y$ and $z = 4y^2 - 250x$ and lies inside the cup of the parabolic cylinder $x = 8 - y^2 - y$.

19. Find the volume of the region common to the interior of the cylinders $3y^2+5z^2 = 84$ and $3x^2 + 5z^2 = 84$.

20. Let D be the region bounded above by the sphere $x^2 + y^2 + z^2 = 900$ and below by the plane $z = 20$. Set up and evaluate the iterated triple integral for the volume of D in (a) rectangular, (b) cylindrical, and (c) spherical coordinates.

21. Find the (exact) volume of the region inside the sphere $x^2 + y^2 + z^2 = 16$, yet outside the cylinder $(x - 2)^2 + y^2 = 4$.

22. Find the mass and center of mass of the region bounded by $z = 200 - 0.03x^2 - 0.04y^2$ and $z = 10 - 0.02x^2 - 0.03y^2$, if the density at the point $P(x, y, z)$ in the region is $\delta(x, y, z) = (40 + 0.18z)^2$. Symmetry suggests that the center of mass should be on the z-axis. Confirm this with your calculations.

23. Let D be the region bounded by $x = 0$ and

$$x = 25\sqrt{1 - \frac{y^2}{16} - \frac{z^2}{9}}$$

where the units are in centimeters. Find the mass and center of mass of D under a uniform mass density of 8 grams per cubic centimeter.

24. A fuel tank has the shape of a circular cylinder with a hemisphere at each end. It lies on its side, and its position is carefully adjusted so that the axis of the cylinder is horizontal. The depth of the fuel level in the tank naturally determines how much fuel remains in the tank. Determine a formula for the volume of the fuel remaining in the tank as a function of the depth h of the fuel level. Let r denote the radius of the cylinder (hence the radius of each hemisphere), and let l denote the length of the cylinder (without the hemispherical ends).

For some values of h the fuel volume is obvious. Verify that your formula holds for these values of h.

25. A mixture of copper and waste rock is located inside an immense hill, and mining engineers are attempting to measure the total mass of the pure copper inside the hill before an attempt is made to buy the mineral rights to the land. A flat map of the region is being studied, and a two-dimensional grid has been placed over the map with units x and y in miles. The copper deposits exist within an area defined roughly by the ellipse

$$1.4x^2 + 3.2y^2 = 5.$$

The elevation H, of the hill, in feet above sea level, at the point (x, y) on or inside the ellipse, is

$$H = (5 - 1.4x^2 - 3.2y^2)(652y^2 + 248y + 350) + 43.$$

The copper/rock mixture begins roughly 20 feet below the surface of the hill, and it continues down to a depth defined by the equation

$$U = 950 - 280x^2 - 640y^2.$$

Let z denote the elevation, in feet above sea level, of a point P above the point (x, y) on the map. Engineers have made a "reasonable" conjecture that the density of pure copper within the rock mixture at a point $P(x, y, z)$ in the ore field is proportional to the product of the vertical distances from P to the upper and lower limits, $H - 20$ and U, of the field. This means, that for some constant of proportionality a, the density formula is

$$\delta = a(H - 20 - z)(z - U).$$

Empirical evidence suggests that a maximum density of 42 pounds per cubic foot (of pure copper) can be expected. Use this to determine a value for the constant a, and then find the total mass of pure copper in the field.

Remember that the elevation z is expressed in feet and that x and y are expressed in miles. The search for the point $P(x_0, y_0, z_0)$ that maximizes δ can be simplified. Surely $(H(x, y) - U(x, y))$ will be a maximum for $x = x_0$, and $y = y_0$. Furthermore, z_0 will be midway between $(H(x_0, y_0) - 20)$ and $U(x_0, y_0)$.

Project: Gravity

Let m denote the mass of a point located at P, and let M denote the mass of a point located at Q. The gravitational force of attraction (exerted at the point P) between the two point masses is

$$\vec{F} = \frac{GmM}{|\vec{r}|^3}\vec{r}$$

where $G =$ is the gravitational constant and $\vec{r}$ is the vector from P to Q. This is one of the three classic *"Inverse Square Laws"* that govern the behavior of not only gravitational forces, but electrical and nuclear forces as well.

If m and M are in *grams*, $\vec{r}$ is in *cm*, and $G = 6.670 \times 10^{-8}$, then $\vec{F}$ is expressed in *dynes*. The problems in this section can be solved, however, without a value for

G. Just carry G along as a constant. Usually, we let $m = 1$ and talk about the gravitational force exerted on a unit point mass located at P.

If a mass-object W (a world) is not a point mass, then the gravitational field created by W is a consequence of the its global shape and size. To determine a formula for the gravitational field for W, we partition W into lots of infinitesimal point masses. At each point mass, the *Inverse Square Law* applies, and the individual forces of each point mass exerted on a unit mass located at the fixed point P can be computed. We simply add up (in the sense of integration) all of the infinitesimal forces.

Force, of course, is a vector quantity. Normally it would have to be decomposed into three real-valued components, and the integration would then be performed on each of the three components. The following two problems, however, have a great deal of symmetry. Use it wisely, and the problems can be greatly simplified.

1. Show that the gravitational field created by a spherical planet of mass M and of constant mass density is the same as the gravitational field created by a point mass of size M located at the center of the sphere. Of course, this makes sense only for points far enough from the center to be outside the sphere.

We may as well suppose that the planet is centered at the origin. Let R denote the radius of the planet, and let δ denote its constant mass density. Then the total mass M of the planet is just δ times the volume of the sphere.

Let $P_0(x_0, y_0, z_0)$ be the location of a unit point mass, and let R_0 be the distance from P0 to the origin. **Maple will have to be told, with the commands assume( ) and additionally( ), that $R > 0$ and that $R_0 > R$.**

2. Have you ever read a science fiction or fantasy book about life inside a hollow planet? Is this possible? Could such a world support a gravitational field that would hold the planet's inhabitants to the "ground?"

Consider a large hollow sphere (planet) of radius R, Suppose that R is quite large (perhaps several thousand kilometers), and that the skin of the sphere (the ground) is made out of a very heavy material of uniform mass density δ (expressed in kilograms per square meter).

Determine a vector-valued formula, that describes the gravitational force of attraction exerted by the entire hollow sphere on a unit point mass located at a point $P_0(x_0, y_0, z_0)$ interior to the sphere.

This problem is meant to be done in spherical coordinates ρ, θ, and ϕ, with ρ fixed at $\rho = R$. Consider a small patch of area on the sphere as a point mass.

A formula for a "small patch of area" on the skin of the sphere will be needed. Recall that the "infinitesimal cube" in spherical coordinates has dimensions $d\rho$ by $\rho\, d\phi$ by $\rho \sin(\phi)\, d\theta$. The area of the face of this cube that lies on the sphere $\rho = R$ is just the product, $\rho^2 \sin(\phi)\, d\theta d\phi$, of the last two dimensions with $\rho = R$. It follows that the expression $R^2 \sin(\phi)$ represents the area of a small patch of area on the surface of the hollow sphere.

Let R_0 be the distance from P_0 to the origin. **Maple will have to be told, with the commands assume( ) and additionally( ), that $R_0 > 0$ and that $R_0 < R$.**

Project: The Atmosphere of a Planet

A new uncharted planet has been discovered, and scientists need to know the total amount of methane in the planet's atmosphere.

The origin of a coordinate system is placed at the planet's center, with units expressed in kilometers. The surface of the planet is defined by the equation

$$\frac{x^2}{5000^2} + \frac{y^2}{5000^2} + \frac{z^2}{4200^2} = 1,$$

and the outer limit of the planet's atmosphere is defined by the equation

$$\frac{x^2}{5030^2} + \frac{y^2}{5030^2} + \frac{z^2}{4230^2} = 1.$$

The density of methane, in grams per cubic meter, is estimated to be

$$\delta = \frac{1.023}{h^2 + 1}$$

at an altitude h meters above the planet's surface. (Altitude is a dimension that is always orthogonal to the planet's surface.) Determine the total amount of methane in the atmosphere.

If you use evalf(Int(Int(..))) or a similar form for triple integrals (notice the capitalized I's) to initiate a numerical approximation process, you will probably have time to enjoy a leisurely dinner while you wait for Maple to calculate an answer to this problem. To speed things up, it could help to take advantage of the comments at the end of Section 12.3. Altitude could be difficult to express in terms of spherical coordinates, and an approximation strategy might be necessary. See the solution to Example 12.13 for some insight.

Index